ASTEROSEISMOLOGY

Our understanding of stars has grown significantly due to recent advances in asteroseismology, the stellar analog of helioseismology, the study of the Sun's acoustic wave oscillations. Using ground-based and satellite observatories to measure the frequency spectra of starlight, researchers are able to probe beneath a star's surface and map its interior structure. This volume provides a wide-ranging and up-to-date overview of the theoretical, experimental, and analytical tools for carrying out front-line research in stellar physics using asteroseismological observations, tools, and inferences. Chapters from seven eminent scientists in residence at the twenty-second Canary Islands Winter School of Astrophysics examine the interior of our Sun relative to data collected from distant stars, how to measure the fundamental parameters of single field stars, diffusion processes, and the effects of rotation on stellar structures. The volume also provides detailed treatments of modeling and computing programs, providing astronomers and graduate students with a practical, methods-based guide.

Pere L. Pallé is a pioneering astrophysicist in the field of helioseismology and head of the Instituto de Astrofísica de Canarias's Research Group for Helioseismology and Asteroseismology. He completed his undergraduate studies and PhD at the University of La Laguna, Spain, and is deeply involved in the deployment, operation, and scientific exploitation of the first Earth-based international helioseismology networks (BiSON, IRIS, GONG), space missions (such as GOLF/SoHO), and asteroseimology networks (SONG). His research is focused on the instrumentation, analysis, and interpretation of solar and stellar internal structures and dynamics by means of seismology tools.

César Esteban is a lecturer in the Astrophysics Department of the University of La Laguna and researcher at the Instituto de Astrofísica de Canarias in Spain. He completed his undergraduate studies and PhD at the University of La Laguna, where he was head of the Astrophysics Department from 2010 to 2011. He researches the physical structure and chemical composition of ionized nebulae, the chemical evolution of galaxies, and the astronomy of ancient cultures. He coordinates different research projects and is a member of various scientific societies and governing boards.

Canary Islands Winter School of Astrophysics

Volume XXII

Series Editor
F. Sánchez, *Instituto de Astrofísica de Canarias*

Previous books in this series

I. Solar Physics
II. Physical and Observational Cosmology
III. Star Formation in Stellar Systems
IV. Infrared Astronomy
V. The Formation of Galaxies
VI. The Structure of the Sun
VII. Instrumentation for Large Telescopes: A Course for Astronomers
VIII. Stellar Astrophysics for the Local Group: A First Step to the Universe
IX. Astrophysics with Large Databases in the Internet Age
X. Globular Clusters
XI. Galaxies at High Redshift
XII. Astrophysical Spectropolarimetry
XIII. Cosmochemistry: The Melting Pot of Elements
XIV. Dark Matter and Dark Energy in the Universe
XV. Payload and Mission Definition in Space Sciences
XVI. Extrasolar Planets
XVII. 3D Spectroscopy in Astronomy
XVIII. The Emission-Line Universe
XIX. The Cosmic Microwave Background: From Quantum Fluctuations to the Present Universe
XX. Local Group Cosmology
XXI. Accretion Processes in Astrophysics

Participants of the XXII Canary Islands Winter School of Astrophysics, in front of the "Casino" (Spanish meaning equivalent to Cultural Club) of La Laguna, located in the historical centre of the town.

Lecturers and scientific organizer of the Winter School in the balcony of the Conference Room at Hotel Nivaria in La Laguna. From left to right: Donald Kurtz, Steven Kawaler, Sarbani Basu, Pere L. Pallé, William Chaplin, Thierry Appourchaux, and Tim Bedding. Not present: Jørgen Christensen-Dalsgaard and Mario J. Monteiro.

ASTEROSEISMOLOGY

XXII Canary Islands Winter School of Astrophysics

Edited by

PERE L. PALLÉ

Instituto de Astrofísica de Canarias, Tenerife

CÉSAR ESTEBAN

Instituto de Astrofísica de Canarias, Tenerife

CAMBRIDGE
UNIVERSITY PRESS

University Printing House, Cambridge CB2 8BS, United Kingdom

One Liberty Plaza, 20th Floor, New York, NY 10006, USA

477 Williamstown Road, Port Melbourne, VIC 3207, Australia

314-321, 3rd Floor, Plot 3, Splendor Forum, Jasola District Centre, New Delhi - 110025, India

79 Anson Road, #06-04/06, Singapore 079906

Cambridge University Press is part of the University of Cambridge.

It furthers the University's mission by disseminating knowledge in the pursuit of education, learning and research at the highest international levels of excellence.

www.cambridge.org
Information on this title: www.cambridge.org/9781107029446

First published 2013

A catalogue record for this publication is available from the British Library

Library of Congress Cataloging in Publication data
Canary Islands Winter School of Astrophysics (22nd : 2010 : La Laguna, Canary Islands)
Asteroseismology : XXII Canary Islands Winter School of Astrophysics / [edited by] Pere L. Pallé, Instituto de Astrofísica de Canarias, Tenerife, César Esteban, Instituto de Astrofísica de Canarias, Tenerife.
pages cm
ISBN 978-1-107-02944-6 (hard covers : alk. paper)
1. Astroseismology – Congresses. I. Pallé, Pere L., editor of compilation. II. Esteban, César, editor of compilation. III. Title.
QB812.C36 2013
523.8–dc23 2013013589

ISBN 978-1-107-02944-6 Hardback

Contents

List of contributors

APPOURCHAUX, THIERRY, Institut d'Astrophysique Spatiale (France)

BASU, SARBANI, Yale University (USA)

BEDDING, TIMOTHY R., University of Sydney (Australia)

CHAPLIN, WILLIAM J., University of Birmingham (UK)

CHRISTENSEN-DALSGAARD, JØRGEN, University of Aarhus (Denmark)

KAWALER, STEVEN D., Iowa State University (USA)

KURTZ, DONALD W., University of Central Lancashire (UK)

List of participants

Alonso Tagle, María Luisa	Universidad Pontificia Católica de Chile (Chile)
Appourchaux, Thierry	Institut d'Astrophysique Spatiale (France)
Basu, Sarbani	Yale University (USA)
Bedding, Timothy R.	University of Sydney (Australia)
Bloemen, Steven	Instituut voor Sterrenkunde, K.U. Leuven (Belgium)
Brandão, Isa	Centro de Astrofísica da Universidade do Porto (Portugal)
Campante, Tiago	Centro de Astrofísica da Universidade do Porto (Portugal)
Chakraborty, Sudeepto	Stanford University (USA)
Chaplin, William J.	University of Birmingham (UK)
Christensen-Dalsgaard, Jørgen	University of Aarhus (Denmark)
Corsaro, Enrico Maria Nicola	INAF-Catania Astrophysycal Observatory (Italy)
Creevey, Orlagh	Instituto de Astrofísica de Canarias (Spain)
Damasso, Mario	University of Padova (Italy)
Díaz Alfaro, Manuel	Instituto de Astrofísica de Canarias (Spain)
Dogan, Gulnur	University of Aarhus (Denmark)
Drobek, Dominik	Instytut Astronomiczny Uniwersytet Wroclawski (Poland)
Esch, Lisa	Yale University (USA)
Escobar, María Eliana	Lab. d'Astrophysique Toulouse-Tarbes (France)
Fernández, Javier	National Tsing Hua University (Taiwan)
Gülmez, Timuçin	University of Johannesburg (South Africa)
Hall, Martin	University of Central Lancashire (UK)
Hambleton, Kelly	University of Central Lancashire (UK)
Kawaler, Steven D.	Iowa State University (USA)
Kurtz, Donald W.	University of Central Lancashire (UK)
Mao, Yongna	National Astronomical Observatories, Chinese Academy of Sciences (China)
Monteiro, Mario J. P. F. G.	Centro de Astrofísica da Universidade do Porto (Portugal)
Murphy, Simon	University of Central Lancashire (UK)
Pallé, Pere L.	Instituto de Astrofísica de Canarias (Spain)
Papics, Peter	Katholieke Universiteit Leuven (Belgium)
Pasek, Mickael	Universite Paul Sabatier (France)
Salmon, Sebastien	University of Liège (Belgium)
Silva Aguirre, Victor	Max Planck Institute for Astrophysics
Sódor, Ádám	Konkoly Observatory (Hungary)
Stott, Jonathan	Vatican Observatory (USA)
Szewczuk, Wojciech	Instytut Astronomiczny Uniwersytet Wroclawski (Poland)
Szymanski, Tomasz	Astronomical Observatory. Jagiellonian Univ. (Poland)
Thygesen, Anders Overaa	Institut for Fysik & Astronomi, Aarhus Universitet (Denmark)
Tognelli, Emanuele	Physics Department – Pisa University (Italy)
Ulusoy, Ceren	University of Johannesburg (South Africa)
Walczak, Przemyslaw	Instytut Astronomiczny Uniwersytet Wroclawski (Poland)

Wang, Huijuan	National Astronomical Observatories, Chinese Academy of Sciences (China)
White, Timothy	University of Sydney (Australia)
Zhang, Chunguang	National Astronomical Observatories, Chinese Academy of Sciences (China)

Preface

Background

The XXII Canary Islands Winter School of Astrophysics, organized by the Instituto de Astrofísica de Canarias (IAC), focuses on the new advances and challenges that asteroseismology provides in the domains of stellar structure, dynamics, and evolution. Every year the Winter School welcomes around 60 Ph.D. students and young postdocs and provides a unique opportunity for them to broaden their knowledge in a key field of astronomy.

Scientific rationale

When oscillations of the Sun were first discovered, a new era of science began. The observed frequencies could be used to probe deep into the stellar interior, the only measurements that could possibly pierce the stellar surface. Today, "helioseismology" has been responsible for some of our deepest understanding of the Sun: we know the radial and longitudinal rotation profile of the interior, we have measured the depth of the outer convection zone, and it has helped solve the so-called neutrino problem when the observations and theory predicted a much hotter central temperature than the observed neutrinos predicted. Today, these seismic observations are not only available in much higher quality, but they are also available for hundreds of other stars. In the last few years, many space missions (CoRoT and Kepler) have produced these data of exquisite quality, and for the first time we are in a position to study the Sun in the context of other stars, measure the fundamental parameters of single field stars to within 2 percent, learn about diffusion processes and the effects of rotation on the stellar structure, and test opacities and equations of state in extreme conditions.

The key objectives of this Winter School are to provide young scientists with knowledge and understanding of observation, instrumentation, theory, modeling, and computing for asteroseismology.

Outline of the school

The primary aim of the XXII Winter School is to provide a wide-ranging and up-to-date overview of the theoretical, experimental, and analytical tools necessary for carrying out front-line research in stellar physics by means of asteroseismological observations, tools, and inferences. The school is particularly designed to offer young researchers tips and guidelines to help them direct their future research toward these themes, which are among the most important in modern astrophysics. To achieve these goals, the Winter School lectures are given by eight eminent and experienced scientists who are actively working on a variety of forefront research projects, and who have played a key role in major advances over the recent years. The list of invited researchers includes leading theoreticians and pioneering observers in each area of the subject. The format of the school also encourages direct interaction between the participating students and lecturers.

The school is primarily intended for doctoral students and recent postdocs in any field of research in Astronomy. Participants of the school will have the opportunity to display their current work by presenting a poster contribution and later to discuss them within a dedicated session.

The Editors

Acknowledgments

The organizers of the XXII Canary Islands Winter School of Astrophysics would like to express their sincere gratitude, first and foremost, to the lecturers, for making it a great scientific and educational event. The careful preparation of the lectures, the attendance and intense interaction with students, and the subsequent writing up of the manuscripts for this book have been a major – but we hope rewarding – commitment in their very busy agendas. The students played a major role in the success of the Winter School: their enormous enthusiasm – maintained throughout the entire two weeks – and outstanding human quality resulted in a really pleasant and fruitful event. Ismael Martínez-Delgado played a major part in the creation of this book, revising in minute detail and with enormous patience and technical editorial skill all the submitted manuscripts.

The Secretary of the Winter School, Mrs. Lourdes González, has been involved in the administration of every school since their inception in 1989. She is therefore the soul and memory of the event. Her great knowledge and diligence before, during, and after each Winter School is a key component in its success. The Web page of the Winter School is a vital tool in the preparation and development of the event, and we thank Jorge Andrés Pérez (SIE/IAC) for its care and maintenance. The press room of the Winter School was the responsibility of Annia Domenech, who did a wonderful job interviewing the lecturers and students. Inés Bonet (Gabinete de Dirección/IAC) complemented the press room coverage with fine and high-quality videos. Ramón Castro and Gabriel Pérez (SMM/IAC) designed the posters and additional multimedia support material for the Winter School, and Diego Sierra and Francisco López (SIC/IAC) set up for us a wonderful wireless system and computer network in the conference room. The visits to the observatories (on Tenerife and La Palma) were possible thanks to the support of their respective managers (Miquel Serra and Juan Carlos Pérez Arencibia) and volunteer guides Antonio-Eff Darwich, Rafael Barrena, and Katrien Uytterhoeven. One of the sessions of the Winter School took place in the Computing Laboratory of the Department of Astrophysics of the Faculty of Physics at the University of La Laguna and we thank Professor Teodoro Roca (Dean of the Faculty) and Dr. Fernando Pérez for their support and invaluable help.

Each year, the Canary Islands Winter School of Astrophysics is a major institutional event at the IAC, whose various departments and support services always actively and enthusiastically contribute to it. We thank all concerned for their support and efficiency.

We are greatly indebted to Spain's Ministerio de Ciencia e Innovación, the Cabildo de Tenerife, and the Ayuntamiento de San Cristóbal de La Laguna for vital financial assistance. The Director of the Museo de Historia y Antropología de Tenerife kindly offered us the use of the Casa Lercaro historical monument to hold the welcoming cocktail party and registration of participants. The Winter School took place in the main conference room of Hotel Nivaria, a sixteenth-century mansion situated in the heart of the city of La Laguna, where all the lecturers and students were lodged. We would like to thank the hotel for its superb facilities and the friendliness of its entire staff.

The Editors

Abbreviations

AAT	Anglo-Australian Telescope
ACRIM	Active Cavity Radiometer Irradiance Monitor
ADIPLS	Aarhus aDIabatic PuLSation code
AMU	Atomic Mass Unit
ASTEC	Aarhus STellar Evolution Code
AT&T	American Telephone and Telegraph
BCE	Before the Common Era
BiSON	Birmingham Solar Oscillations Network
BJD	Barycentric Julian Date
BJED	Barycentric Julian Ephemeris Date
CCD	Charge Coupled Device
C-D	Christensen-Dalsgaard diagram
CEFF	Coulomb corrections Eggleton-Faulkner-Flannery equation
CNES	Centre National d'Estudes Spatiales
CNO	Carbon-Nitrogen-Oxygen cycle
CORALIE	Echelle Spectrograph on the 1.2-metre Leonard Euler Swiss telescope
COROT	COnvection ROtation et Transits planétaires
CP	Chemically Peculiar
CTIO	Cerro Tololo Inter-American Observatory
CZ	Convective Zone
DFT	Discrete Fourier Transform
DSP	Digital Signal Processing
EFF	Eggleton-Faulkner-Flannery equation
ESA	European Space Agency
ET	Ephemeris Time
EVRIS	Etude de la Variabilité et de la Rotation des Intérieurs Stellaire
FFT	Fast Fourier Transform
FWHM	Full Width at Height Maximum
GK	Gough and Kosovichev set
GMST	Greenwich Mean Sidereal Time
GOLF	Global Oscillations at Low Frequency
GONG	Global Oscillations Network Group
GPS	Global Positioning System
HARPS	High Accuracy Radial Velocity Planet Searcher
HJD	Heliocentric Julian Date
HR	Hertzsprung Russell diagram
HST	Hubble Space Telescope
IAC	Instituto de Astrofísica de Canarias
IAU	International Astronomical Union
IPHIR	InterPlanetary Helioseismology by IRadiance measurements
JD	Julian Date
KIC	Keppler Input Catalogue
KOI	Keppler Object of Interest
LBV	Luminous Blue Variable
LHS	Left Hand Side
LOWL	Low-l
LS	Lomb-Scargle periodogram
MAP	Maximum A Posteriori
MDI	Michelson Doppler Imager

MH	Metropolis-Hastings algorithm
MHD	Mihalas, Hummer and Däppen equation
MIX	Core Mixing Model
MJD	Modified Julian Date
MLE	Maximum Likelihood Estimation
MOLA	Multiplicative Optimally Localised Averages
MONS	Measuring Oscillations in Nearby Stars
MOST	Microvariability and Oscillations of STars
NACRE	Nuclear Astrophysics Compilation of REaction rates
NASA	National Aeronautics and Space Administration
NIST	National Institute of Standards and Technology
NODIF	No Diffusion
OGLE	Optical GRavitational Lensing Experiment
OLA	Optimally Localised Averages
OPAL	Opacity Project Los ALamos
PLATO	PLAnetary Transits and Oscillations of stars
PSD	Power Spectral Density
PRISMA	Probing Rotation and Interior of Stars: Microvariability and Activity
PVSG	Periodically Variable Super Giant star
RHS	Right Hand Side
RJD	Reduced Julian Date
RLS	Regularized List Squares
SARG	Spettrografo Alta Risoluzione Galileo
sdB	subdwarf B
SDSS	Sloan Digital Sky Survey
SMEI	Solar Mass Ejection Experiment
SMM	Solar Maximum Mission
SOHO	SOlar and Heliospheric Observatory
SOI	Solar Oscillations Investigations
SOLA	Subtractive Optimally Localised Averages
SONG	Stellar Observations Network Group
SPB	Slowly Pulsating B stars
STD	Standard Model
SVD	Singular Value Decomposition
TAI	Temps Atomique International (International Atomic Time)
TCB	Temps Coordonnée Barycentrique (Barycentric Coordinate Time)
TCG	Temps Coordonnée Géocentrique (Geocentric Coordenate Time)
TDB	Temps Dynamique Barycentrique (Barycentric Dynamical Time)
TDT	Terrestrial Dynamical Time
TJD	Truncated Julian Date
TT	Terrestrial Time
UCLES	University College London Echelle Spectrograph
UK	United Kingdom
US	United States
USA	United States of America
UT	Universal Time
UTC	Coordinated Universal Time
UVES	Ultraviolet and Visual Echelle Spectrograph
VIRGO	Variability of solar IRadiance and Gravity Oscillations
VLBI	Very Long Baseline radio Interferometry
WIRE	Wide-field Infra Red Explorer
YREC	Yale Rotating Stellar Evolution Code

1. Sounding the solar cycle with helioseismology: Implications for asteroseismology

WILLIAM J. CHAPLIN

1.1 Introduction

My brief for the IAC Winter School was to cover observational results on helioseismology, flagging where possible implications of those results for the asteroseismic study of solar-type stars. My desire to make such links meant that I concentrated largely on results for low angular-degree (low-l) solar p modes, in particular results derived from "Sun-as-a-star" observations (which are, of course, most instructive for the transfer of experience from helioseismology to asteroseismology). The lectures covered many aspects of helioseismology – modern helioseismology is a diverse field. In these notes, rather than discuss each aspect to a moderate level of detail, I have instead made the decision to concentrate on one theme, that of "sounding" the solar activity cycle with helioseismology. I cover the topics from the lectures and I also include some new material, relating both to the lecture topics and to other aspects I did not have time to cover. Implications for asteroseismology are developed and discussed throughout.

The availability of long time series data on solar-type stars, courtesy of the NASA Kepler Mission (Chaplin *et al.*, 2010; Gilliland *et al.*, 2010) and the French-led CoRoT satellite (Appourchaux *et al.*, 2008), is now making it possible to "sound" stellar cycles with asteroseismology. The prospects for such studies have been considered in some depth (Chaplin *et al.*, 2007a, 2008a; Metcalfe *et al.*, 2007; Karoff *et al.*, 2009, e.g.), and in the last year the first convincing results on stellar-cycle variations of the p-mode frequencies of a solar-type star (the F-type star HD49933) were reported by García *et al.* (2010). This result is important for two reasons: first, the obvious one of being the first such result, thereby demonstrating the feasibility of such studies; and second, the period of the stellar cycle was evidently significantly shorter than the 11-year period of the Sun (probably between 1 and 2 years). If other similar stars show similar short-length cycles, there is the prospect of being able to "sound" perhaps two or more complete cycles of such stars with Kepler (assuming the mission is extended, as expected, to 6.5 year or more). The results on HD49933 may be consistent with the paradigm that stars divide into two groups, activity-wise, with stars in each group displaying a similar number of rotation periods per cycle period (e.g., see Böhm-Vitense, 2007), meaning solar-type stars with short rotation periods – HD49933 has a surface rotation period of about 3 days – tend to have short cycle periods. We note that Metcalfe *et al.* (2010) recently found another F-type star with a short (1.6-year) cycle period (using chromospheric H & K data). Extension of the Kepler Mission will, of course, also open the possibility of detecting full swings in activity in stars with cycles having periods up to approximately the length of the solar cycle.

The rest of my notes break down as follows. Section 1.2 gives an introductory overview of the solar cycle, as seen in helioseismic data. A brief history of observations of solar cycle changes in low-angular degree (low-l) solar p modes is given in Section 1.2.1. Then, in Section 1.2.2, we consider the causes of the observed changes in the mode frequencies; and in Section 1.2.3, we discuss variations in the mode powers and damping rates, and what the relative sizes of those changes imply for the underlying cause.

Section 1.3 considers several subtle ways in which the stellar activity cycles can affect values of, and inferences made from, the mode parameters. Concepts are introduced using the example of the solar cycle, and the impact it has on low-l p modes observed in

Sun-as-a-star data. Implications for asteroseismic observations of solar-type stars are then developed. We start in Section 1.3.1 with a discussion of the impact of stellar activity on estimates of the mode frequencies. This is followed in Section 1.3.2 by a similar discussion for frequency separation ratios. Finally, Section 1.3.3 explains how mode peaks in the frequency-power spectrum can be "distorted" by stellar cycles, rendering commonly used fitting models inappropriate.

In Section 1.4 we develop a simple model to illustrate the impact on the mode frequencies of the range in latitudes covered by near-surface magnetic activity on solar-type stars, and discuss how the angle of inclination can affect significantly the observed frequency shifts (because that angle affects which mode components are visible in the observations). We also show how measurements of the frequency shifts of modes having different angular and azimuthal degrees may be used to make inferences on the spatial distribution of the near-surface magnetic activity on solar-type stars.

We end in Section 1.5 by thinking somewhat longer-term, and consider how much low-l data would be needed to measure evolutionary changes of the solar p-mode frequencies.

1.2 The seismic solar cycle: overview

A rich and diverse body of observational data is now available on temporal variations of the properties of the global solar p modes. The signatures of these variations are correlated strongly with the well-known 11-year cycle of surface activity. The search for temporal variations of the p-mode properties began in the early 1980s, following accumulation of several years of global seismic data. The first positive result was reported by Woodard and Noyes (1985), who found evidence in observations made by the Active Cavity Radiometer Irradiance Monitor (ACRIM) instrument, on board the Solar Maximum Mission (SMM) satellite, for a systematic decrease of the frequencies of low-l p modes between 1980 and 1984. The first year coincided with high levels of global surface activity, while during the latter period activity levels were much lower. The modes appeared to be responding to the Sun's 11-year cycle of magnetic activity. Woodard and Noyes (1985) found that the frequencies of the most prominent modes had decreased by roughly 1 part in 10,000 between the activity maximum and minimum of the cycle. By the late 1980s, an in-depth study of frequency variations of global p modes, observed in the Big Bear data, had demonstrated that the agent of change was confined to the outer layers of the interior (Libbrecht and Woodard, 1990).

Accumulation of data from the new networks and instruments has made it possible to study the frequency variations to unprecedented levels of detail, and has revealed signatures of subtle, structural change in the subsurface layers. The discovery of solar-cycle variations in mode parameters associated with the excitation and damping (e.g., power, damping rate, and peak asymmetry) followed. Patterns of flow that penetrate a substantial fraction of the convection zone have also been uncovered as well as possibly signatures of changes in the rotation rate of the layers that straddle the tachocline, and much more recently evidence for a quasi-periodic 2-year signal, superimposed on the solar-cycle variations of the mode frequencies.

The modern seismic data give unprecedented precision on measurements of frequency shifts. Examples of average frequency shifts for low-l data are shown in Fig. 1.1, for Sun-as-a-star data collected by the ground-based Birmingham Solar-Oscillations Network (BiSON) and the Global Oscillations at Low Frequency (GOLF) instrument on board the ESA/NASA Solar and Heliospheric Observatory (SOHO). From observations of the medium-l frequency shifts it is possible to produce surface maps showing the strength of the solar-cycle shifts as a function of latitude and time (Howe *et al.*, 2002), like the example shown in Fig. 1.2 (which is made from Global Oscillations Network Group (GONG) data). These maps bear a striking resemblance to the butterfly diagrams that

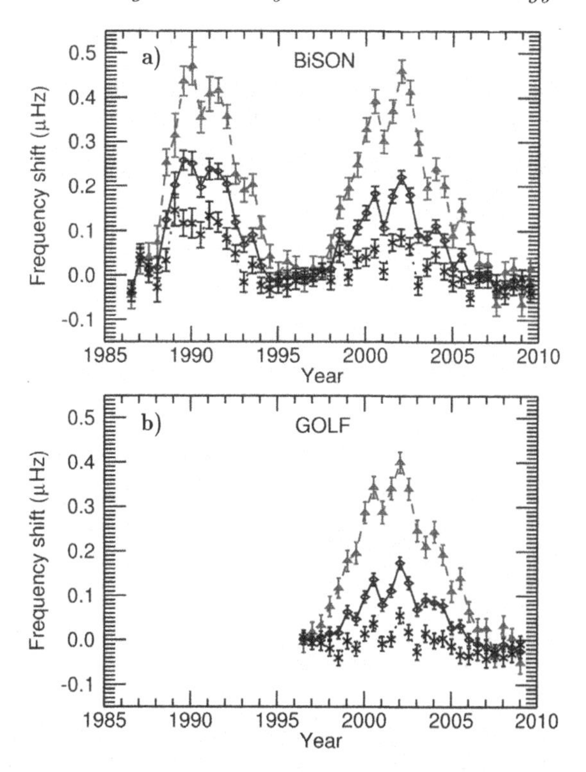

FIG. 1.1. Average frequency shifts of most prominent low-l solar p modes, as measured in: *a)* BiSON data and *b)* GOLF data over the last two 11-year activity cycles. The three curves show results for averages made over different ranges in frequency: 1,880 to 3,710 μHz (diamonds, joined by solid line); 1,880 to 2,770 μHz (crosses, joined by dotted line); 2,820 to 3,710 μHz (triangles, joined by dashed line). (From Fletcher *et al.*, 2010.)

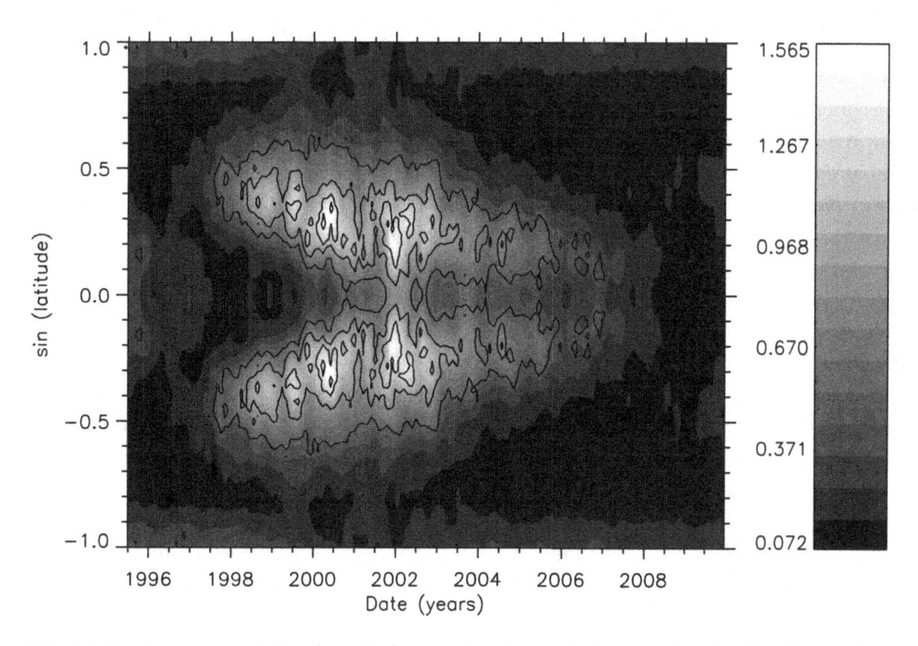

FIG. 1.2. Mode frequency shifts (in μHz) as a function of time and latitude. The values come from analysis of GONG data. The contour lines indicate the surface magnetic activity. (Figure courtesy of R. Howe.)

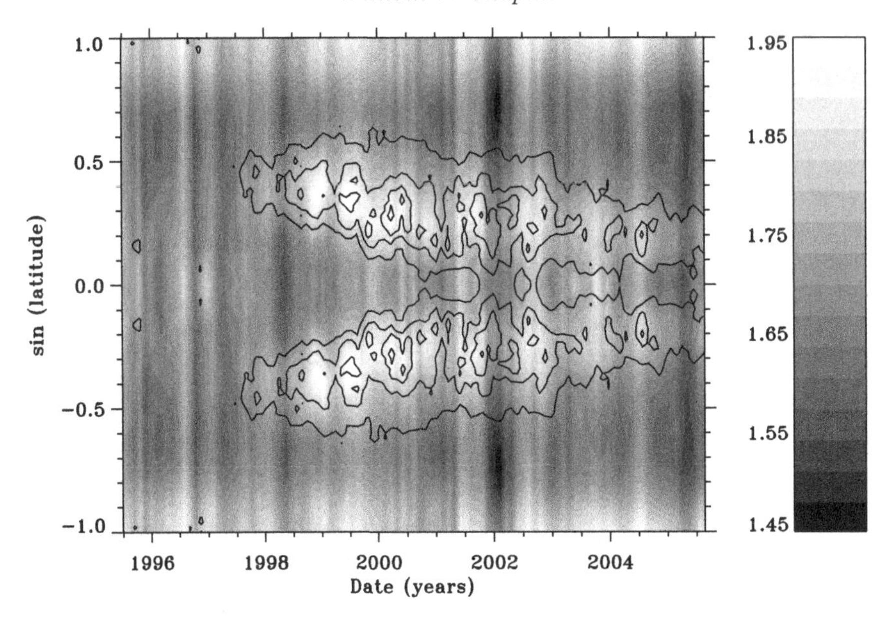

FIG. 1.3. Mode line width (in μHz) as a function of time and latitude. The values come from analysis of GONG data. The contour lines indicate the surface magnetic activity. (Courtesy of R. Howe.)

show spatial variations in the strength of the surface magnetic field over time. The implication is that the frequency shift of a given mode depends on the strength of that component of the surface magnetic field that has the same spherical harmonic projection on the surface. Similar maps may also be made for variations observed in the p-mode powers and damping rates (Komm *et al.*, 2002), which, like the frequency maps, show a close spatial and temporal correspondence with the evolution of active-region field (Fig. 1.3).

1.2.1 *The seismic solar cycle at low angular degree*

As noted above, the first evidence for activity-related changes to the low-l p modes of the Sun was reported by Woodard and Noyes (1985). These changes were soon confirmed (and the results extended) by Pallé *et al.* (1989) and Elsworth *et al.* (1990). Almost two decades on, there now exists an extensive literature devoted to studies of low-l mode parameter variations. Since the frequencies are by far the most precisely determined parameters, it is hardly surprising that analyses of the frequencies threw up the first positive results for solar-cycle changes. Evidence for changes in mode power followed next (Pallé *et al.*, 1990a; Anguera Gubau *et al.*, 1992; Elsworth *et al.*, 1994). The first tentative claims for changes in mode line width (damping) were made by Pallé *et al.* (1990b). Subsequently, Toutain and Wehrli (1997) and Appourchaux (1998) provided stronger evidence in support of an increase in damping with activity, and these claims were confirmed beyond all doubt by Chaplin *et al.* (2000). (Komm *et al.*, 2000, did likewise for medium l modes at about the same time.) Peak asymmetry is the most recent addition to the list of parameters that show solar-cycle variations (Jiménez-Reyes *et al.*, 2007). Careful measurement of variations in the powers, damping rates, and peak asymmetries – all parameters associated with the excitation and damping – are allowing studies to be made of the impact of the solar cycle on the convection properties in the near-surface layers.

Frequencies of modes below $\approx$4,000 μHz are observed to increase with increasing activity. For the most prominent modes at around 3,000 μHz, the size of the shift is about 0.4 μHz between activity minimum and maximum. Furthermore, higher-frequency modes experience a larger shift than their lower-frequency counterparts. This frequency dependence suggests that the perturbations responsible for the frequency shifts are located very close to the solar surface. The upper turning points of the modes – which for low-degree modes are effectively independent of l – lie deeper in the Sun for low-frequency modes than they do for high-frequency modes: higher-frequency modes are as such more sensitive to surface perturbations.

That the shifts do not scale like ν_{nl}/L, where ν_{nl} is mode frequency and $L = \sqrt{l(l+1)}$, rules out the possibility that the perturbation is spread throughout a significant fraction of the solar interior. This may be understood by thinking classically, in terms of ray paths followed by the acoustic waves. When a wave reaches the lower turning point of its cavity it will by definition be moving horizontally, and its phase speed will be equal to

$$c = \frac{\omega_{nl}}{k} = \frac{2\pi\nu_{nl}R}{\sqrt{l(l+1)}} \propto \nu_{nl}/L, \tag{1.1}$$

where k is the horizontal wavenumber and R the outer cavity radius. The ratio ν_{nl}/L therefore maps to the location of the lower boundary of the cavity, and hence the cavity size. Since the shifts do not scale like this ratio, we conclude that the perturbation must be confined within a narrow layer, and not spread so widely that it covers the entirety of the cavities of many of the modes.

At frequencies above 4,000 μHz the size of the shift decreases, and also changes sign above $\simeq$4,500 μHz, meaning that these very high-frequency modes suffer a reduction in frequency as activity levels rise (Anguera Gubau *et al.*, 1992; Chaplin *et al.*, 1998; Jiménez-Reyes *et al.*, 2001; Gelly *et al.*, 2002; Salabert *et al.*, 2004).

Detailed comparison of the low-l frequency shifts with changes in various disc-averaged proxies of global surface activity provides further tangible input to the solar cycle studies. This is because different proxies show differing sensitivity to various components of the surface activity. While the changes in frequency are observed to correlate fairly well with contemporaneous changes in global proxies, the match is far from perfect. Jiménez-Reyes *et al.* (1998) were the first to show that the relation of the frequency shifts to variations in the proxies was markedly different on the rising and falling parts of the 11-year Schwabe cycle. Since the magnitudes of the shifts should reflect the different spatial sensitivities of modes of different angular and azimuthal degree to the time-dependent variation of the surface distribution of the activity, a better choice for the activity proxy would clearly be one that has been decomposed to have a similar spatial distribution as the mode under study. Chaplin *et al.* (2004a) and Jiménez-Reyes *et al.* (2004) have shown that the sizes of the low-l shifts do indeed scale better with activity proxies that have the same spherical harmonic projection as the modes.

Chaplin *et al.* (2007b) compared frequency changes in 30 years of BiSON data with variations in six well-known activity proxies. Interestingly, they found that only activity proxies having good sensitivity to the effects of weak-component magnetic flux – which is more widely distributed in latitude than the strong flux in the active regions – were able to follow the frequency shifts consistently over the three cycles.

The unusual behavior during the most recent solar minimum (straddling solar cycles 23 and 24) of many diagnostics and probes of solar activity has raised considerable interest and debate in the scientific community (e.g., see the summary by Sheeley, 2010). The minimum was unusually, and unexpectedly, extended and deep. Polar magnetic fields were very weak, and the open flux was diminished compared to other preceding minima.

Helioseismology has been used to probe the behavior of sub-surface flows during the minimum. Howe *et al.* (2009) found that the equatorward progression of the lower branches of the so-called torsional oscillations (east–west flows) was late in starting compared to previous cycles. They flagged this delayed migration as a possible precursor of the delayed onset of cycle 24. The meridional (north–south) flow also carries a signature of the solar cycle, which converges toward the active-region latitudes and also intensifies

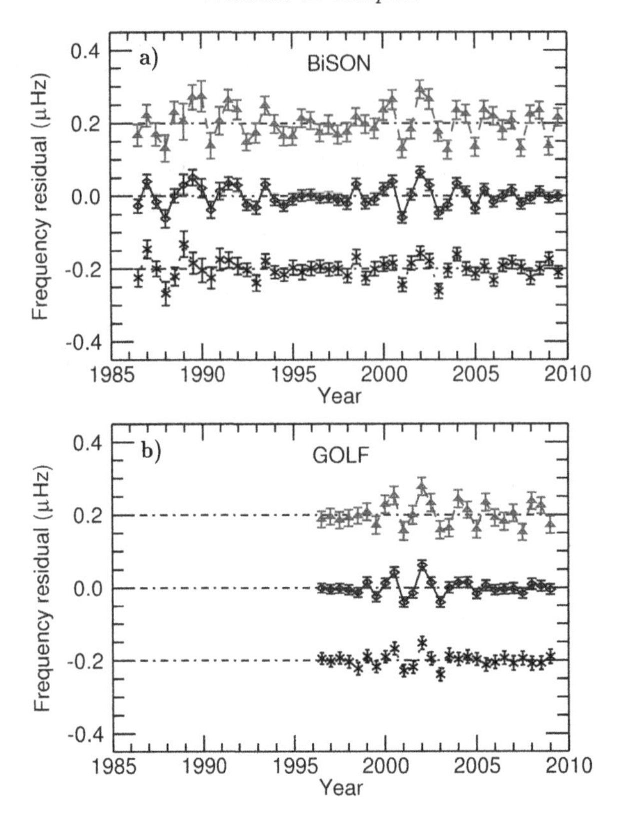

FIG. 1.4. Frequency residuals that remain after the long-term solar-cycle variation have been removed from the average frequency shifts of: *a)* BiSON data and *b)* GOLF data. The three curves in each panel show results for averages made over different ranges in frequency: 1,880 to 3,710 μHz (diamonds, joined by solid line); 1,880 to 2,770 μHz (crosses, joined by dotted line); 2,820 to 3,710 μHz (triangles, joined by dashed line). (Plot from Fletcher *et al.*, 2010.)

in strength as activity increases: González Hernández *et al.* (2010) found that during the current minimum this component had developed to detectable levels even before the visual onset of magnetic activity on the solar surface.

The globally coherent acoustic properties of the recent solar minimum have been studied extensively with low-degree p modes (e.g., Broomhall *et al.*, 2009; Salabert *et al.*, 2009) and medium-degree p modes (Tripathy *et al.*, 2010). These studies have shown that while the surface proxies of activity (e.g., the 10.7-cm radio flux) were quiescent and very stable during the minimum, the p-mode frequencies showed much more variability. Tripathy *et al.* (2010) noted further surprising behavior compared to the previous cycle 22–23 minimum, that is, an apparent anticorrelation of the p-mode frequency shifts and the surface proxies of activity.

Broomhall *et al.* (2009) had suggested the possible presence of a quasi-biennial modulation of the frequencies of the low-degree modes, superimposed upon the well-established ~11-year variation of the frequencies. This has since been confirmed by further in-depth analysis, which reveals a signature that is consistent in the frequencies extracted from BiSON and GOLF data (Fletcher *et al.*, 2010), as shown in Fig. 1.4. This plots the frequency residuals that remain after the long-term solar-cycle variation has been removed from the average frequency shifts. The residuals show significant variability, with a period of around 2 years, that variability being more pronounced at times of higher surface activity. The fact that this biennial signature has similar amplitude in the low-frequency *and* the high-frequency modes used in this analysis suggests that its origins lie deeper than the very superficial layers responsible for the 11-year shifts.

Finally in this section on the frequencies, we note that from appropriate combinations of low-l frequencies Verner *et al.* (2006) uncovered apparent solar-cycle variations in the amplitude of the depression in the adiabatic index, Γ_1, in the He II zone. These variations presumably reflect the impact of the changing activity on the equation of state of the gas in the layer, and confirmed the findings of Basu and Mandel (2004), who used the more numerous data available from medium-l frequencies.

1.2.2 *What is the cause of the frequency shifts?*

Broadly speaking, the magnetic fields can affect the modes in two ways. They can do so directly, by the action of the Lorentz force on the gas. This provides an additional restoring force, the result being an increase of frequency, and the appearance of new modes. Magnetic fields can also influence matters indirectly, by affecting the physical properties in the mode cavities and, as a result, the propagation of the acoustic waves within them. This indirect effect can act both ways, to either increase or decrease the frequencies.

We begin this section with a back-of-the-envelope calculation, which makes an (admittedly) approximate prediction of the impact on the frequency of a typical low-l mode of changes in stratification brought about by the near-surface magnetic field in sunspots. The calculation is instructive, in that the size of the predicted frequency shift it gives is broadly in line with the observations, providing further evidence in support of the shifts being caused by near-surface perturbations due to the presence of magnetic field.

As we have seen, frequencies of the most prominent modes increase with increasing magnetic activity. Here, we consider the indirect effect of the near-surface magnetic field on the near-surface properties, and hence the frequencies of these modes. If magnetic fields modify the surface properties in such a way as to reduce the effective size of the mode cavity, the required increase of frequency will result.

In regions of strong magnetic field – such as those occupied by sunspots – there will be a gas pressure deficit (assuming those regions to be in pressure equilibrium with their field-free surroundings). This is because gas and magnetic pressure combine within the magnetic regions, while only gas pressure acts in the field-free regions. Sunspots are also characterized by a reduced temperature relative to the surroundings. The central part of the spot therefore exhibits a lower pressure, temperature, and density than the surroundings, resulting in the so-called Wilson Depression. Values of pressure, temperature, and density found at the surface in the field-free plasma are only reached at some depth beneath the surface in strong-field regions. Recent measurements suggest that for a sunspot the typical size of this depression is about 1,000 km (Watson *et al.*, 2009).

Let us assume that 1,000 km corresponds approximately to the amount, δR, by which the mode cavities are reduced in size beneath sunspots. The fractional area of the solar surface occupied by sunspots reaches $\sim$0.5 % at modern cycle maxima. We therefore obtain a net, surface-averaged estimate of δR via

$$\langle \delta R \rangle \approx 0.5\,\% \times 1000\,\mathrm{km} \approx 5\,\mathrm{km}. \tag{1.2}$$

The sound speed, c, at the solar surface is approximately $10\,\mathrm{km\,s^{-1}}$, implying a reduction in the travel time in the mode cavity, due to the "shrinkage" at the surface, of

$$\delta T \approx 5\,\mathrm{km}/10\,\mathrm{km\,s^{-1}} \approx 0.5\,\mathrm{s}. \tag{1.3}$$

The travel time across the cavity, T, is related to the large frequency separation, $\Delta\nu$, via:

$$\Delta\nu \simeq \left(2 \int_0^R \frac{dr}{c} \right)^{-1} \simeq (2T)^{-1}. \tag{1.4}$$

For the Sun, $\Delta\nu = 135\,\mu$Hz, implying $T \approx 3700$ s. The fractional change (reduction) in T is therefore:

$$\delta T/T \simeq 0.5/3700 \simeq 1.4 \times 10^{-4}. \quad (1.5)$$

Since $\delta\nu/\nu = -\delta T/T$, the predicted increase in frequency of a mode at $\approx 3{,}000\,\mu$Hz will be

$$\delta\nu = 3000 \times 1.4 \times 10^{-4} \simeq 0.4\,\mu\text{Hz}. \quad (1.6)$$

This estimate is of a very similar size to the observed frequency shifts.

What does detailed modelling suggest? Perhaps the most significant contribution of recent years in this area is that of Dziembowski and Goode (2005). Their results suggest that the indirect effects dominate the perturbations, and that the magnetic fields are too weak in the near-surface layers for the direct effect to contribute significantly to the observed frequency shifts. However, Dziembowski and Goode (2005) also found some evidence to suggest that the direct effect may play a more important rôle for low-frequency modes, at depths beneath the surface where the magnetic field is strong enough to give a significant direct contribution to the frequency shifts. We now go on to explain how dependence of the frequency shifts on mode inertia and mode frequency can tell us something about the location and nature of the perturbations.

We begin by noting that when the frequency shifts are multiplied by the mode inertia, and then normalized by the inertia of a radial mode of the same frequency, the modified shifts are found to be a function of frequency alone. This in effect removes any l dependence of the shifts (at fixed frequency, the higher the l, the larger is the observed frequency shift). The observed l dependence may be understood in terms of, for example, a physical interpretation of the mode inertia. The normalized inertia, I_{nl}, may be defined according to (Christensen-Dalsgaard and Berthomieu, 1991):

$$I_{nl} = \text{M}_\odot^{-1} \int_V |\xi|^2 \rho dV = 4\pi \text{M}_\odot^{-1} \int_0^R |\xi|^2 \rho r^2 dr = M_{nl}/\text{M}_\odot \quad (1.7)$$

where ξ is the (surface-normalized) displacement associated with the mode, and the integration is performed over the volume V of the Sun, which has mass $\text{M}_\odot$. The mode mass M_{nl} is therefore the interior mass affected by the perturbations associated with the mode. As l increases, so M_{nl} decreases, and the more sensitive a mode will be to a near-surface perturbation of a given size (giving a larger frequency shift). One may therefore render the shifts l independent by multiplying them by the inertia ratio Q_{nl} (Christensen-Dalsgaard and Berthomieu, 1991), which is given by:

$$Q_{nl} = I_{nl}/\bar{I}(\nu_{nl}). \quad (1.8)$$

Multiplication of the raw shifts by Q_{nl} is indeed seen to collapse the shifts of different l onto a single curve. As shown in Fig. 1.5, this then allows one to combine data spanning a range in l, which reduces errors, giving tighter constraints on frequency dependence of the shifts.

Chaplin *et al.* (2001) studied in detail the frequency dependence of the inertia-ratio-corrected shifts, $\delta\nu_{nl}Q_{nl}$, of both low-l modes and medium-l modes up to $l = 150$. They fitted these data to a power law of the form

$$\delta\nu_{nl}Q_{nl} = \frac{c}{I_{nl}}\nu_{nl}^{\alpha}, \quad (1.9)$$

where the power-law index is α and c is a constant. They found that $\alpha \simeq 2$ for $\nu \geq 2{,}500\,\mu$Hz, while α is approximately zero for $\nu < 2{,}500\,\mu$Hz. Rabello-Soares *et al.* (2008) repeated the exercise for medium-l and high-l modes. They extracted very similar behavior, and, because they had more medium- and high-l data at low frequencies, they

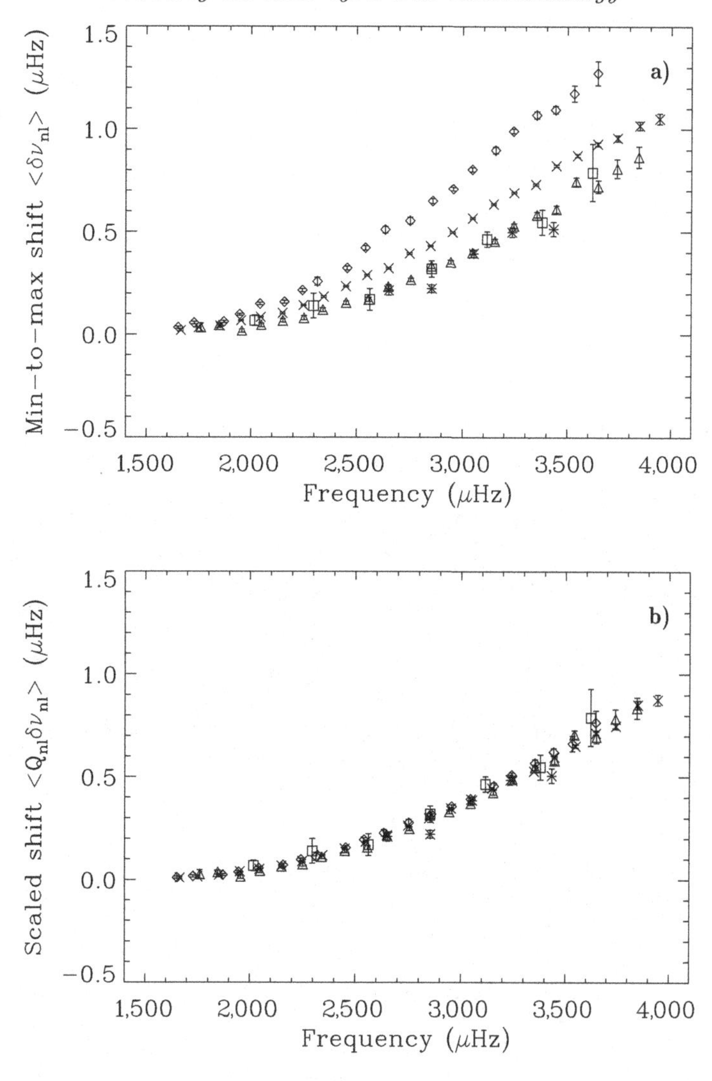

FIG. 1.5. *a)* Frequency shifts, $\delta\nu$, plotted as a function of mode frequency, for results from different instruments over different ranges in l (diamonds: GONG, l = 100 to 150; crosses: GONG, l = 30 to 90; triangles: GONG, l = 4 to 25; squares: BiSON, l = 0 to 2; asterisks: LOI, l = 0 to 8). *b)* Scaled frequency shifts, after multiplication by the inertia ratio Q_{nl}. (From Chaplin *et al.*, 2001.)

were able to extend the analysis to $\nu < 2{,}000\,\mu$Hz where they found that α appears to change sign and go negative. The high-l modes provide important information, as they are confined in the layers close to the surface where the physical changes responsible for the frequency shifts are also located.

What do the values of α imply? If the perturbation is located within the photosphere (but is confined to extend over an extent no longer than one pressure scale height), then we might expect $\alpha \simeq 3$ (e.g., Libbrecht and Woodard, 1990). If instead the perturbation extends beneath the surface, the frequency dependence will be weaker and α will get smaller (Gough, 1990). The observed values of α in the abovementioned frequency regimes are both less than 3, suggesting that the perturbations cannot arise solely in the photosphere. Since α is smaller for the lower-frequency modes, this suggests that the perturbation extends to greater depths, the lower in frequency one goes. This is consistent with the inferences made by Dziembowski and Goode (2005).

1.2.3 *Solar-cycle variations of mode power and mode damping*

Variations in mode damping have now been uncovered in all the major low-l datasets (see Chaplin, 2004, and references therein). The trend found is an increase of about 20 percent between activity minimum and maximum, with some suggestion of the variations being peaked in size at about $\simeq$3,000 μHz. Mode powers are at the same time observed to decrease by about the same fractional amount, while the mode heights decrease by twice the amount. Use of the analogy of a damped, randomly forced oscillator is particularly instructive for understanding these results.

We define our oscillator according to the equation:

$$\ddot{x}(t) + 2\eta\dot{x}(t) + (2\pi\nu_0)^2 x(t) = K\delta(t - t_0), \tag{1.10}$$

where $x(t)$ is the displacement, ν_0 the natural frequency of the oscillator, η the linear damping constant, and K is the amplitude of the forcing function (assumed to be a random Gaussian variable), with "kicks" applied at times t_0, $\delta(t - t_0)$ being the delta function.

Provided that $\mathcal{F}(\nu)$ – the frequency spectrum of the forcing function – is a slowly varying function of ν, and $\eta \ll 2\pi\nu_0$, the power spectral density (PSD) in the frequency domain will be a Lorentzian profile, that is,

$$\mathrm{PSD}(\nu) \propto \mathcal{F}(\nu)\left(1 + \left(\frac{\nu - \nu_0}{\eta/2\pi}\right)^2\right)^{-1}. \tag{1.11}$$

This holds for both the spectrum of the displacement, $x(t)$, and the velocity, $\dot{x}(t)$. The FWHM of the peak in cyclic frequency is

$$\Delta = \frac{\eta}{\pi}. \tag{1.12}$$

The maximum power spectral density, or height, of the resonant peak is:

$$H \propto \frac{\mathcal{F}(\nu)}{\eta^2}. \tag{1.13}$$

The total mean-square power (variance in the time domain) is proportional to peak height times width, that is, $P \propto H\Delta$, so that

$$P \propto \frac{\mathcal{F}(\nu)}{\eta}. \tag{1.14}$$

The energy (kinetic plus potential) of a resonant mode with associated inertia I is given by:

$$E = MP. \tag{1.15}$$

The rate at which energy is supplied to (and dissipated by) the modes, dE/dt, is readily derived by again having recourse to the oscillator analogy. The amplitude of the oscillator is attenuated in time by the factor $\exp(-\eta t)$, and its energy is proportional to the amplitude squared. Hence, we may write

$$E = (\mathrm{constant}) \times \exp(-2\eta t). \tag{1.16}$$

It follows that:

$$\log E = -2\eta t + \log(\mathrm{constant}), \tag{1.17}$$

and taking derivatives:

$$dE/dt = -2\eta E. \tag{1.18}$$

If we combine Equations 1.14, 1.15, and 1.18, we have:

$$dE/dt = \dot{E} \propto -\mathcal{F}(\nu)I. \tag{1.19}$$

We may attempt to measure solar-cycle variations in Δ, H, P, E, and $\dot{E}$. The parameters that are usually extracted directly by the observers are the widths (Δ) and heights (H) of the peaks in the frequency power spectrum. The mode powers, P, are then estimated from

$$P = \frac{\pi}{2}\Delta H \tag{1.20}$$

(i.e., area under Lorentzian). The E may be computed from the P using model computed inertias, I. We note that fractional changes in E will be the same as those in P (assuming the I do not show any significant changes). Finally, $\dot{E}$ follows from

$$\dot{E} = -\pi E\Delta = -\pi IP\Delta. \tag{1.21}$$

If solar-cycle variations are uncovered in the above parameters, what do they imply for the underlying cause (or causes)? If we think in terms of the basic oscillator parameters – the damping rate η and the forcing function $\mathcal{F}(\nu)$ – Equations 1.12, 1.13, 1.14, 1.15, and 1.19 then imply the following:

$$\delta\Delta\nu/\Delta\nu = \delta\eta/\eta, \tag{1.22}$$

$$\delta H/H = \delta\mathcal{F}(\nu)/\mathcal{F}(\nu) - 2\delta\eta/\eta, \tag{1.23}$$

$$\delta P/P = \delta E/E = \delta\mathcal{F}(\nu)/\mathcal{F}(\nu) - \delta\eta/\eta, \tag{1.24}$$

and

$$\delta\dot{E}/\dot{E} = \delta\mathcal{F}(\nu)/\mathcal{F}(\nu). \tag{1.25}$$

Observations of p modes at low l are consistent with $\delta\dot{E}/\dot{E} = 0$, which in turn implies that $\delta\mathcal{F}(\nu)/\mathcal{F}(\nu) = 0$: the acoustic forcing of the low-l modes remains, on average, constant over the cycle. Any changes that are observed in the other parameters must therefore arise from $\delta\eta/\eta$ alone, and we have that:

$$\delta\Delta\nu/\Delta\nu = -\delta P/P = -\delta H/2H, \tag{1.26}$$

The observed changes in Δ, P, and H follow these ratios. We may therefore infer that changes to these parameters are in all likelihood the result of changes to the damping of the low-l modes. Houdek *et al.* (2001) noted that the high Rayleigh number of the magnetic field could modify the preferred horizontal length scale of the convection, which in turn could modify the damping rates. Their numerical calculations of the implied changes for low-l modes show reasonable agreement with the observations.

1.3 Subtle signatures of stellar activity and stellar-cycle variations

Asteroseismic observations of stars will, for the foreseeable future, be limited to obtaining data on low-l modes because of the cancellation effects resulting from unresolved observations of stellar discs. Moreover, the angle of inclination, i, offered by a star also determines which of the low-l components will have nonnegligible visibility in the data. The visibility of the radial ($l = 0$) modes is not affected by i, but for the non-radial modes the visibility, $\mathcal{E}(i)_{lm}$ (in power) of a given (l, m) component is assumed to be (e.g., Gizon and Solanki, 2003):

$$\mathcal{E}(i)_{lm} \propto \frac{(l-|m|)!}{(l+|m|)!}\left(P_l^{|m|}(\cos i)\right)^2, \tag{1.27}$$

where $P_l^{|m|}$ are Legendre polynomials. The fact that some components may in effect be missing from the data has important implications for the analysis. There are some obvious implications, for example, it will not be possible to measure frequency splittings of non-radial modes if i is close to 90 degrees – so that the star is viewed rotation-pole on – since only the zonal components ($m = 0$) will have nonnegligible visibility. In this section we highlight a more subtle aspect, which relates to the fact that estimates of the frequencies of non-radial modes are affected by the interplay between the effects of near-surface magnetic activity and the relative visibility (or absence) of the mode components. This has consequences for inferences made on solar-type stars. We first introduce and explain the problems using "Sun-as-a-star" helioseismic observations as a test case, before considering the wider implications for asteroseismology. We return, in Section 1.4 to discuss the implications for estimation of average frequency shifts of solar-type stars.

1.3.1 *Explanation and description of frequency bias*

Extant Sun-as-a-star observations are made from a perspective in which the plane of the rotation axis of the Sun is nearly perpendicular to the line-of-sight direction (i.e., i is always close to 90 degrees). This means that only components with $l + m$ even have nonnegligible visibility. As we shall explain below, the impact of surface magnetic activity on the arrangement of the components in frequency means it may not then be possible to estimate the true frequency centroid of the multiplet. The frequency centroids carry information on the spherically symmetric component of the internal structure, and are the input data that are required, for example, for hydrostatic structure inversions.

In the complete absence of the near-surface activity, the $l + m$ odd components that are "missing" from the Sun-as-a-star data would be an irrelevance. All mode components would be arranged symmetrically in frequency, meaning centroids could be estimated accurately from the subset of visible components. A near-symmetric arrangement is found at the epochs of modern solar-cycle minima (Chaplin *et al.*, 2004b). However, when the observations span a period having medium to high levels of activity – as a long dataset by necessity must – the arrangement of components will no longer be symmetric. The frequencies given by fitting the Sun-as-a-star data will then differ from the true centroids by an amount that is sensitive to l. The l dependence arises because in the Sun-as-a-star data, modes of different l comprise visible components having different combinations of l and m; and these different combinations show different responses (in amplitude and phase) to the spatially nonhomogeneous surface activity (i.e., as observed in the active bands of latitude for solar-type stars).

These effects are illustrated in Figs. 1.6 and 1.7. Figure 1.6 is a schematic plot of the arrangement in frequency of the components of an $l = 2$ mode, under circumstances where activity levels are assumed to be negligible. The top panel shows the arrangement in frequency of the various m (as labeled on the ordinate), assuming a rotational frequency splitting of 0.4 μHz between adjacent m. The dotted line marks the frequency centroid. The middle panel shows schematically both the placement in frequency (abscissa) and relative visibility in the frequency power spectrum (height on ordinate) of the three mode components that would have nonnegligible visibility in the Sun-as-a-star data. The bottom panel shows the Lorentzian peak profiles expected of a mode at the center of the p-mode spectrum (where peak line widths are around 1 μHz).

Figure 1.7 shows the situation at an epoch coinciding with high levels of solar activity. This activity is most prominent in the active latitudes. The $|m| = 2$ components are more sensitive to acoustic perturbations due to activity that arises in these latitudes than is the $m = 0$ component (as a result of the spatial distribution of the relevant spherical harmonic components). The result is a distortion of the symmetric arrangement in frequency of the multiplet, because the different $|m|$ experience a frequency shift of different size.

The nonsymmetric arrangement of the components is very evident in Fig. 1.7. Two lines have been plotted in the middle panel, which were not shown in Fig. 1.6. The solid line marks the (unweighted) average frequency of the components showing nonnegligible

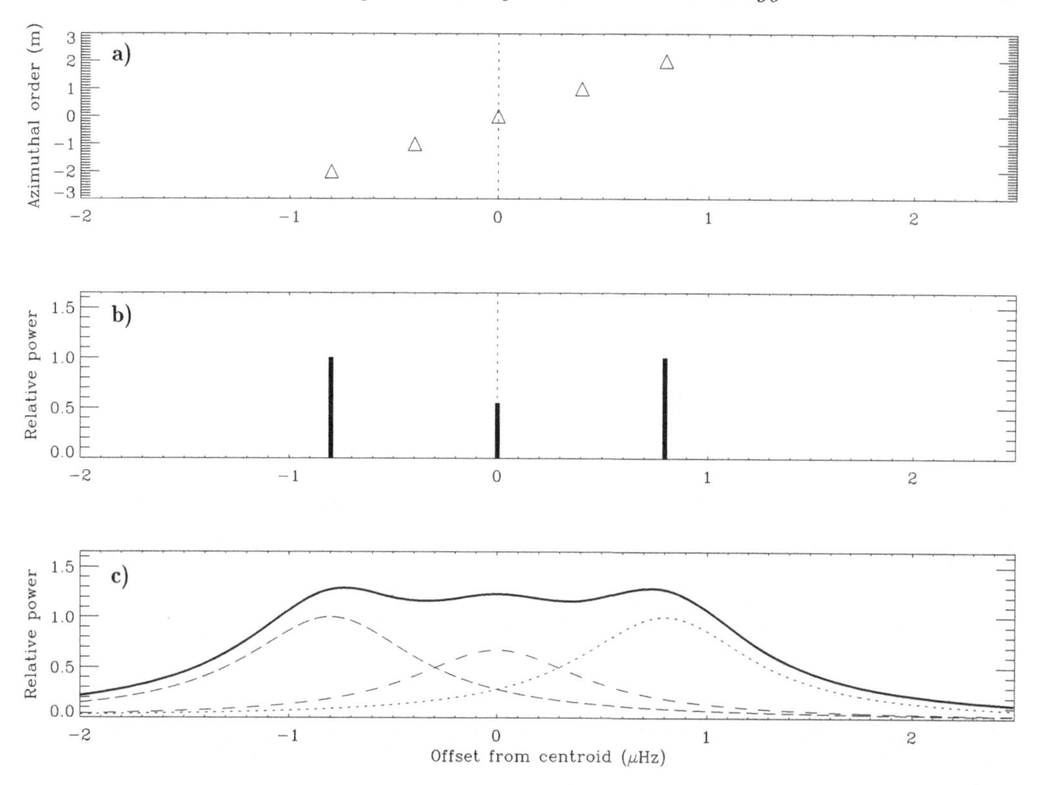

FIG. 1.6. *a)* Placement in frequency, relative to the frequency centroid of all $2l + 1$ components (abscissa), of the various m (labeled on ordinate) of an $l = 2$ mode during an epoch of negligible solar surface activity. The solid vertical line marks the location of the centroid. *b)* Placement in frequency (abscissa) and relative visibility in the frequency power spectrum (height on ordinate) of the three mode components that would have nonnegligible visibility in the Sun-as-a-star data. *c)* Lorentzian peak profiles expected of an $l = 2$ mode at the center of the p-mode spectrum (where peak line widths are around $1\,\mu$Hz).

visibility. It is clearly offset, or biased, by about $0.07\,\mu$Hz, from the centroid (dotted line). The dashed line marks the actual, expected location of the Sun-as-a-star frequency (offset from the centroid by more than $0.1\,\mu$Hz). It differs from the unweighted average of the three visible components because the "peak bagging" procedures used to estimate the mode frequencies are influenced by the relative heights of the different m. Here, the outer $|m| = 2$ components carry greater weight in determining the fitted $l = 2$ frequency than does the weaker $m = 0$ component, thereby shifting the estimated frequency from the unweighted average.

It can be quite hard to estimate, to any reasonable accuracy and precision, the locations in frequency of each component individually (which, as noted earlier, would help to reduce the offset between the estimated frequency and the true centroid because it would be possible to compute the unweighted average frequency). Frequency estimation must contend with the fact that within the non-radial multiplets the various m lie in such close frequency proximity to one another that suitable models that seek to describe the characteristics of the m present must usually be fitted to the components simultaneously.

Is there any way to get around the problem, and in some way "correct" the Sun-as-a-star frequencies for the frequency bias? Two methods have been applied successfully. The first relies on the availability of contemporaneous resolved-Sun helioseismic data – which are sensitive to all m components – to in effect map the strength and spatial distribution of the acoustic asphericity arising from the surface activity over the epoch in question. Chaplin *et al.* (2004b) and Appourchaux and Chaplin (2007) show how

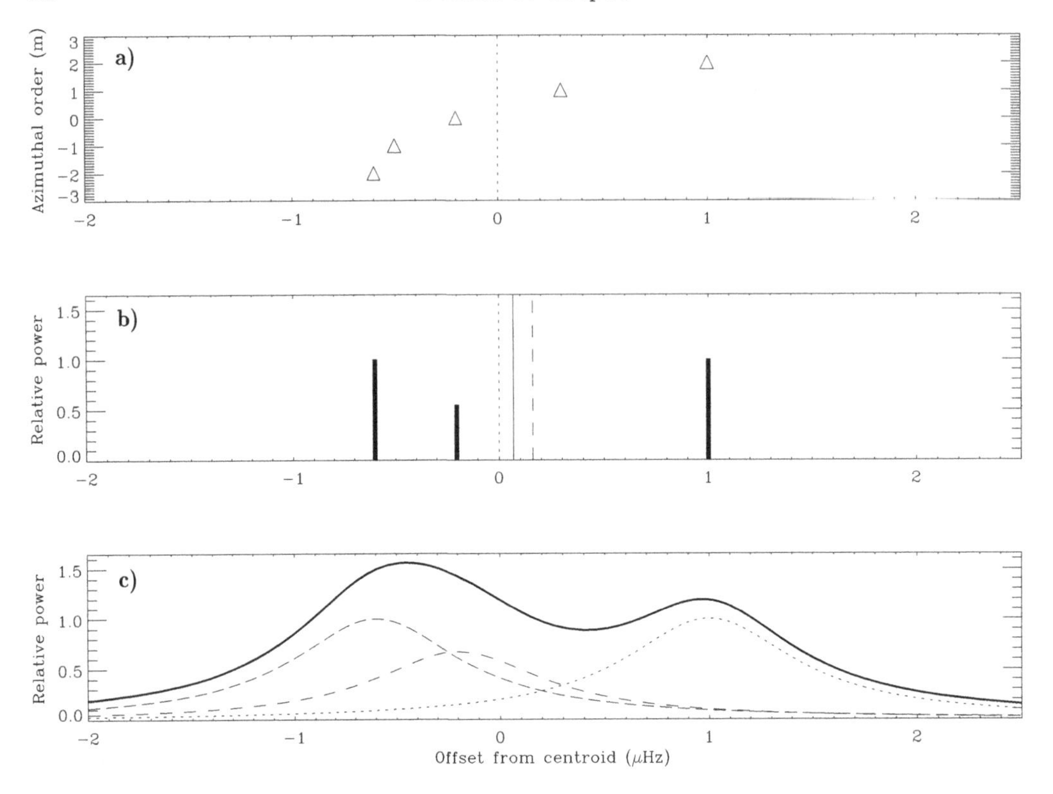

FIG. 1.7. Same as Fig. 1.6, but during an epoch with activity levels corresponding to the maximum of a typical (modern) solar cycle. The solid and dashed lines in the middle panel mark estimates of the $l = 2$ frequency that might be derived from the Sun-as-a-star data.

to make the correction, using the so-called "even a coefficients" from fits to the resolved-Sun frequencies (the a coefficients are discussed later in this section). This approach is, of course, currently not viable for other solar-type stars. Fortunately, the second method in principle is: here, the same dataset that is used to produce the frequency estimates is also used to help calibrate the corrections that are needed to "clean" the frequencies. Implicit in the procedure is the assumption that at epochs coinciding with cycle minima, any frequency asymmetry in the mode multiplets is negligible. This assumption turns out to be reasonable for the Sun, as noted previously (Chaplin *et al.*, 2004b).

The second method is described at length in Broomhall *et al.* (2009). In sum, the long dataset is divided into smaller subsets, and the frequencies of each subset are estimated. Tried and trusted procedures are then used to measure the frequency shifts of different modes from one subset to another. The frequency shift of any given mode may be measured relative to its average frequency over the long dataset, or, for example, its frequency in the subsets (or subset) that coincide(s) with the cycle minimum (the choice has subtle consequences for propagation of the frequency uncertainties, e.g., see Chaplin *et al.*, 2007b). It is common practice, at least for the low-l data, to average shifts over several modes to reduce uncertainties in the shifts. By making an average over many orders in n, one can track the average frequency shift from subset to subset. By making an average over a few (say, two or three) orders in n it is possible to constrain the dependence of the shifts on frequency (or n).

In the next stage of the procedure the frequency shifts are parametrized. Sun-as-a-star procedures have parametrized the shifts as a linear function of one of the global proxies of magnetic activity, the 10.7-cm radio flux (Tapping and Detracey, 1990) being a favored choice. The mode-averaged frequency shifts are fitted to a linear function of the 10.7-cm

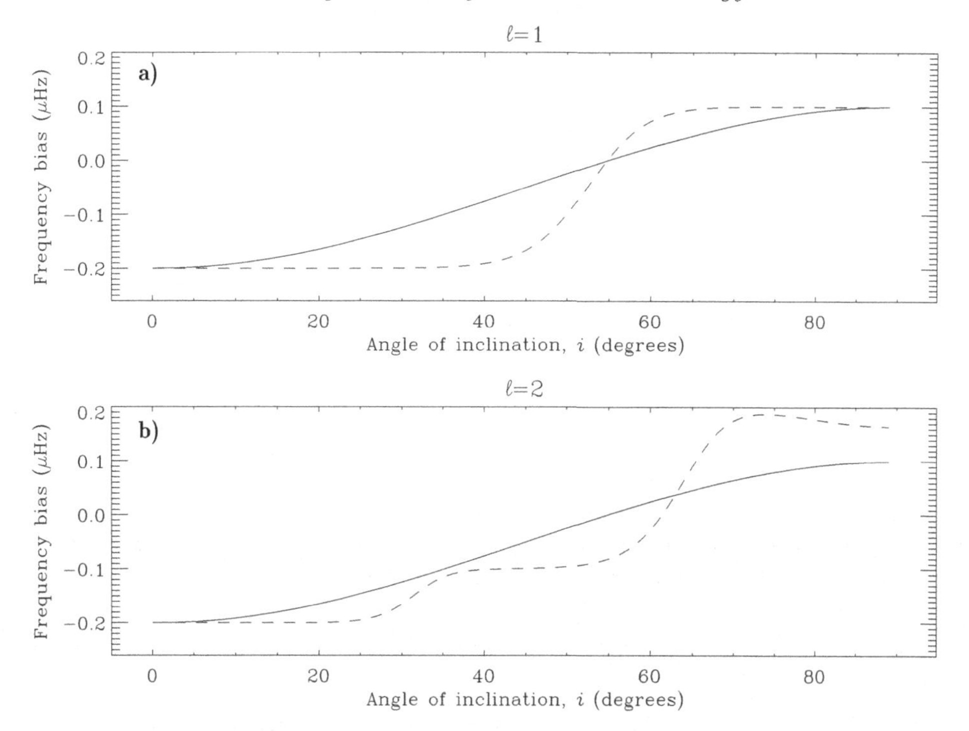

FIG. 1.8. Expected frequency bias for Sun-as-a-star observations made for different angles of inclination, i (see text for explanations of solid and dashed lines).

flux (averaged over the same periods as the seismic data). This linear model then suffices to predict the expected average frequency shift for any epoch, given the 10.7-cm flux for that period. Expected shifts for individual modes are then computed using a best-fitting polynomial, which describes the measured frequency shifts as a function of frequency, relative to the average frequency shift. The predicted frequency shifts may then be subtracted from the raw frequencies (the latter measured from the full time series) to yield a set of corrected frequencies, commensurate with negligible surface activity. The procedure yields frequencies that are not subject to the bias discussed above, since that bias is assumed, *by definition*, to be missing at low activity.

Application of the procedure, verbatim, to other stars would, of course, demand that contemporaneous activity data be available (e.g., chromospheric H & K data). Such data may be hard to obtain in sufficient quantities on a large number of stars. One might instead use measures of variability in the time series – arising from rotational modulation of surface activity – to form a suitable proxy (Basri *et al.*, 2010). It may instead be possible to parameterize the frequency shifts by fitting nonlinear functions in time (i.e., to describe the observed shifts as functions of time), hence circumventing the need to have complementary activity data.

How large might the frequency bias be for other stars? The magnitude of the bias is governed to first order by the size of the acoustic asphericity, arising from the magnetic activity, and the angle of inclination of the star. Second-order effects come in through the interplay between the mode properties and the peak-bagging procedures used to estimate the frequencies (e.g., see discussion in Ballot *et al.*, 2008, on correlations between different parameters).

Figure 1.8 shows the predicted bias from the first-order effects. Here, we assumed an acoustic asphericity similar to that observed on the Sun, as experienced by $l = 1$ and $l = 2$ modes at the center of the solar p-mode spectrum. The predictions therefore in principle show the bias that would be expected in the most prominent low-l modes if the

Sun were to be observed at different angles of inclination, i. Two predictions are plotted in each panel. We first explain how we derived the solid-line prediction (using the $l = 2$ case as an example).

The full set of m frequencies of a mode may be described by a polynomial expansion of the form

$$\nu_{nlm} = \nu_{\text{cen}} + \sum_{j=1}^{j=2l} a_j(n,l) l \mathcal{P}_l^j(m), \tag{1.28}$$

where ν_{cen} is the frequency centroid of the mode, and $\mathcal{P}_l^j(m)$ are polynomials related to the Clebsch-Gordan coefficients (Ritzwoller and Lavely, 1991). An expansion of this type is commonly used to fit the frequencies observed in resolved-Sun data, where all m components are available. The even a_j describe perturbations that are nonspherically symmetric in nature, for example, the acoustic asphericity from activity, while the odd a_j coefficients describe spherically symmetric contributions to the frequency splittings (rotation). Here, for simplicity, we assume that for the low-l modes the a_2 term dominates other terms in the description of the asphericity, and that the a_1 term dominates other terms in the description of the rotation. The required $\mathcal{P}_l^j(m)$ are:

$$\mathcal{P}_l^1(m) = m/l \tag{1.29}$$

and

$$\mathcal{P}_l^2(m) = \frac{6m^2 - 2l(l+1)}{6l^2 - 2l(l+1)}. \tag{1.30}$$

We may then write the frequencies of each of the components explicitly as:

$$\nu_{n20} = \nu_{\text{cen}} - 2a_2, \tag{1.31}$$

and

$$\nu_{n21} = \nu_{\text{cen}} + a_1 - a_2, \tag{1.32}$$

$$\nu_{n2-1} = \nu_{\text{cen}} - a_1 - a_2, \tag{1.33}$$

and

$$\nu_{n22} = \nu_{\text{cen}} + 2a_1 + 2a_2, \tag{1.34}$$

$$\nu_{n2-2} = \nu_{\text{cen}} - 2a_1 + 2a_2. \tag{1.35}$$

The solid line prediction in Fig. 1.8 assumes that the estimated frequency of the mode is the weighted combination of the individual frequencies, where the weights are proportional to the relative heights of the components, $\mathcal{E}(i)_{lm}$ (see Equation 1.27). We weight in this way because the estimated frequency is influenced by the relative heights of the different m. We then have:

$$\nu_{n2} = \left(\sum_{m=-2}^{m=+2} \mathcal{E}(i)_{2m} \times \nu_{n2m} \right) \bigg/ \left(\sum_{m=-2}^{m=+2} \mathcal{E}(i)_{2m} \right). \tag{1.36}$$

The solid lines in Fig. 1.8 show $\nu_{nl} - \nu_{\text{cen}}$, that is, the predicted frequency bias, as a function of i. To make the prediction we adopt $a_2 = 0.1\,\mu$Hz (typical value at solar maximum). (The results do not depend upon a_1, which cancels when ν_{nl} is computed.)

Results on Sun-as-a-star data, and simulated asteroseismic data on solar-type stars, suggest that the weights are actually nonlinear functions in $\mathcal{E}(i)_{lm}$ (e.g., see Chaplin *et al.*, 2004b). A better description is one for which the weights are proportional to

$\mathcal{E}(i)_{lm}$ raised to some positive power (four is probably a reasonable value). Here, we therefore also make predictions based

$$\nu_{n2} = \left(\sum_{m=-2}^{m=+2} \mathcal{E}(i)^4_{2m} \times \nu_{n2m} \right) \Bigg/ \left(\sum_{m=-2}^{m=+2} \mathcal{E}(i)^4_{2m} \right), \tag{1.37}$$

which gives the dashed lines in Fig. 1.8.

Equations 1.36 and 1.37 may be written in a more general form (i.e., for any given l) as:

$$\nu_{nl} = \left(\sum_{m=-l}^{m=+l} \mathcal{E}(i)^\gamma_{lm} \times \nu_{nlm} \right) \Bigg/ \left(\sum_{m=-l}^{m=+l} \mathcal{E}(i)^\gamma_{lm} \right), \tag{1.38}$$

where γ is a constant.

The results indicate that the bias is least severe at intermediate values of i. The worst case is when $i = 0$ degrees. The nonlinear weighting in $\mathcal{E}(i)_{lm}$ also changes slightly the functional form of the bias in i, for example, in the case of $l = 1$ it extends the range of i over which the bias is more important.

When we obtain long datasets on solar-type stars, frequency uncertainties will drop to levels well below $0.1\,\mu$Hz. In this context, the bias at some angles is by no means insignificant. Moreover, the bias would, of course, be larger on a star with higher-than-solar acoustic asphericity, that is, one might quite reasonably expect to encounter solar-type stars where the bias may be a significant fraction of $1\,\mu$Hz or more. In sum, this source of bias in the frequencies will need to be taken into account in the not-too-distant future, if we are to avoid errors on the inferences drawn on target stars.

1.3.2 *Impact on frequency separation ratios*

Roxburgh and Vorontsov (2003) proposed the use for asteroseismology of the ratio of separations in frequency between low-l p-modes: *Frequency separation ratios* are formed from the large and small frequency separations of the p modes,

$$\Delta\nu_l(n) = \nu_{nl} - \nu_{n-1,l}, \tag{1.39}$$

and

$$d_{l\,l+2}(n) = \nu_{n,l} - \nu_{n-1,l+2}. \tag{1.40}$$

Data on modes in the range $0 \leq l \leq 3$ may be used to make ratios, according to:

$$r_{02}(n) = \frac{d_{02}(n)}{\Delta\nu_1(n)}, \qquad r_{13}(n) = \frac{d_{13}(n)}{\Delta\nu_0(n+1)}. \tag{1.41}$$

Roxburgh and Vorontsov (2003) used an asymptotic method to show that the ratios are very insensitive to conditions in the near-surface layers of a solar-type star. This is an attractive property, as the outer layers are uncertain and hard to model accurately. As such, the ratios in principle offer a clean diagnostic of the internal properties of a star (see also Otí Floranes *et al.*, 2005; Roxburgh, 2005). They are nevertheless potentially sensitive to the acoustic asphericity – as was pointed out by Otí Floranes *et al.* (2005), and discussed at length by Chaplin *et al.* (2005) – which is clearly undesirable if one wishes to suppress any signatures from the near-surface layers.

That the ratios are sensitive to the solar cycle may be understood as follows. For the Sun-as-a-star data, measured frequency shifts of the $l = 2$ modes are larger than those of the neighboring $l = 0$ modes (because the frequencies of these modes are dominated by the $|m| = 2$ components, which are more sensitive to the acoustic perturbations arising in the active latitudes than are the $l = 0$ modes). The measured $d_{l\,l+2}(n)$ will therefore carry a residual signature of the solar cycle, being smaller at higher levels of activity.

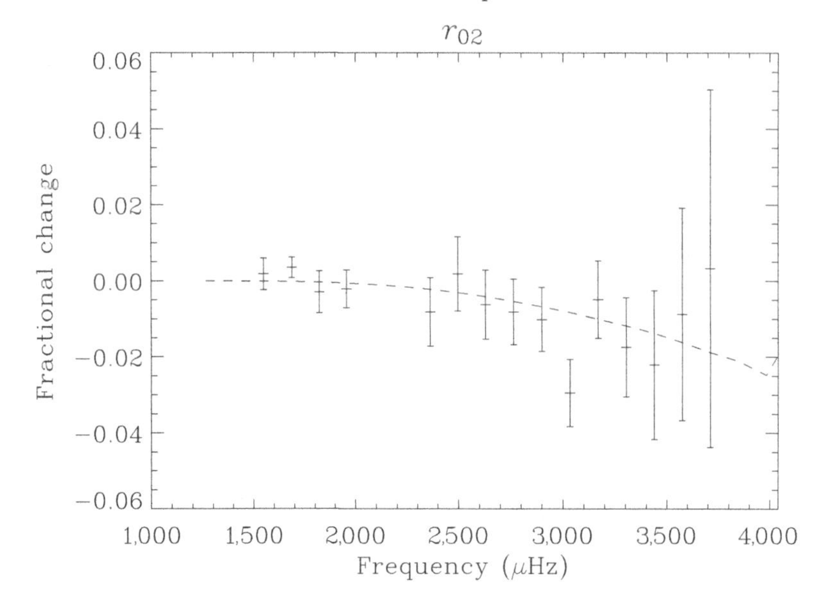

FIG. 1.9. Fractional change in r_{02} – as averaged over the most prominent low-l modes – between two BiSON Sun-as-a-star datasets, one around solar maximum and one around solar minimum (differences in sense maximum minus minimum). (From Chaplin *et al.*, 2005).

Residual solar-cycle changes in $\Delta\nu_l(n)$ are, fractionally, much smaller (but nevertheless measurable; see Broomhall *et al.*, 2011), so that in the case of, for example, $r_{02}(n)$:

$$\delta r_{02}(n)/r_{02}(n) \simeq \delta d_{02}(n)/d_{02}(n). \tag{1.42}$$

For the Sun-as-a-star data, $\delta d_{02}(n)/d_{02}(n) \approx -0.01$ for modes at the center of the p-mode spectrum observed at solar maximum, hence, we would expect the ratios to decrease between solar minimum and solar maximum by about this amount. Figure 1.9 plots measured fractional differences in r_{02}, as averaged over the most prominent low-l modes. The differences were computed between two BiSON Sun-as-a-star datasets, one formed of data from around solar maximum, the other formed of data from around solar minimum.

The potential impact on observations of other solar-type stars will depend on both the nature of the near-surface magnetic fields, and also the angles of inclination offered by the stars. In Section 1.4, we shall develop a simple model of the frequency shifts that will allow us to make these predictions for other stars. We therefore return in that section to the frequency separation ratios.

1.3.3 *Impact on shapes of resonant peaks*

What effect will stellar-cycle shifts in frequency through a long time series have on the underlying shapes of the p-mode peaks, when those peaks are observed in frequency-power spectra made from the full time series? Given a significant shift, one would expect the Lorentzian-like peak shapes to be distorted, with the mode power being spread out in frequency thereby flattening the profiles. Whether or not a frequency shift $\delta\nu$ is significant in this context is determined by the ratio

$$\epsilon = \delta\nu/\Delta, \tag{1.43}$$

that is, the ratio of the shift to the line width of the mode. The larger this ratio, the more distorted a mode peak will be.

Chaplin *et al.* (2008b) derived analytical descriptions of the distorted peak profiles. Since for solar-type stars we expect the modes to be excited and damped on timescales much shorter than that on which any significant change of the mode frequencies is observed, the profiles may be assumed to correspond to the average of all the instantaneous profiles taken at any time t within the full period of observation T. Each instantaneous profile may be described as, for example, a Lorentzian with a central frequency $\nu(t)t$, such that the time-averaged profile is

$$\langle P(\nu)\rangle = \frac{1}{T}\int_0^T \frac{H}{1+\left(\frac{\nu-\nu(t)}{\Delta/2}\right)^2}\,dt, \tag{1.44}$$

where the angled brackets indicate an average over time, and H and Δ are the mode height (maximum power spectral density) and line width, respectively.

Chaplin *et al.* (2008b) derived profiles given by two functions describing the frequency shifts in time: first, the simplest possible function, this being a linear variation over time (which would be relevant for describing the profiles given by observations made on the steepest parts of the rising or falling phases of a stellar cycle); and second, a co-sinusoidal variation, to mimic a full stellar cycle. They neglected the effects of the solar-cycle variations in mode power, height, and width.

A simple linear variation in time may be described by

$$\nu(t) = \nu_0 + \delta\nu\frac{t}{T}, \tag{1.45}$$

where ν_0 is the unperturbed frequency, and the frequency is shifted by an amount $\delta\nu$ from the start ($t = 0$) to the end ($t = T$) of the time series. Substitution of Equation 1.45 into Equation 1.44, followed by solution of the integral, gives the predicted mode profile:

$$\langle P(\nu)\rangle = \frac{H}{2\epsilon}\mathrm{atan}\left(\frac{2\epsilon}{1-\epsilon^2+X^2}\right), \tag{1.46}$$

where ϵ is the shift-to-line width ratio (Equation 1.43), and

$$X = \frac{\nu-(\nu_0+\delta\nu/2)}{\Delta/2}. \tag{1.47}$$

The top panel of Fig. 1.10 shows profiles given by Equation 1.46. The unperturbed profile (solid line) is for a mode having an unperturbed frequency of $\nu_0 = 3{,}000\,\mu$Hz, an unperturbed line width of $\Delta = 1\,\mu$Hz, and an unperturbed height of $H = 100$ units. The other curves show the profiles that result when the frequency shift, $\delta\nu$, is: 0.15 μHz (dotted line); 0.40 μHz (dashed line); 1.50 μHz (dot-dashed line); and 3.0 μHz (dot-dot-dot-dashed line). Since $\Delta = 1\,\mu$Hz, the $\delta\nu$ also correspond to the shift-to-line width ratios, ϵ. To put the values in context, we once more recall that low-l solar p modes at $\approx 3000\,\mu$Hz, which also have width $\approx 1\,\mu$Hz, show a frequency shift of approximately $0.40\,\mu$Hz from the minimum to the maximum of the solar activity cycle.

Only at the two largest shifts (dot-dashed and dot-dot-dot-dashed lines) do the profiles depart appreciably from the Lorentzian form. Closer inspection of the profiles does reveal some modest distortion at the two, smaller, Sun-like shifts. These have $\epsilon = 0.15$ and 0.40 respectively.

A co-sinusoidal variation of the mode frequency may be described by

$$\nu(t) = \nu_0 + \frac{\Delta\nu}{2}\left[1-\cos\left(\frac{2\pi t}{P_{\mathrm{cyc}}}\right)\right], \tag{1.48}$$

where ν_0 is again the unperturbed frequency (i.e., the frequency at minimum activity), while $\delta\nu$ is now the full amplitude (from minimum to maximum) of the cyclic frequency

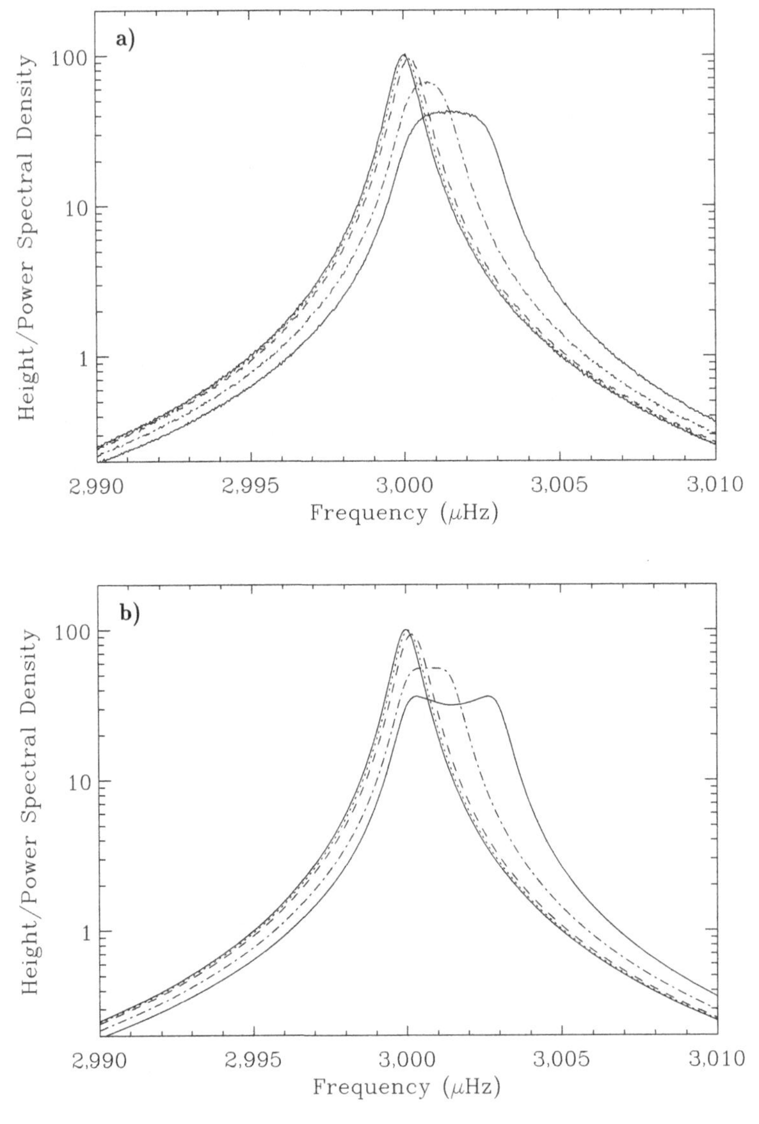

FIG. 1.10. Top panel: Peak profiles expected for a single mode of width $\Delta = 1\,\mu$Hz in the frequency-power spectrum of a time series within which the frequency was varied in a linear manner by total amount $\delta\nu$. Various line styles are: no shift (solid line); $\delta\nu = 0.15\,\mu$Hz (dotted line); $0.40\,\mu$Hz (dashed line); $1.50\,\mu$Hz (dot-dashed line); and $3.0\,\mu$Hz (dot-dot-dot-dashed line). Bottom panel: Peak profiles expected for a single mode of the same width, but where the frequency has instead been subjected to a co-sinusoidal variation in time. The time series is assumed to have length equal to one cycle period; while the amplitude of the cycle is $\delta\nu/2$ (giving a total minimum-to-maximum shift in frequency of $\delta\nu$). line styles are as per the upper panel. (From Chaplin *et al.*, 2008b.)

shift (not the *total* shift, as in the linear model). When the length of observation, T, equals the cycle period $P_{\rm cyc}$ (or one-half of the period), the average profile resulting from Equation 1.48 is:

$$\langle P(\nu)\rangle = \frac{\sqrt{L_1 \cdot L_2}}{\sqrt{1 - \epsilon^2 \cdot F(L_1, L_2)}}, \tag{1.49}$$

where

$$F(L_1, L_2) = \frac{4L_1L_2}{H\left(\sqrt{L_1} + \sqrt{L_2}\right)^2},$$
$$L_1 = \frac{H}{1 + (X - \epsilon)^2},$$
$$L_2 = \frac{H}{1 + (X + \epsilon)^2},$$
$$X = \frac{\nu - (\nu_0 + \frac{\delta\nu}{2})}{\Delta/2}.$$

Since the frequency spends more time around its maximum and minimum values, power near these extreme frequencies will have more weight in the time-averaged profile, giving the profile a double-humped appearance. This is reflected in the analytical expression through the two Lorentzians L_1 and L_2. It is obvious from Equation 1.49 that when $\epsilon \ll 1$ the profile tends to a single Lorentzian.

The bottom panel of Fig. 1.10 shows predicted profiles from Equation 1.49,[1] assuming observations made over a complete activity cycle, and with the same shifts $\delta\nu$ as were applied in the linear-model case (top panel of Fig. 1.10). As for the simpler linear variation, it is only at the two largest $\delta\nu$ that the profiles depart appreciably from the unperturbed (Lorentzian) form, here showing the predicted "humps" at the extreme frequencies of the cycle. However, closer inspection again reveals some distortion of the Lorentzian shapes at the small Sun-like shifts. For a given shift, this distortion appears to be slightly larger than in the simpler, linear case.

While the distortion seen at the Sun-like shifts is evidently modest, Chaplin *et al.* (2008b) showed that it is nevertheless just detectable in long, multi-year Sun-as-a-star datasets. By fitting modes to the usual Lorentzian-like models – which do not allow for the distortion – rather than modified models, like the ones shown above, Chaplin *et al.* (2008b) showed that an overestimation (underestimation) of the line width (height) parameter results. This bias is estimated to be of size comparable to the observational uncertainties given by datasets with a length of several years. Bias in the frequency parameter is much less of an issue.

The distortion may, of course, be more important for asteroseismic datasets on some solar-type stars, for example, those for which the ratio of the stellar-cycle frequency shifts to the mode line widths is larger than for the Sun. The shifts need only be about twice as strong as those on the Sun before significant distortion of the peaks results. Visible distortion of the mode profiles in asteroseismic data may as such provide an initial diagnostic of strong stellar-cycle signatures over the duration of the observations.

1.4 Estimated frequency shifts, and inference on spatial distribution of surface activity

We have seen that near-surface activity that is distributed in a nonspherically symmetric manner (i.e., the aforementioned acoustic asphericity) will give rise to differences in the magnitudes of the frequency shifts of modes of different $(l, |m|)$. For a scenario like that on the Sun, where bands of strong magnetic activity are located at lower latitudes, it will be the sectoral components of the non-radial modes that show the largest shifts. As was noted in Section 1.3.1, these components, for which $l = |m|$, have more

[1] Versions of Equations 1.46 and 1.49, which allow for the small amounts of peak asymmetry observed in real solar p modes, are presented in Chaplin *et al.* (2008b). We note here that this asymmetry is negligible for modes at the center of the solar p-mode spectrum.

sensitivity to changes at lower latitudes than do their zonal counterparts. It is possible to measure differences in the sizes of the frequency shifts of the different low-l modes (e.g., see Chaplin *et al.*, 2004a; Jiménez-Reyes *et al.*, 2004). If such measurements could be made on other stars, might it not then be possible to make some inference on the surface distribution in latitude of the activity, from measurement of the relative sizes of the frequency shifts (assuming, as mentioned elsewhere, that it is near-surface perturbations that are the dominant cause of the observed shifts)?

This question was considered by Chaplin *et al.* (2007a). They used a very simple model of the spatial distribution of the surface activity on a solar-type star, in which the activity was given a uniform amplitude between some lower and upper bounds in latitude, λ. As should be evident from what follows, it would be very straightforward to incorporate a more sophisticated model of the activity. They used this model to predict expected frequency shifts of modes of different $(l, |m|)$. Here, we reproduce their model, and use it to show explicitly how it might be used with results on observed frequency shifts to make inference on the active latitudes of a star. Before that, we use its predictions to show how estimated *average* frequency shifts, and estimated frequency separation ratios, of solar-type stars are affected by the angle of inclination.

Implicit in the calculation of the expected frequency shifts is the assumption that contributions to the fractional change in the sound speed are nonzero only very close to the surface. Consideration of the radial dependence of the sound speed is neglected (again, to simplify the description). The expected shifts, $\delta\nu_{lm}$, are then proportional to (e.g., Moreno-Insertis and Solanki, 2000)

$$\delta\nu_{lm} \propto \left(l + \frac{1}{2}\right) \frac{(l-m)!}{(l+m)!} \int_0^{\pi} |P_l^m(\cos\theta)|^2 B(\theta) \sin\theta \, \mathrm{d}\theta, \tag{1.50}$$

where the $P_l^m(\cos\theta)$ are Legendre polynomials, and $B(\theta)$ describes the distribution in co-latitude θ of the activity. (Note that $\lambda = 90\,\text{degrees} - \theta$.) Chaplin *et al.* (2007a) calibrated the model so that for a Sun-like configuration the predicted $\delta\nu_{00}$ was equal to the observed average shift of the solar $l = 0$ modes. This calibration was subsumed within the $B(\theta)$, as will be discussed. For all surface configurations tested, the activity at cycle minimum was set to $B(\theta) = 0$. Values of $\delta\nu_{lm}$ determined at the simulated cycle maxima therefore corresponded to the sought-for stellar cycle shifts.

In a first sequence of models $B(\theta)$ was set to some constant value at cycle maximum. This value was the same for *all* model configurations. The only parameter that was changed from model to model, and therefore the only factor that could alter the observed shifts, was the maximum latitude of the activity, $\lambda_{\max}$. The minimum latitude was fixed for all models at $\lambda_{\min} = 5\,\text{degrees}$. In summary:

$$B(\theta) = \begin{cases} \text{const} & \text{for } \lambda_{\min} \le |\lambda| \le \lambda_{\max}, \\ 0 & \text{otherwise.} \end{cases} \tag{1.51}$$

In the scenario above, the total surface magnetic flux is proportional to $\lambda_{\max} - \lambda_{\min}$, and therefore varied by over an order of magnitude for the range of computations made ($10 \le \lambda_{\max} \le 70\,\text{degrees}$). Chaplin *et al.* (2007a) also made a second sequence of models in which the total flux was conserved for all values of $\lambda_{\max}$. This model is described by:

$$B(\theta) = \begin{cases} \text{const}/\left(\lambda_{\max} - \lambda_{\min}\right) & \text{for } \lambda_{\min} \le |\lambda| \le \lambda_{\max}, \\ 0 & \text{otherwise.} \end{cases} \tag{1.52}$$

Chaplin *et al.* (2007a) used results from Knaack *et al.* (2001) to help guide their calibration of Equations 1.51 and 1.52. Knaack *et al.* noted that observations of the surface activity during the cycle 22 maximum showed sunspots confined to bands from 5 to 30 degrees, and faculae in bands from 5 to 40 degrees. On the Sun, faculae occupy a

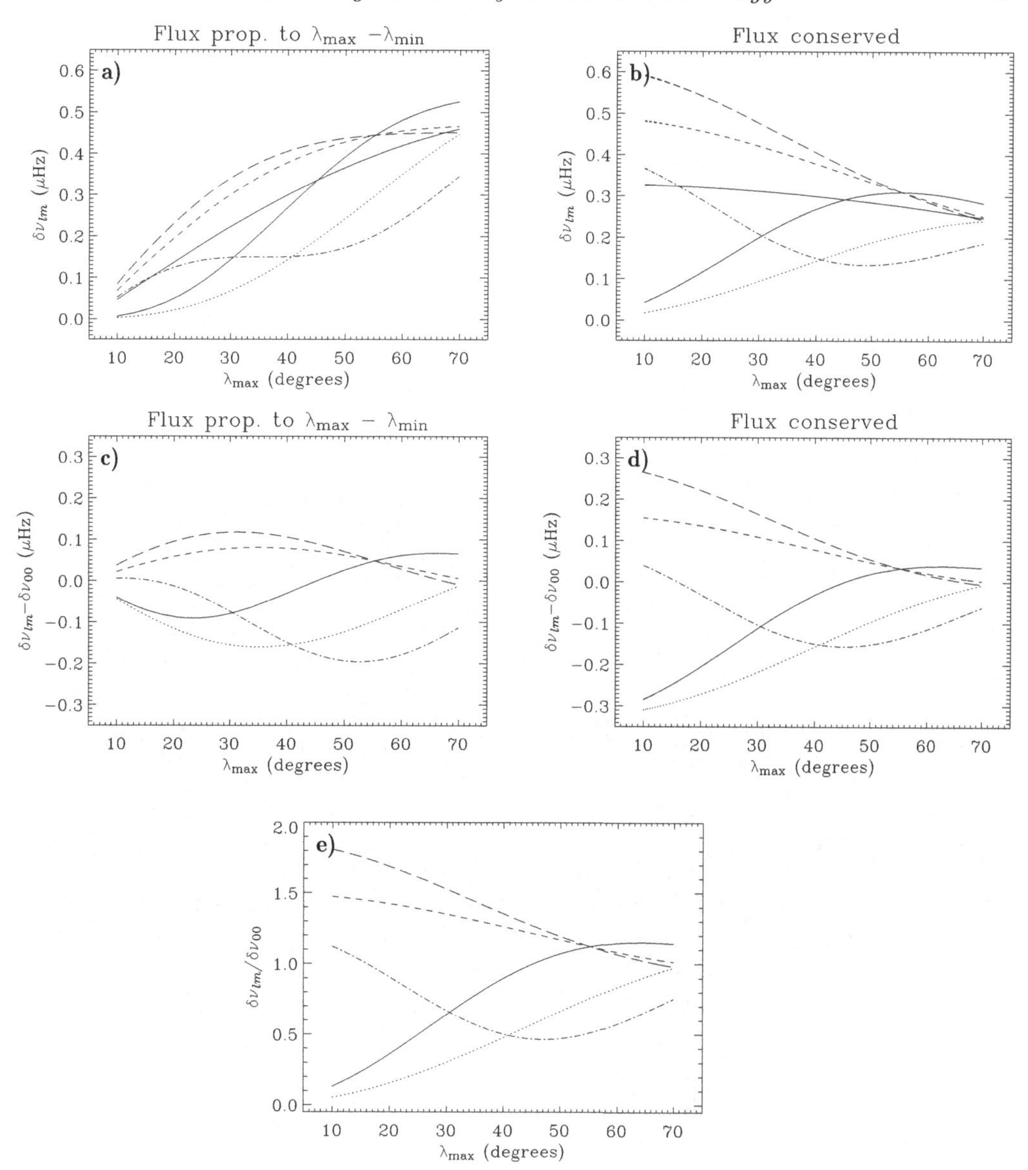

FIG. 1.11. Frequency shifts given by a simple model of stellar surface magnetic activity (magnetic flux). line styles, which are the same in all panels, show results for different $(l,|m|)$ components as follows: solid for (0,0); dotted for (1,0); short dashes for (1,1); dot-dashed for (2,0); dot-dot-dot-dashed for (2,1); and long dashes for (2,2). *a)* shifts calculated from models based on Equation 1.51 and *b)* based on Equation 1.52. *c)–d)* Residuals given by subtracting the (0,0) shifts from the other component shifts. *e)* Ratio of component shifts to (0,0) shifts (these ratios are the same for both sequences of models). (From Chaplin *et al.*, 2007a.)

much larger surface area than the spots (see also de Toma *et al.*, 2004). The constant in Equations 1.51 and 1.52 was therefore calibrated so that when $\lambda_{max} = 40$ degrees, the calculated shift of the radial modes was $\delta\nu_{00} \sim 0.3\,\mu$Hz. This is the value observed for the most prominent low-l radial modes of the Sun.

The top two panels of Fig. 1.11 show the calibrated shifts from both sequences of models, based on Equation 1.51 (left-hand panel) and Equation 1.52 (right-hand panel). The middle panels show residuals given by subtracting the $(l, |m|)=(0,0)$ shifts from the other component shifts. The bottom panel shows the ratio of the component shifts and

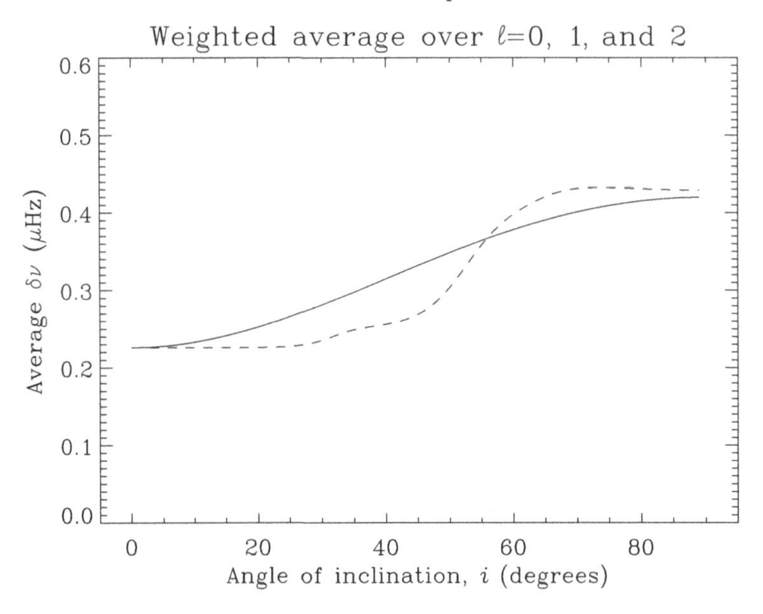

FIG. 1.12. Predicted frequency shift, averaged over $l = 0$, 1, and 2 modes, as a function of angle of inclination, i, assuming a solar-like spatial distribution of surface activity (see text for details).

the (0,0) shifts: since $B(\theta) =$ const. for both sequences of models, the ratios are also the same (hence only one plot).

Let us first consider the impact of the angle of inclination, i, on estimated *average* frequency shifts for solar-type stars. We again adopt the approach of Section 1.3.1, to take into account the effect of the relative visibility of components within any given l (Equations 1.36 to 1.38). We take the frequency shifts $\delta\nu_{lm}$ predicted by the simple model in this section, for different angles of inclination, i (as computed assuming the solar calibration). Since the $\delta\nu_{lm}$ are functions of i, we write $\delta\nu_{lm}(i)$. We weight the shifts of each component within a multiplet, to give the effective (weighted average) shift at that l:

$$\delta\nu_l = \left(\sum_{m=-l}^{m=+l} \mathcal{E}(i)^{\gamma}_{lm} \times \delta\nu_{lm} \right) \Big/ \left(\sum_{m=-l}^{m=+l} \mathcal{E}(i)^{\gamma}_{lm} \right), \tag{1.53}$$

Because one may then choose to average results over different l, to reduce errors, we compute a final, weighted average shift according to:

$$\langle \delta\nu \rangle = \left(\sum_{l=0}^{l=2} \mathcal{E}'_l \times \delta\nu_{lm} \right) \Big/ \left(\sum_{l=0}^{l=2} \mathcal{E}'_l \right), \tag{1.54}$$

using the predictions for $l = 0$, 1, and 2. Here, the $\mathcal{E}'_l$ are the relative total l visibilities, which we take to be 1, 1.5, and 0.5 (for $l = 0$, 1, and 2, respectively).

Figure 1.12 plots the $\langle \delta\nu \rangle$ as a function of i, for $\gamma = 1$ (solid line) and $\gamma = 4$ (dashed lines). When i is low, the $m = 0$ components have the highest visibility in the $l = 1$ and $l = 2$ modes, which reduce the size of the average frequency shift (since these components show smaller shifts than the other components when $\lambda_{\rm max} = 40$ degrees). The plot shows clearly that for a solar-like spatial distribution of surface activity, estimated average frequency shifts could differ by up to a factor of two, depending upon the angle i. It will clearly be important to have good constraints on i in order to properly interpret measured frequency shifts of solar-type stars (e.g., see Ballot *et al.*, 2006, 2008).

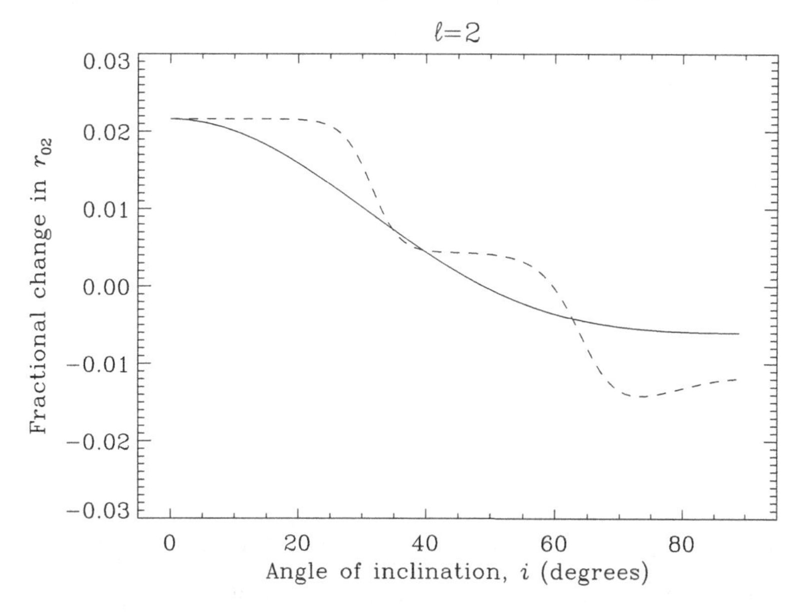

FIG. 1.13. Predicted fractional change (from solar minimum to solar maximum) in frequency separation ratios r_{02} of the most prominent solar p modes, as a function of angle of inclination, i, assuming a solar-like spatial distribution of surface activity (see text for details).

We also use predictions from Equation 1.53 above to show the impact of i on the observed frequency separation ratios. Figure 1.13 plots the implied fractional change, between solar minimum and maximum, in the frequency separation ratios r_{02} of the most prominent solar p modes. Recall from Section 1.3.2 that results from BiSON data give a fractional change of -0.01; this is consistent with the prediction shown here for $i = 90$ degrees. Figure 1.13 shows that for the Sun the largest offset occurs at $i = 0$ degrees, when the predicted fractional change is 0.02 (positive). Since we should expect to observe solar-type stars with larger acoustic asphericity than the Sun, it is possible that bias in the frequency separation ratios could reach levels of several percent.

Let us now use the frequency shift *ratios* plotted in the bottom panel of Fig. 1.11, together with results on measured frequency shifts of low-l solar p modes, to make an estimate of $\lambda_{\max}$ for the Sun. By using the ratios as opposed to the absolute values of the shifts we remove any dependence of the result on the absolute calibration of the model shifts. Let us denote the shift ratios by $k_{lm} = \delta\nu_{lm}/\delta\nu_{00}$. From the BiSON Sun-as-a-star data in Chaplin *et al.* (2004a), we may infer observed ratios of $k_{11} = 1.23 \pm 0.14$ and $k_{22} = 1.45 \pm 016$. (The observed $l = 2$ Sun-as-a-star shifts will include a contribution from the (2,0) components, however, the contribution is relatively weak, and the observed shifts may be regarded, to very good approximation, as being those of the (2,2) components.) To infer $\lambda_{\max}$, we use the appropriate curves from Fig. 1.11. Uncertainties in $\lambda_{\max}$ follow from

$$\delta\lambda_{\max} = \frac{\Delta\lambda_{\max}}{\Delta k_{lm}}\delta k_{lm}, \tag{1.55}$$

where δk_{lm} are the observed uncertainties in the k_{lm}, and $\Delta\lambda_{\max}/\Delta k_{lm}$ are the gradients of the respective curves in the vicinity of k_{lm}. From the $l = 1$ shift ratio we find $\lambda_{\max} = 43 \pm 16$ degrees, and from the $l = 2$ shift ratio we find $\lambda_{\max} = 34 \pm 10$ degrees. These values are of course reasonable estimates for the Sun.

One may adopt a similar procedure using asteroseismic data on other solar-type stars. One would again need to have good constraints on i, which may be obtained from peak-bagging of the modes in the frequency-power spectrum.

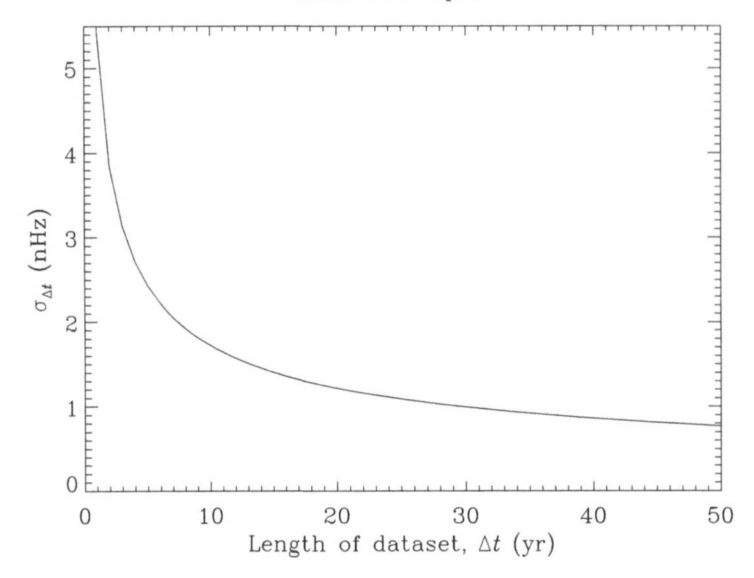

FIG. 1.14. Combined frequency precision, $\sigma_{\Delta t}$, for low-l observations, extrapolated to different dataset lengths, Δt (see text for more details).

1.5 How long will it take to detect evolutionary changes?

We finish by thinking in the longer term, and ask whether it might be possible to measure evolutionary changes in solar p-mode frequencies, using low-l data (the choice of data again made with other solar-type stars in mind).

First, consider the precision in the estimated frequencies. We take the estimated uncertainties, σ_{nl}, from Broomhall *et al.* (2009), who measured low-l frequencies in 8,640 d (23.7 year) of BiSON Sun-as-a-star data. The published list included 81 frequencies, covering $l = 0$ to 3. An estimate of the *combined* precision in *all* the measured frequencies, $\langle\sigma\rangle$, is given by:

$$\langle\sigma\rangle = \left(\sum_{nl} 1/\sigma_{nl}^2\right)^{-1/2} \simeq 1.1\,\mathrm{nHz}. \tag{1.56}$$

We may then estimate the combined precision, $\sigma_{\Delta t}$, for any dataset length, Δt (in year) from:

$$\sigma_{\Delta t} = \left(\frac{8640}{365}\right)^{1/2} \langle\sigma\rangle\, \Delta t^{-1/2} \simeq 5.4\, \Delta t^{-1/2}\,\mathrm{nHz}, \tag{1.57}$$

because the frequency uncertainties of the observed modes are expected to scale with the square root of time. Figure 1.14 plots $\sigma_{\Delta t}$ as a function of dataset length, Δt.

How do the $\sigma_{\Delta t}$ compare to the expected frequency shifts from evolutionary effects? Those shifts should, to good approximation, be dominated by the slow expansion in radius of the Sun as it evolves on the main sequence. Since, to first order, the frequencies scale like the square root of the mean density of the star, we have:

$$\left(\frac{\delta\nu}{\nu}\right) \simeq \frac{1}{2}\left(\frac{\delta\rho}{\rho}\right) \simeq -\frac{3}{2}\left(\frac{\delta R}{R}\right). \tag{1.58}$$

This implies that

$$\frac{1}{\nu}\left(\frac{\delta\nu}{\delta t}\right) \simeq -\frac{3}{2R}\left(\frac{\delta R}{\delta t}\right). \tag{1.59}$$

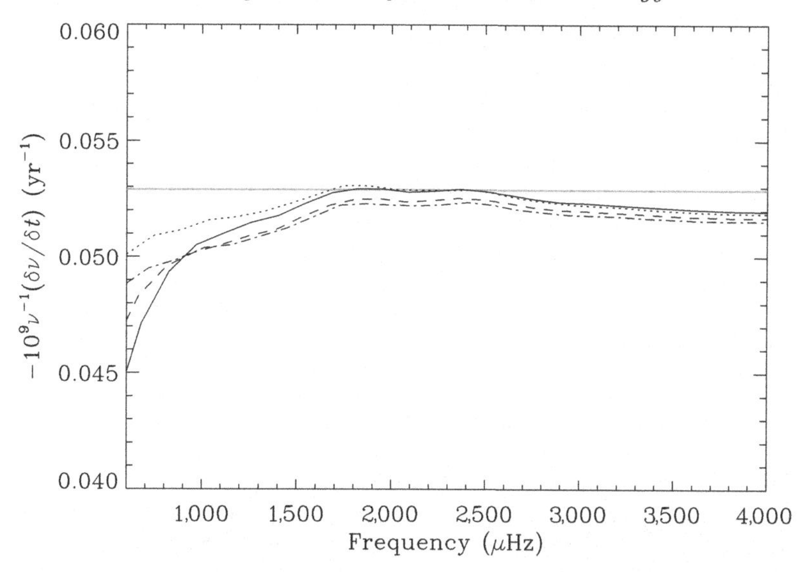

FIG. 1.15. $\nu(\delta\nu/\nu)$, as determined from the frequencies of several stellar evolutionary models made with solar mass and solar composition, but with varying ages, ranging ± 0.02 Gyr about the accepted solar age. line styles show data for $l = 0$ (thin solid line), $l = 1$ (dotted line), $l = 2$ (dashed line), and $l = 3$ (dot-dashed line). The thick gray line plots $-(3/2R)(\delta R/R)$, as determined from the radii of the stellar models.

Figure 1.15 plots $1/\nu(\delta\nu/\nu)$, as determined from the frequencies of several stellar evolutionary models (Yale-Yonsei models) made with solar mass and solar composition, but with varying ages ranging ± 0.02 Gyr about the accepted solar age. The different line styles show results for different l ($l = 0$ as a thin solid line, $l = 1$ as a dotted line, $l = 2$ as a dashed line, and $l = 3$ as a dot-dashed line). The thick gray line plots $-(3/2R)(\delta R/R)$, as determined from the radii of the stellar models. It has a value $-5.3 \times 10^{-11}\,\mathrm{yr}^{-1}$, which agrees quite well with the estimated gradient from the model-computed frequencies, verifying that the frequencies scale in an approximately homologous fashion.

We may then estimate to good approximation the expected evolutionary frequency shift $\Delta\nu_{nl}$ of a mode over a length of time Δt (in year):

$$\Delta\nu_{nl} \simeq -5.3 \times 10^{-11} \nu_{nl} \Delta t. \tag{1.60}$$

For a mode at the center of the solar p-mode spectrum, having a frequency $\approx$3,000 μHz, we obtain -1.6×10^{-4} nHz in 1 year, and -1.6×10^{-2} nHz in 100 year. These expected shifts are significantly smaller than the combined uncertainties expected for each length, being just over 5 nHz in 1 year, and about 0.5 nHz in 100 year. The shift and combined uncertainty have similar sizes only when $\Delta t \approx 1,100$ yr.

Acknowledgements

The author would like to express his thanks to P. Pallé and colleagues at the IAC, for their invitation to lecture at the Winter School and for the generous hospitality shown throughout the meeting. He thanks the attendees and his fellow lecturers for making the School such an enjoyable experience. He also acknowledges S. Basu for computing stellar model frequencies, A.-M. Broomhall and R. Howe for help with figures, and Y. Elsworth for useful discussions.

REFERENCES

Anguera Gubau, M., Palle, P. L., Perez Hernandez, F., Regulo, C., and Roca Cortes, T. 1992. The low L solar p-mode spectrum at maximum and minimum solar activity. *A&A*, **255**(Feb.), 363–72.

Appourchaux, T. 1998. The structure of the solar core: an observer's point of view. Pages 37–46 of: S. Korzennik (ed.), *Structure and Dynamics of the Interior of the Sun and Sun-like Stars*. ESA Special Publication, vol. 418.

Appourchaux, T., and Chaplin, W. J. 2007. On understanding the meaning of l = 2 and 3 p-mode frequencies as measured by various helioseismic instruments. *A&A*, **469**(July), 1151–4.

Appourchaux, T., Michel, E., Auvergne, M., Baglin, A., Toutain, T., Baudin, F., Benomar, O., Chaplin, W. J., Deheuvels, S., Samadi, R., Verner, G. A., Boumier, P., García, R. A., Mosser, B., Hulot, J.-C., Ballot, J., Barban, C., Elsworth, Y., Jiménez-Reyes, S. J., Kjeldsen, H., Régulo, C., and Roxburgh, I. W. 2008. CoRoT sounds the stars: p-mode parameters of Sun-like oscillations on HD 49933. *A&A*, **488**(Sept.), 705–14.

Ballot, J., Appourchaux, T., Toutain, T., and Guittet, M. 2008. On deriving p-mode parameters for inclined solar-like stars. *A&A*, **486**(Aug.), 867–75.

Ballot, J., García, R. A., and Lambert, P. 2006. Rotation speed and stellar axis inclination from p modes: how CoRoT would see other suns. *MNRAS*, **369**(July), 1281–6.

Basri, G., Walkowicz, L. M., Batalha, N., Gilliland, R. L., Jenkins, J., Borucki, W. J., Koch, D., Caldwell, D., Dupree, A. K., Latham, D. W., Meibom, S., Howell, S., and Brown, T. 2010. Photometric variability in kepler target stars: the sun among stars–a first look. *ApJ*, **713**(Apr.), L155–L159.

Basu, S., and Mandel, A. 2004. Does solar structure vary with solar magnetic activity? *ApJ*, **617**(Dec.), L155–L158.

Böhm-Vitense, E. 2007. Chromospheric activity in G and K main-sequence stars, and what it tells us about stellar dynamos. *ApJ*, **657**(Mar.), 486–93.

Broomhall, A.-M., Chaplin, W. J., Elsworth, Y., and New, R. 2011. Solar-cycle variations of large frequency separations of acoustic modes: implications for asteroseismology. *MNRAS*, **413**(June), 2978–86.

Broomhall, A.-M., Chaplin, W. J., Elsworth, Y., Fletcher, S. T., and New, R. 2009. Is the current lack of solar activity only skin deep? *ApJ*, **700**(Aug.), L162–L165.

Chaplin, W. J. 2004 (Oct.). Low-degree helioseismology: the state of play on its silver anniversary. Pages 34–43 of: D. Danesy (ed), *SOHO 14 Helio- and Asteroseismology: toward a Golden Future*. ESA Special Publication, vol. 559.

Chaplin, W. J., Appourchaux, T., Elsworth, Y., García, R. A., Houdek, G., Karoff, C., Metcalfe, T. S., Molenda-Żakowicz, J., Monteiro, M. J. P. F. G., Thompson, M. J., Brown, T. M., Christensen-Dalsgaard, J., Gilliland, R. L., Kjeldsen, H., Borucki, W. J., Koch, D., Jenkins, J. M., Ballot, J., Basu, S., Bazot, M., Bedding, T. R., Benomar, O., Bonanno, A., Brandão, I. M., Bruntt, H., Campante, T. L., Creevey, O. L., Di Mauro, M. P., Doğan, G., Dreizler, S., Eggenberger, P., Esch, L., Fletcher, S. T., Frandsen, S., Gai, N., Gaulme, P., Handberg, R., Hekker, S., Howe, R., Huber, D., Korzennik, S. G., Lebrun, J. C., Leccia, S., Martic, M., Mathur, S., Mosser, B., New, R., Quirion, P.-O., Régulo, C., Roxburgh, I. W., Salabert, D., Schou, J., Sousa, S. G., Stello, D., Verner, G. A., Arentoft, T., Barban, C., Belkacem, K., Benatti, S., Biazzo, K., Boumier, P., Bradley, P. A., Broomhall, A.-M., Buzasi, D. L., Claudi, R. U., Cunha, M. S., D'Antona, F., Deheuvels, S., Derekas, A., García Hernández, A., Giampapa, M. S., Goupil, M. J., Gruberbauer, M., Guzik, J. A., Hale, S. J., Ireland, M. J., Kiss, L. L., Kitiashvili, I. N., Kolenberg, K., Korhonen, H., Kosovichev, A. G., Kupka, F., Lebreton, Y., Leroy, B., Ludwig, H.-G., Mathis, S., Michel, E., Miglio, A., Montalbán, J., Moya, A., Noels, A., Noyes, R. W., Pallé, P. L., Piau, L., Preston, H. L., Roca Cortés, T., Roth, M., Sato, K. H., Schmitt, J., Serenelli, A. M., Silva Aguirre, V., Stevens, I. R., Suárez, J. C., Suran, M. D., Trampedach, R., Turck-Chièze, S., Uytterhoeven, K., Ventura, R., and Wilson, P. A. 2010. The asteroseismic potential of kepler: first results for solar-type stars. *ApJ*, **713**(Apr.), L169–L175.

Chaplin, W. J., Elsworth, Y., Isaak, G. R., Lines, R., McLeod, C. P., Miller, B. A., and New, R. 1998. An analysis of solar p-mode frequencies extracted from BiSON data: 1991–1996. *MNRAS*, **300**(Nov.), 1077–90.

Chaplin, W. J., Appourchaux, T., Elsworth, Y., Isaak, G. R., Miller, B. A., and New, R. 2000. Source of excitation of low-l solar p modes: characteristics and solar-cycle variations. *MNRAS*, **314**(May), 75–86.

Chaplin, W. J., Appourchaux, T., Elsworth, Y., Isaak, G. R., and New, R. 2001. The phenomenology of solar-cycle-induced acoustic eigenfrequency variations: a comparative and complementary analysis of GONG, BiSON and VIRGO/LOI data. *MNRAS*, **324**(July), 910–16.

Chaplin, W. J., Appourchaux, T., Elsworth, Y., Isaak, G. R., Miller, B. A., and New, R. 2004a. On comparing estimates of low-l solar p-mode frequencies from Sun-as-a-star and resolved observations. *A&A*, **424**(Sept.), 713–17.

Chaplin, W. J., Elsworth, Y., Isaak, G. R., Miller, B. A., and New, R. 2004b. The solar cycle as seen by low-l p-mode frequencies: comparison with global and decomposed activity proxies. *MNRAS*, **352**, 1102–08.

Chaplin, W. J., Appourchaux, T., Elsworth, Y., Isaak, G. R., Miller, B. A., New, R., and Toutain, T. 2004c. Solar p-mode frequencies at $\ell = 2$: what do analyses of unresolved observations actually measure? *A&A*, **416**, 341–51.

Chaplin, W. J., Elsworth, Y., Miller, B. A., New, R., and Verner, G. A. 2005. Impact of the solar activity cycle on frequency separation ratios in helioseismology. *ApJ*, **635**(Dec.), L105–L108.

Chaplin, W. J., Elsworth, Y., Houdek, G., and New, R. 2007a. On prospects for sounding activity cycles of Sun-like stars with acoustic modes. *MNRAS*, **377**, 17–29.

Chaplin, W. J., Elsworth, Y., Miller, B. A., Verner, G. A., and New, R. 2007b. Solar p-mode frequencies over three solar cycles. *ApJ*, 659, 1749–60.

Chaplin, W. J., Houdek, G., Appourchaux, T., Elsworth, Y., New, R., and Toutain, T. 2008a. Challenges for asteroseismic analysis of Sun-like stars. *A&A*, 485, 813–22.

Chaplin, W. J., Elsworth, Y., New, R., and Toutain, T. 2008b. Distortion of the p-mode peak profiles by the solar-cycle frequency shifts: do we need to worry? *MNRAS*, 384, 1668–74.

Christensen-Dalsgaard, J., and Berthomieu, G. 1991. *Theory of solar oscillations*, 401–78.

de Toma, G., White, O. R., Chapman, G. A., Walton, S. R., Preminger, D. G., and Cookson, A. M. 2004. Solar cycle 23: an anomalous cycle? *ApJ*, **609**(July), 1140–52.

Dziembowski, W. A., and Goode, P. R. 2005. Sources of oscillation frequency increase with rising solar activity. *ApJ*, **625**(May), 548–55.

Elsworth, Y., Howe, R., Isaak, G. R., McLeod, C. P., and New, R. 1990. Variation of low-order acoustic solar oscillations over the solar cycle. *Nature*, **345**(May), 322–4.

Elsworth, Y., Howe, R., Isaak, G. R., McLeod, C. P., Miller, B. A., New, R., Speake, C. C., and Wheeler, S. J. 1994. Solar p-mode frequencies and their dependence on solar activity recent results from the BISON network. *ApJ*, **434**(Oct.), 801–06.

Fletcher, S. T., Broomhall, A.-M., Salabert, D., Basu, S., Chaplin, W. J., Elsworth, Y., Garcia, R. A., and New, R. 2010. A seismic signature of a second dynamo? *ApJ*, **718**(July), L19–L22.

García, R. A., Mathur, S., Salabert, D., Ballot, J., Régulo, C., Metcalfe, T. S., and Baglin, A. 2010. CoRoT reveals a magnetic activity cycle in a sun-like star. *Science*, **329**(Aug.), 1032.

Gelly, B., Lazrek, M., Grec, G., Ayad, A., Schmider, F. X., Renaud, C., Salabert, D., and Fossat, E. 2002. Solar p-modes from 1979 days of the GOLF experiment. *A&A*, **394**(Oct.), 285–97.

Gilliland, R. L., Brown, T. M., Christensen-Dalsgaard, J., Kjeldsen, H., Aerts, C., Appourchaux, T., Basu, S., Bedding, T. R., Chaplin, W. J., Cunha, M. S., De Cat, P., De Ridder, J., Guzik, J. A., Handler, G., Kawaler, S., Kiss, L., Kolenberg, K., Kurtz, D. W., Metcalfe, T. S., Monteiro, M. J. P. F. G., Szabó, R., Arentoft, T., Balona, L., Debosscher, J., Elsworth, Y. P., Quirion, P.-O., Stello, D., Suárez, J. C., Borucki, W. J., Jenkins, J. M., Koch, D., Kondo, Y., Latham, D. W., Rowe, J. F., and Steffen, J. H. 2010. Kepler asteroseismology program: introduction and first results. *PASP*, **122**(Feb.), 131–43.

Gizon, L., and Solanki, S. K. 2003. Determining the inclination of the rotation axis of a sun-like star. *ApJ*, **589**(June), 1009–19.

González Hernández, I., Howe, R., Komm, R., and Hill, F. 2010. Meridional circulation during the extended solar minimum: another component of the torsional oscillation? *ApJ*, **713**(Apr.), L16–L20.

Gough, D. O. 1990. Comments on helioseismic inference. Pages 283–318 of: Y. Osaki and H. Shibahashi (ed), *Progress of Seismology of the Sun and Stars.* Lecture Notes in Physics, vol. 367, Springer Verlag, Berlin.

Houdek, G., Chaplin, W. J., Appourchaux, T., Christensen-Dalsgaard, J., Däppen, W., Elsworth, Y., Gough, D. O., Isaak, G. R., New, R., and Rabello-Soares, M. C. 2001. Changes in convective properties over the solar cycle: effect on p-mode damping rates. *MNRAS*, **327**(Oct.), 483–7.

Howe, R., Christensen-Dalsgaard, J., Hill, F., Komm, R., Schou, J., and Thompson, M. J. 2009. A note on the torsional oscillation at solar minimum. *ApJ*, **701**(Aug.), L87–L90.

Howe, R., Komm, R. W., and Hill, F. 2002. Localizing the solar cycle frequency shifts in global p-modes. *ApJ*, **580**(Dec.), 1172–87.

Jiménez-Reyes, S. J., Regulo, C., Palle, P. L., and Roca Cortes, T. 1998. Solar activity cycle frequency shifts of low-degree p-modes. *A&A*, **329**(Jan.), 1119–24.

Jiménez-Reyes, S. J., Corbard, T., Pallé, P. L., Roca Cortés, T., and Tomczyk, S. 2001. Analysis of the solar cycle and core rotation using 15 years of Mark-I observations: 1984–1999. I. The solar cycle. *A&A*, **379**(Nov.), 622–33.

Jiménez-Reyes, S. J., García, R. A., Chaplin, W. J., and Korzennik, S. G. 2004. On the spatial dependence of low-degree solar p-mode frequency shifts from full-disk and resolved-Sun observations. *ApJ*, **610**(July), L65–L68.

Jiménez-Reyes, S. J., Chaplin, W. J., Elsworth, Y., García, R. A., Howe, R., Socas-Navarro, H., and Toutain, T. 2007. On the Variation of the Peak Asymmetry of Low-l Solar p Modes. *ApJ*, **654**(Jan.), 1135–45.

Karoff, C., Metcalfe, T. S., Chaplin, W. J., Elsworth, Y., Kjeldsen, H., Arentoft, T., and Buzasi, D. 2009. Sounding stellar cycles with Kepler. I. Strategy for selecting targets. *MNRAS*, **399**(Oct.), 914–23.

Knaack, R., Fligge, M., Solanki, S. K., and Unruh, Y. C. 2001. The influence of an inclined rotation axis on solar irradiance variations. *A&A*, **376**(Sept.), 1080–1089.

Komm, R. W., Howe, R., and Hill, F. 2000. Solar-cycle changes in gong p-mode widths and amplitudes 1995–1998. *ApJ*, **531**(Mar.), 1094–1108.

Komm, R., Howe, R., and Hill, F. 2002. Localizing width and energy of solar global p-modes. *ApJ*, **572**(June), 663–73.

Libbrecht, K. G., and Woodard, M. F. 1990. Solar-cycle effects on solar oscillation frequencies. *Nature*, **345**(June), 779–82.

Metcalfe, T. S., Dziembowski, W. A., Judge, P. G., and Snow, M. 2007. Asteroseismic signatures of stellar magnetic activity cycles. *MNRAS*, **379**(July), L16–L20.

Metcalfe, T. S., Basu, S., Henry, T. J., Soderblom, D. R., Judge, P. G., Knölker, M., Mathur, S., and Rempel, M. 2010. Discovery of a 1.6 year magnetic activity cycle in the exoplanet host star ι horologii. *ApJ*, **723**(Nov.), L213–17.

Moreno-Insertis, F., and Solanki, S. K. 2000. Distribution of magnetic flux on the solar surface and low-degree p-modes. *MNRAS*, **313**(Apr.), 411–22.

Otí Floranes, H., Christensen-Dalsgaard, J., and Thompson, M. J. 2005. The use of frequency-separation ratios for asteroseismology. *MNRAS*, **356**(Jan.), 671–9.

Pallé, P. L., Régulo, C., and Roca Cortés, T. 1989. Solar cycle induced variations of the low L solar acoustic spectrum. *A&A*, **224**(Oct.), 253–8.

Pallé, P. L., Régulo, C., and Roca Cortés, T. 1990a. Frequencies, line widths, and Splittings of Low-Degree Solar p-Modes. Pages 189–195 of: Y. Osaki and H. Shibahashi (ed), *Progress of Seismology of the Sun and Stars*. Lecture Notes in Physics. Springer Verlag, Berlin.

Pallé, P. L., Régulo, C., and Roca Cortés, T. 1990b. The Spectrum of Solar p-Modes and the Solar Activity Cycle. Pages 129–134 of Y. Osaki and H. Shibahashi (ed), *Progress of Seismology of the Sun and Stars*. Lecture Notes in Physics. Springer Verlag, Berlin.

Rabello-Soares, M. C., Korzennik, S. G., and Schou, J. 2008. Variations of the solar acoustic high-degree mode frequencies over solar cycle 23. *Advances in Space Research*, **41**, 861–867.

Ritzwoller, M. H., and Lavely, E. M. 1991. A unified approach to the helioseismic forward and inverse problems of differential rotation. *ApJ*, **369**(Mar.), 557–66.

Roxburgh, I. W. 2005. The ratio of small to large separations of stellar p-modes. *A&A*, **434**(May), 665–9.

Roxburgh, I. W., and Vorontsov, S. V. 2003. The ratio of small to large separations of acoustic oscillations as a diagnostic of the interior of solar-like stars. *A&A*, **411**(Nov.), 215–20.

Salabert, D., Fossat, E., Gelly, B., Kholikov, S., Grec, G., Lazrek, M., and Schmider, F. X. 2004. Solar p modes in 10 years of the IRIS network. *A&A*, **413**(Jan.), 1135–42.

Salabert, D., García, R. A., Pallé, P. L., and Jiménez-Reyes, S. J. 2009. The onset of solar cycle 24: what global acoustic modes are telling us. *A&A*, **504**(Sept.), L1–L4.

Sheeley, Jr., N. R. 2010 (June). What's so peculiar about the cycle 23/24 solar minimum? Pages 3–13 of: S. R. Cranmer, J. T. Hoeksema, and J. L. Kohl (ed), *SOHO-23: Understanding a Peculiar Solar Minimum*. Astronomical Society of the Pacific Conference Series, vol. 428.

Tapping, K. F., and Detracey, B. 1990. The origin of the 10.7 CM flux. *Sol. Phys.*, **127**(June), 321–32.

Toutain, T., and Wehrli, C. 1997. First Results from the SPM of the VIRGO/SOHO Experiment. Pages 254–258 of: B. Schmieder, J. C. del Toro Iniesta, and M. Vazquez (ed), *1st Advances in Solar Physics Euroconference. Advances in Physics of Sunspots*. Astronomical Society of the Pacific Conference Series, vol. 118.

Tripathy, S. C., Jain, K., Hill, F., and Leibacher, J. W. 2010. Unusual trends in solar p-mode frequencies during the current extended minimum. *ApJ*, **711**(Mar.), L84–L88.

Verner, G. A., Chaplin, W. J., and Elsworth, Y. 2006. BiSON data show change in solar structure with magnetic activity. *ApJ*, **640**(Mar.), L95–L98.

Watson, F., Fletcher, L., Dalla, S., and Marshall, S. 2009. Modelling the longitudinal asymmetry in sunspot emergence: the role of the Wilson Depression. *Sol. Phys.*, **260**(Nov.), 5–19.

Woodard, M. F., and Noyes, R. W. 1985. Change of solar oscillation eigenfrequencies with the solar cycle. *Nature*, **318**(Dec.), 449–50.

2. Learning physics from the stars: It's all in the coefficients

STEVEN D. KAWALER

2.1 Overview and basic discussion of equations of stellar structure

This section is intended merely as a reminder of things you have already learned (or will soon learn) in a first course in stellar structure and evolution. Here, we will introduce, and in some cases derive, the four basic equations of stellar structure. If you would like additional details, by all means consult one of the standard texts in this subject – I recommend without hesitation the text by Hansen *et al.* (2004).

2.1.1 *Dependent and independent variables*

In solving the equations that describe the conditions within stars, we need to decide on a coordinate system. A reasonable simplification of the equations arises if we assume that stars are spherically symmetric. With that assumption, a single positional variable will suffice as the independent variable. Our choice could be r, the distance from the center of the star, or m_r, the mass contained within a shell of radius r. For some purposes, we may use r and for others, m_r. Their values are related through the equation of mass conservation (or continuity):

$$\frac{dm_r}{dr} = 4\pi r^2 \rho(r), \tag{2.1}$$

where $\rho(r)$ is the mass density at position r.

As for the quantities that we would like to use to describe the physical conditions within the star at each position, we will define them as follows:

- velocity: v (= 0 for a hydrostatic star)
- density: ρ or $n = N_A\rho/\mu$
- pressure: P
- temperature: T
- chemical composition (mass fraction): X_i
- ion/charge balance: y_i, n_e
- internal energy: U
- entropy: S
- heat flow parameters/cross sections: $\kappa_{\rm rad}, \kappa_{\rm cond}$
- energy flow: $L_{\rm r}, F_{\rm conv}$
- energy generation/loss: $\epsilon_{\rm nuc}, \epsilon_\nu$

These quantities provide a fairly complete description of the state of the stellar interior, but they clearly are not independent of one another. In fact, one can show that only four dependent variables at each point in a model suffice to fully describe the conditions there. For computation of stellar structure and evolution, those quantities are usually chosen as r (or m_r), P, T, and L_r.

Of course, the other quantities are important, but they can be derived or computed given the main four quantities. That said, the computation of a stellar model requires knowledge of the values of all of the above quantities. They appear in the *coefficients* of the equations, and contain all of the physics of relevance to stellar interiors. Thus they provide the direct connection between fundamental physics and the appearance and behavior of the stars.

Their appearance in the coefficients of the equations of stellar structure allows us to further categorize what we need in terms of the primary *mechanical* quantities r (or m_r) and P, and the primary *thermal* quantities T and L_r. Necessary extra information is given by the composition (X_i). Various branches of physics provide the tools to determine important *derived* quantities, including equation–of–state parameters ρ, μ, U, and S, from which we can calculate $\nabla_{\rm ad}$, specific heats, and other thermodynamic quantities. From atomic physics we can compute y_i and n_e, and the opacities $\kappa_{\rm rad}$ and $\kappa_{\rm cond}$. Nuclear physics provides the tools for computing $\epsilon_{\rm nuc}$ and $\epsilon_{\rm neut}$, and hydrodynamics gives us hope for computing $F_{\rm conv}$.

2.1.2 *The equations of stellar structure*

Let's now see how this set of parameters appears when we write down the equations of stellar structure. This is possible to do in a compact form only if we expect that the assumption of spherical symmetry everywhere is a good one (so that all quantities are functions only of r), and that the star that we are modeling is in hydrostatic equilibrium (ensuring that time derivatives are zero). Then and only then do the equations take on the form that is used almost universally to produce stellar models for asteroseismological (and indeed most other) purposes.

2.1.2.1 Mass conservation

The first equation has already been introduced above; Equation 2.1 ensures that the mass and radius are consistent with the density from the center to the surface. Equation 2.1 also serves as a coordinate transformation between using m_r and r as independent variables. Consider a general quantity Z (which could stand for temperature, pressure, or something else). Then the differential equation describing the dependence of Z on r can be written down in terms of the dependence of Z on m_r as follows:

$$\frac{dZ}{dr} = 4\pi r^2 \rho(r) \frac{dZ}{dm_r}. \tag{2.2}$$

2.1.2.2 Mechanical equilibrium

We're operating under the assumption of nothing moving, that is, there is mechanical (hydrostatic) equilibrium within the entire stellar model. This means that the downward force of gravity at any position r,

$$\rho(r) g(r) = \rho(r) \frac{Gm_r}{r^2} \tag{2.3}$$

must be balanced by the balance of the pressure upward:

$$P(r) - P(r + dr) = -\left(\frac{dP}{dr}\right) dr. \tag{2.4}$$

If they do indeed balance, then this equilibrium requires that

$$\frac{dP}{dr} = -\frac{Gm_r}{r^2}\rho(r) \quad \text{or} \quad \frac{dP}{dm_r} = -\frac{Gm_r}{4\pi r^4}. \tag{2.5}$$

Note, in passing, that Equations 2.1 and 2.5 contain only r, m_r, P, and ρ. So, under circumstances where one can write the pressure in terms of the density only (i.e., without an explicit temperature dependence), then these two equations, plus the $P(\rho)$ relation, suffice to completely describe the mechanical structure of the model. As an example, consider a polytropic equation of state such as:

$$P(r) = K\rho^{\gamma}(r). \tag{2.6}$$

Under this condition, the two equations can be written in terms of the independent variable (r) and two dependent variables (P and m_r):

$$\frac{dP}{dr} = \frac{Gm_r}{r^2}\frac{1}{K^{1/\gamma}}P^{1/\gamma}(r) \tag{2.7}$$

$$\frac{dm_r}{dr} = 4\pi r^2 \frac{1}{K^{1/\gamma}}P^{1/\gamma}(r) \tag{2.8}$$

That is, two equations, two unknowns – so just add a pair of boundary conditions, and you have a complete stellar model, in hydrostatic equilibrium!

2.1.2.3 Energy generation

In the more common case where we do care about the thermal content of stellar material, we need to determine the energy balance within each zone of the star. To this end, we appeal to conservation of energy: that is, the energy flowing into a zone must be balanced by the energy flowing out of that same zone, possibly affected by energy production or consumption within the zone. If no energy is gained or lost, then clearly

$$L_r = L_{r+dr}. \tag{2.9}$$

But, more generally,

$$L_{r+dr} = L_r + 4\pi r^2 \rho(r)\left(-\frac{dQ}{dt} + \epsilon\right)dr, \tag{2.10}$$

where ϵ is the net energy generation rate per unit mass, and Q is the heat content per unit mass. The quantity ϵ generally denotes nuclear processes (i.e., energy production via fusion) but also includes energy losses through neutrino emission. In those cases, ϵ can be computed as a function of the thermodynamic state and the composition, that is, $\epsilon(\rho, T, X_i)$.

The term involving the time derivative of Q describes the rate of heat gain or loss, per gram, of material. Generally, we have

$$\frac{dQ}{dt} = \frac{dE}{dt} - P\frac{\partial V}{\partial t} = \frac{dE}{dt} + \frac{P}{\rho^2}\frac{\partial \rho}{\partial t} \tag{2.11}$$

where the first term on the right is the rate of change of internal energy, and the second term accounts for any PdV work done on (or by) the zone. Equation 2.11 can be recast using the definition of entropy,

$$\frac{dQ}{dt} = T\frac{\partial S}{\partial t}, \tag{2.12}$$

which leads to the more familiar form for the energy conservation equation

$$\frac{dL_r}{dr} = 4\pi r^2 \rho\left(\epsilon - T\frac{\partial S}{\partial t}\right). \tag{2.13}$$

We note that the time derivative in Equation 2.13 is the only place where time explicitly appears in the equations of stellar structure. More on that later.

2.1.2.4 Energy transport

Finally, we require an equation that tells us how temperature changes with position in a stellar model. To get there, we invoke thermal equilibrium – the assumption that since energy is flowing out of the "top" of a given region, energy must also be flowing in from the bottom. We model this transport as a diffusive process driven by the fact that the energy density has to change through the region. This in turn is a result of the continuously increasing radius of each zone along with any temperature change. Schematically, we can write the radiant flux F_r as follows:

$$F_r = \text{energy-density gradient} \times \text{speed} \times \text{mean-free-path}.$$

In a less schematic form, we have

$$F_r = -\frac{d}{dr}\left(\frac{aT^4}{3}\right) \times c \times \lambda \tag{2.14}$$

where λ is the mean free path between photon scatterings and c is the speed of light. Recognizing that for radiation, the mean free path $\lambda = (\kappa\rho)^{-1}$ where κ is the opacity per gram, we can rewrite this equation as

$$F_r = \frac{4ac}{3\kappa\rho} T^3 \frac{dT}{dr} \tag{2.15}$$

for the energy flux carried by radiation. Multiplying by the surface area of the zone and rearranging yields an expression for the gradient in the temperature when radiative diffusion carries the flux:

$$\frac{dT}{dr} = -\frac{3\kappa\rho}{16\pi acr^2}\frac{L_r}{T^3} \tag{2.16}$$

We can also express the temperature gradient in more general terms as a function of the pressure gradient:

$$\frac{dT}{dr} = \frac{dT}{dP}\frac{dP}{dr} = \frac{T}{P}\frac{d\ln T}{d\ln P}\frac{dP}{dr}. \tag{2.17}$$

If we define

$$\nabla \equiv \frac{d\ln T}{d\ln P} \tag{2.18}$$

then

$$\frac{dT}{dr} = -\nabla \frac{GM_r}{r^2}\frac{\rho T}{P}. \tag{2.19}$$

This is a general expression for the temperature gradient because the mechanism of heat transport is not specified, but is contained in the way ∇ is calculated.

Equation 2.16, in fact, can easily be transformed to look like Equation 2.19. Looking at Equations 2.16 and 2.19, we can define a "del" for the case when radiation carries the flux:

$$\nabla_{\rm rad} \equiv \frac{3\kappa_r}{16\pi ac}\frac{L_r}{T^4}\frac{P}{GM_r}. \tag{2.20}$$

In turn, we can now write the temperature gradient in the model as

$$\frac{dT}{dr} = -\nabla_{\rm rad}\frac{GM_r}{r^2}\frac{\rho T}{P} \tag{2.21}$$

for the case where ∇ is determined by radiative diffusion (that is, $\nabla = \nabla_{\rm rad}$). Later, we will consider other forms of heat transport that can provide values for ∇ under a variety of physical conditions (e.g., when the material is convective or conductive).

2.1.3 *The constitutive relations: where the physics is*

The previous section summarized the four differential equations of stellar structure and the general background from where the came. Each equation relates the dependent variables to the independent variable, but each also includes some other factors and terms, within which the physics that govern stellar structure reside. In this section, we outline these quantities and describe the "constitutive relations" that provide the route to evaluating the values of these terms.

2.1.3.1 The equation of state

The dependent variables P and T, along with the compositional mix of the stellar material (mass fractions X_i for each atomic species i) suffice to determine the density ρ, the relevant thermal quantities such as internal energy, and the ionization state of the material. The route to calculating these quantities can be a difficult one in complete generality, but with a few simplifying assumptions we can make progress. The principal assumption is that the material in a stellar interior is everywhere in local thermodynamic equilibrium, and that all quantities are isotropic. The fact that the photon mean free path is much smaller than the length scales of interest (the pressure scale height, stellar radius, etc.) ensures that this approximation is a relevant one.

With that assumption (and assuming isotropy) one can show that for an unionized gas, the pressure and density are related through

$$P = \sum_i n_i kT \quad \text{where} \quad n_i = N_A \rho X_i / A_i. \tag{2.22}$$

We now define the "mean molecular weight" μ as

$$\mu^{-1} \equiv \sum_i X_i / A_i \tag{2.23}$$

so that Equation 2.22 becomes

$$P = \frac{\rho}{\mu} N_A kT \tag{2.24}$$

and also the internal energy per gram is

$$E = \frac{3}{2}\frac{P}{\rho}. \tag{2.25}$$

Of course, the interiors of stars are mostly ionized, so that the above equations need some modification to account for the free electrons in addition to the nuclei. Thus, we have

$$P = P_e + P_I = P_e + \frac{\rho}{\mu_I} N_A kT \tag{2.26}$$

and further note that the right-hand term for P_I remains valid even in the case of electron degeneracy when dealing with most stars that have asteroseismic potential. On the other hand, for the electron pressure under non-degenerate conditions, we have

$$P_e = \frac{\rho}{\mu_e} N_A kT \tag{2.27}$$

where μ_e depends on the ionization state of the material.

As the reader might expect, there are complications beyond simply computing the ionization state. Departures from ideal gas behavior can be significant under conditions found within the Sun and the stars, and we will discuss those later. In massive stars, radiation pressure ($P_{\rm rad} = aT^3/3$) becomes increasingly important in the outer layers as the mass increases.

2.1.3.2 Energy generation rates

Stars are, through much of their visible lifetime, gigantic nuclear furnaces. To ensure energy conservation as per Equation 2.13 we need an accurate determination of the net energy generation rate per gram, ϵ. For the nuclear fusion process, this means computing the rate of interaction between a target and a projectile. We express the probability of these interactions in terms of a cross section σ for a given process. Then the rate of the interaction is proportional to $n\sigma v$, where n is the number density of targets and v is the relative velocity of the reactants. The velocity will depend on energy (and scale with $\sqrt{kT}$).

The reaction cross section σ depends on energy as well, and theory can provide a reasonable functional form of that energy dependence, but the scaling of that relation requires laboratory measurement. Since the energy dependence of the velocity and cross-section can be understood, common practice is to tabulate $\langle\sigma v\rangle$, averaged over the energy distribution corresponding to a temperature T.

For resonant reactions, your stellar structure class should have taught you that

$$\langle\sigma v\rangle_{ij} = K_1\, g\frac{\Gamma_i\Gamma_j}{\Gamma}T^{-3/2}e^{-K_2/T} \tag{2.28}$$

where K_1 and K_2 are constants that depend on the properties of the interacting particles, and the lifetimes of the incoming state and outgoing state are indicated by the energy widths Γ_i and Γ_j, with the total lifetime given by $1/\Gamma$. The quantity g is a statistical weight factor. In the case where a reaction does not have a resonance in the energy range of interest, the $\langle\sigma v\rangle$ value is given instead by

$$\langle\sigma v\rangle_{ij} = \frac{K_0 S(0)}{Z_i Z_j}T^{-2/3}e^{-K_3 T^{-1/3}}. \tag{2.29}$$

Here, $S(0)$ is the "astrophysical S factor" which is the value, at zero energy, of a slowly varying function of energy that helps isolate a part of the energy dependence of the reaction. Once we have determined these cross sections, the rate of energy production per gram of material can be computed with knowledge of the energy yield per reaction and the abundances of projectiles and targets:

$$\epsilon_{ij}(\rho, T, X) = Q_{ij}\,\rho N_A^2 \frac{X_i}{A_i}\frac{X_j}{A_j}\langle\sigma v\rangle_{ij}. \tag{2.30}$$

2.1.3.3 Radiation, conduction, and convection

(i) Radiative Transport

Section 2.1.2.4 expressed the temperature gradient in terms of ∇, which is the local power law slope of the $T(P)$ relation. The value of ∇ will reflect the dominant energy transport mechanism. The example shown in that section was for energy transport by photon diffusion, which defined a $\nabla_{\rm rad}$. That quantity depends in turn on the conditions at that point (P, T, M_r, L), some constants, and the radiative opacity $\kappa_r(\rho, T, X_i)$. The radiative opacity represents scattering photons through a variety of atomic processes involving electronic transitions (bound-free, bound-bond) and free particles (free-free and electron scattering). Though the scattering cross-section will be a sharp function of frequency (we live in a quantum world), the radiative opacity to be used in computations of isotropic stellar interiors may be simplified using a suitable average opacity over the expected frequency or wavelength distribution of the flux at the local temperature T. A favorite frequency average is the Rosseland mean opacity

$$\frac{1}{\bar{\kappa}_r} = \frac{\int_0^\infty \frac{1}{\kappa_\nu}\frac{\partial B_\nu}{\partial T}d\nu}{\int_0^\infty \frac{\partial B_\nu}{\partial T}d\nu} \tag{2.31}$$

where the average is weighted by the local flux derivative (and suitably normalized). Note that this is an average of the reciprocal of the opacity, since the radiative flux is inversely proportional to the opacity (see Equation 2.15) and it is F that is the quantity that is driving the temperature gradient.

(ii) Conductive Transport

Under certain circumstances in stellar matter (as well as the much more common household applications), heat can be transported by conduction. Under the constraints of the formalism adopted in Section 2.1.2.4, we must seek a *conductive* opacity κ_c that encapsulates the physics of conduction. Thus, we write

$$\kappa_c \equiv \frac{4acT^3}{3\rho D_e} \tag{2.32}$$

where the diffusion coefficient for conductive heat transport is

$$D_e \approx \frac{c_v v_e \lambda}{3}. \tag{2.33}$$

In Equation 2.33, c_v is the specific heat at constant density, v_e is the thermal velocity of the electrons, and λ is again a mean free path. Skipping several steps, we can eventually show that

$$\kappa_c \propto \frac{\mu_e^2}{\mu_I} Z_I^2 \left(\frac{T}{\rho}\right)^2 \tag{2.34}$$

where the constant of proportionality is the hard part. Finally, we note that when both conductive and radiative opacities are comparable, the combined opacity adds in parallel; that is, because

$$F = F_r + F_c = -\frac{4acT^3}{3\rho}\frac{dT}{dr}\left(\frac{1}{\kappa_r} + \frac{1}{\kappa_c}\right) \tag{2.35}$$

the net opacity is

$$\kappa = \left(\frac{1}{\kappa_r} + \frac{1}{\kappa_c}\right)^{-1}. \tag{2.36}$$

(iii) Convection

Finally, we come to one of the continuing perplexities of computational stellar evolution, and that is the treatment of the convective flux. Again, as the energy transport problem has been posed, the problem reduces to determining the value of ∇ when convection carries the flux – that is, when there is heat transport via bulk turbulent motion, we employ ∇_{conv} in Equation 2.19.

Convection is required should the radiative temperature gradient exceed the adiabatic temperature gradient. To demonstrate this, consider an ephemeral blob of material where the temperature exceeds the ambient surrounding temperature. Such a blob will have a lower density than the surroundings since pressure equilibrium is enforced. Archimedes taught us that material with a lower density than its surroundings experiences a buoyancy force upward; this blob begins to rise. If it does so without losing its identity by exchanging heat with its surroundings, then the temperature will drop as it moves upward to lower pressures. The temperature will fall along an adiabat. The surroundings, however, need not be along the same adiabat; the degree to which the surrounding temperature falls is determined by the local value of ∇ and is presumably ∇_r. If, after a displacement upward, the blob finds that it is warmer than its surroundings, it will continue to experience an upward force. That can only happen if the surrounding temperature drops faster than the adiabatic (blob) value – this occurs when $\nabla_{ad} < \nabla_r$. If not, and the blob is cooler than its surrounding after an upward displacement ($\nabla_{ad} > \nabla_r$), then the blob's density will become higher than the surroundings, and it will sink back to its equilibrium value.

Eventually, a blob that is buoyantly unstable (and therefore has a higher heat content than its surroundings) will eventually dissolve, resulting in a net transport of heat from the inside of the star to the outside. This excess heat transport will lower the overall gradient somewhat; a cascade of such blobs therefore can globally affect the temperature gradient in the star, if ∇_r is larger than ∇_{ad} at the start.

This convective instability, therefore, is generated in regions of the star where ∇_r is large when compared to the adiabatic value – from Equation 2.20, that is, where κ is large or where L/m_r is large. In the first case (large κ_r), we often see the effects of convection in cooler regions of stellar envelopes where partial ionization of a dominant species can drive up the radiative opacity. In the latter case, if energy generation is confined to a small region so that the luminosity rises rapidly over a small mass range, then convection can be required to carry all of this locally generated flux. This is the situation in the cores of stars, and in helium shell burning on the asymptotic giant branch.

In the extreme, convection suppresses the temperature gradient in unstable regions, bringing it down to ∇_{ad} in the case of perfectly efficient convection. Of course, modeling this process as resulting in neutrally stratified material is a simplification of the turbulent and dynamic motions expected where convection is present. Approximating $\nabla_{\rm conv}$ as ∇_{ad} in regions that are unstable, and mixing the material completely to simulate these dynamical processes, may be good enough for some purposes (such as core convection). And we have the added benefit that ∇_{ad} is a thermodynamic quantity, meaning that it comes for "free" from the equation of state constitutive relation.

But in reality we should allow for imperfect convection, and seek a truer treatment of convection that can yield a value for ∇ that is intermediate between ∇_r and ∇_{ad}. One such treatment, with a small number of free parameters, is the famous mixing-length theory. Tuning the mixing length to pressure scale height ratio ($\alpha \equiv \ell/\lambda_P$) so that models reproduce the solar radius and luminosity at the solar age, for example, provides a touchstone to explore how convection might work in other stars. Many variants of the mixing length have been proposed and employed over the 52+ years since it was introduced into astrophysics by Erika Böhm-Vitense (Böhm-Vitense, 1958).

2.1.4 *Conclusions: room for improvement?*

Given the importance of convective energy transport across stellar astrophysics, and the relatively crude model we have been using for it for decades, it is not surprising that astrophysics is being held back by our ignorance of the convection process. Computationally hydrodynamics simulations of conditions found in stars are being brought to bear on the subject, but are (still) computational expensive and time consuming. Implementation into stellar evolution codes is a relatively young endeavor, but an important one that is relying on seismic diagnostics of the Sun and stars.

As we will see in the remaining portions of this chapter, the coefficients of the equations of stellar structure rely on basic physics that, in some cases, is well understood. However, like convection, many of the processes that go into these coefficients are not very tightly constrained by experiment or by theory. With asteroseismic probes, we may be able to place new and strong constraints on a variety of physics problems.

2.2 The physics behind the coefficients

In the previous section, we reviewed how the basic equations of stellar structure result from various aspects of equilibrium in a self-gravitating (mostly) gaseous sphere. The coefficients within those equations connect with some basic physics: the equation of state of material under extreme conditions, nuclear reaction cross-sections, and the interaction between matter and radiation. In this section we review the relevant constitutive relations and point out areas where the physical inputs could be improved through the indirect probes provided by asteroseismology.

2.2.1 *The equation of state: $\rho(P, T, X_i)$*

2.2.1.1 Basic elements

The "easy" form of the equation of state involves the assumption of a perfect gas (perhaps undergoing ionization). Complicating factors include the role of electron degeneracy, but the computational aspects of that are well known. Less manageable aspects include non-ideal effects, where some (electromagnetic) interaction between particles leads to effects such as pressure ionization, Coulomb interactions, and state changes such as crystallization.

To begin with, though, the ideal gas relation leads to an equation of state of the form shown in Equations 2.26 and 2.27, where μ_e depends on the ionization state of the material. Ionization balance, in the ideal case, is provided by recursive solution of the Saha equation

$$\frac{n_{k+1}}{n_k} n_e = \frac{u_{k+1}(T)}{u_k(T)} \frac{2(2\pi k T m_e)^{3/2}}{h^3} e^{-\chi_k/(kT)}, \tag{2.37}$$

where the index k denotes the ionization level, χ_k is the ionization potential of level k, and u_k is a suitable partition function.

These all work well enough when the density is "sufficiently" low. But as the density increases, the quantum statistical nature of the electrons begins to overwhelm the simple ideal, perfect gas assumption. For Fermi–Dirac particles such as electrons (with available spins $\pm 1/2$), the number density is a function of particle momentum:

$$n(p) = \frac{2}{h^3} \frac{1}{\exp\left[-\mu + mc^2 + E(p)\right]/kT + 1} \tag{2.38}$$

where, generally,

$$E(p) = (p^2c^2 + m^2c^4)^{1/2} - mc^2, \tag{2.39}$$

and the chemical potential is μ.

From these, one can obtain the pressure and internal energy from

$$P = \frac{1}{3} \int_p n(p) p v 4\pi p^2 dp \quad - \text{and} - \quad E = \int_p n(p) E(p) 4\pi p^2 dp\,. \tag{2.40}$$

The interesting part of this is the energy (and temperature) dependence, the left-hand term of Equation 2.38, which allows us to write

$$n = \frac{8\pi}{h^3} \int_0^\infty F(E) p^2 dp. \tag{2.41}$$

If we now define the Fermi energy in terms of the Fermi potential μ_F:

$$E_F = \mu_F - mc^2 \tag{2.42}$$

then in the limiting case of low T (that is, $kT \ll E_F$) is $F(E) = 1$ if $E \leq E_F$ and $F(E) = 0$ if $E > E_F$. At these low temperatures, then, P and E become independent of T and depend only on the Fermi energy (which in turn depends only on the density). This is the well-known degeneracy pressure for a $T = 0$ system. At higher temperatures, thermal energy can bump particles to $E > E_F$ at the top of the energy range only, eroding the "edge" of the $F(E)$ distribution at the $E = E_F$ edge. Computation of the electron pressure under this intermediate-degeneracy range is well understood (but does involve some messier integrals).

2.2.1.2 Trouble in an ideal world: charged particles do interact

In Equation 2.37, the relative populations depend on the ionization energies χ_k. However, those energies are valid for the ion in isolation, and are measured relative to the continuum. At densities relevant to stellar interiors, however, the electric field from surrounding ions can influence the ionization level by effectively suppressing the continuum, making ionization easier at a given temperature. The separation between ions, a, depends on the density:

$$a = 7.3 \times 10^{19}\text{cm} \times \left(\rho \frac{X}{A}\right)^{-1/3} \tag{2.43}$$

so that the surrounding ions lead to a depression of the continuum of

$$\Delta\chi \approx \frac{Z^2 e^2}{2a} = 9.8\text{eV} \times Z^2 \left(\rho \frac{X}{A}\right)^{1/3}. \tag{2.44}$$

For hydrogen, at densities of a few g/cm^3, the effective change in χ_H is comparable to the ionization energy itself – that is, hydrogen should be completely ionized independent of the temperature. It is this "pressure ionization" that is responsible for the interiors of stars being fully ionized. Without it, the high densities within stellar interiors could lead to recombination if the Saha equation remains unmodified – that is, if the unadulterated ionization potential is used. Another consequence of these proximity effects is that the internal energy is reduced below perfect gas levels. Simply, the ion–electron interactions result in a binding energy that would not exist otherwise, adding a negative component to the total internal energy.

If the ions and electrons "feel" one another, then at some level, there are ion–ion interactions as well. This is because the shielding of ions by surrounding electrons can become imperfect if the density is high enough. Under those circumstances, we must consider Coulomb repulsion between the imperfectly shielded ions. The energy of these Coulomb effects is proportional to Z^2e^2/a. Scaling this to kT gives us the Coulomb parameter Γ_C:

$$\Gamma_C \equiv \frac{Z^2 e^2}{akT} = 2.27\ Z^2 \left(\frac{\rho}{10^6}\right)^{1/3} \left(\frac{T}{10^7}\right)^{-1} \left(\frac{A}{12}\right)^{-1/3} \tag{2.45}$$

using cgs units for scaling the density, Kelvin for the temperature, and AMU for the atomic weight A. When $\Gamma_C = 1$, Coulomb effects begin to be significant; in the solar interior, $\Gamma_C = 0.1$. Although they are still small, the sensitivity of helioseismic analysis demands that Coulomb effects be taken into account. For other stars, such as red giant cores and white dwarfs, Γ_C can be much greater than 1, and mutual ion repulsion becomes an important element of the equation of state budget.

In particular, as first proposed by Salpeter (1961), in white dwarfs where Γ_C can exceed 100, this mutual ion repulsion could result in crystallization of white dwarf interiors. In the modern formulation, the onset of crystallization is believed to occur when Γ_C exceeds 175. In that case, when a white dwarf interior cools to below a threshold temperature

$$T_{\text{xtal}} \approx 3.4 \times 10^6 \left(\frac{Z}{8}\right)^2 \left(\frac{A}{16}\right)^{-1/3} \left(\frac{\rho}{10^6 \text{g cm}^{-3}}\right)^{1/3} K \tag{2.46}$$

the core solidifies. Note that the crystallization varies as with $Z^2/A^{-1/3}$, meaning that as a white dwarf cools, heavier elements crystallize at higher temperatures, so in the course of white dwarf cooling, they crystallize *before* lighter elements. The order of crystallization remains a difficult issue, because of rotation and other mass motions in the degenerate interior. Whether the core of the white dwarf solidifies as some sort of alloy (principally of oxygen and carbon) or if the oxygen precipitates out prior to reaching the lower temperature of carbon crystallization remains an interesting and difficult question.

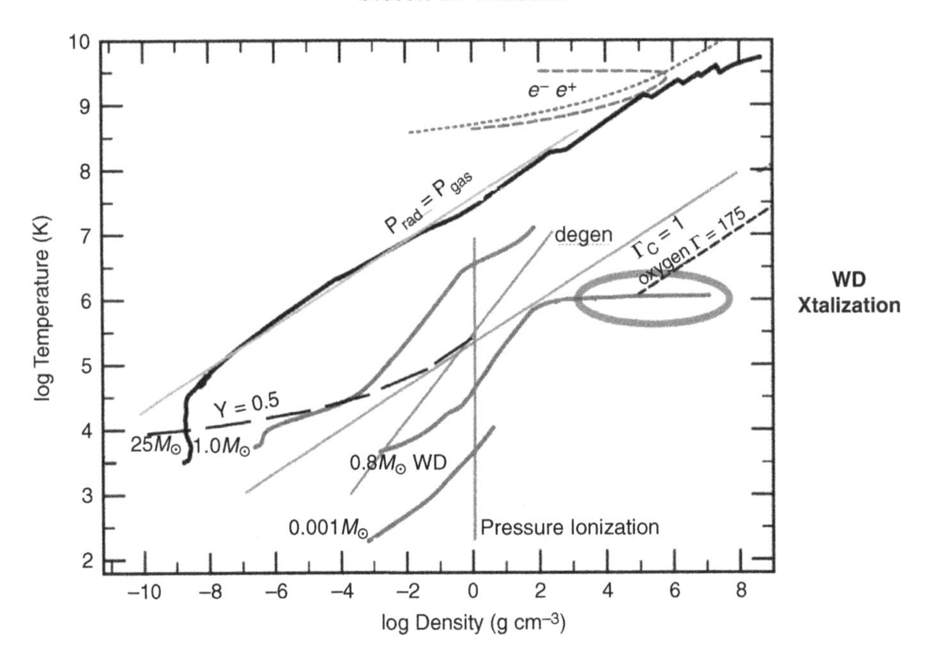

FIG. 2.1. Equation of state regions in the $\rho - T$ plane, adapted from Paxton *et al.* (2011). The labeled lines show borders where various effects are relevant as described in the text. The solid lines also denote the run of density and temperature for stellar models on the main sequence and for a 0.8 $M_\odot$ white dwarf.

2.2.1.3 Full-up equation of state calculations

Figure 2.1 summarizes the various places in phase space where these effects become important. Putting together the elements of Sections 2.2.1.1 and 2.2.1.2, the computation of a complete, realistic equation of state is not trivial. One must address ionization balance using a full mixture of elements and realistic partition functions for the bound states, while also accounting for degeneracy and nonideal effects. Most modern treatments use variations of the free-energy minimization approach to compute extensive tables with equations of state parameters for mixtures of interest at a range of pressures (or densities) and temperatures.

As an example, the OPAL effort (see Rogers *et al.*, 1996, and references therein) covers conditions relevant to the Sun and solar-type stars. For lower temperatures and relevant densities, Saumon *et al.* (1995) produced widely used tables. Under conditions relevant to white dwarfs and red giant cores, the most recent useful algorithm (with analytic fits to the full calculations) is by Potekhin and Chabrier (2010), which allows for Coulomb interactions as well as crystallization.

2.2.2 *Energy generation and loss rates:* $\epsilon(\rho, T, X_i)$

2.2.2.1 Nuclear cross sections

Developing expressions for the rate of energy generation through nuclear fusion (and loss through neutrino emission) is at the crossroads between experimental and theoretical nuclear physics and computational astrophysics. The essential scheme is a collision–physics one: this is a "rate $= n\langle\sigma v\rangle$" problem at its core. The relative velocity v depends on the mean kinetic energy per particle ($v(E) = \sqrt{(}kT/m)$) assuming a Boltzmann distribution for particle velocities, and $\sigma(E)$ is the "cross section" for the given interaction.

Nuclear physics provides a functional form for the cross section $\sigma(E)$, but laboratory measurements are essential for an absolute determination for use in stellar models.

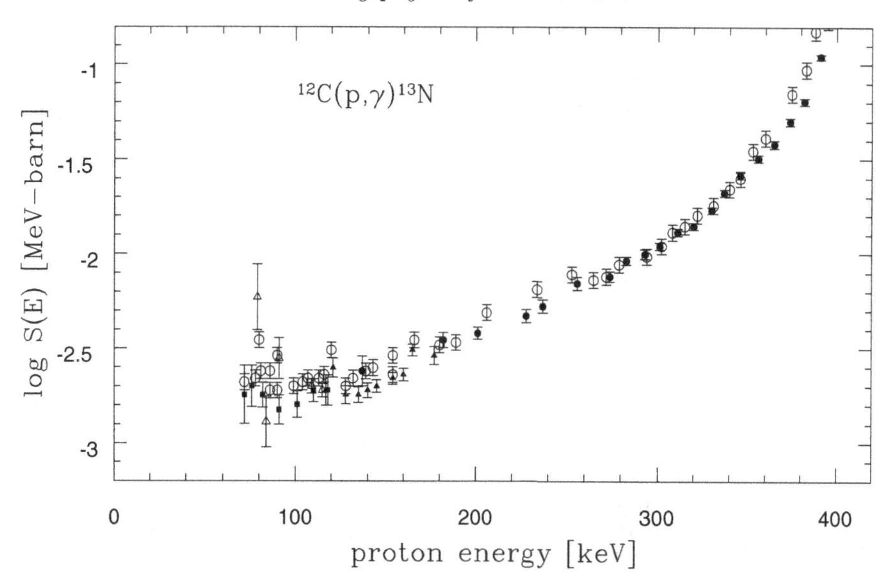

FIG. 2.2. Astrophysical S factor for a representative nuclear reaction as measured. Data from the NACRE compilation (http://pntpm3.ulb.ac.be/Nacre/barre_database.htm) for the $^{12}C(p,\gamma)^{13}N$ reaction. Typical relevant stellar energies are approximately 1-2 keV, well below the experimentally accessible energies.

Generally, these semi-empirical cross sections are tabulated as $\langle \sigma v \rangle_{ij}$ for interaction between a projectile particle i and target nucleus j, as described in Section 2.1.3.2.

The nonresonant form of $\langle \sigma v \rangle_{ij}$, Equation 2.29, contains the quantity $S(0)$, which is the astrophysical S factor at zero energy. Why zero energy? Simply put, the temperatures that we find in the centers of stars range from 10^7 K for hydrogen burning to a few times 10^8 K for helium burning, or roughly 1-20 keV. Though this represents the mean thermal energy of the particles, it is an order of magnitude or more *below* the energies available to nuclear physics experimentation. In Fig. 2.2 we show representative experimental efforts to measure the astrophysical S factor for a nonresonant reaction. Clearly, the extrapolation to zero energy needs to be done with some care given the experimental uncertainties alone, added to the analytic approximations. For more precision, additional terms allowing for the value of S at stellar energies can be derived from the experimental data for some reactions.

For *resonant* reactions, Equation 2.28 will be effective if the resonance lies within the range of energies of the interactions (or, more precisely, near the Gamow peak) provided that the energy width Γ, and its position $E_{\rm res}$, is known. In practice, though, we see from Fig. 2.2 that the relevant energies are poorly sampled by experiment in many cases. Thus, a "hidden" resonance can affect the behavior of $S(E)$ significantly, resulting in poor estimates of $\langle \sigma v \rangle_{ij}$. Perhaps the most important example of the complicating role of resonances in nuclear fusion rates, in terms of stellar astrophysics, is the $^{12}C(\alpha,\gamma)^{16}O$ reaction that is mostly responsible for setting the C/O ratio in the galaxy. There are low-energy resonances (at about 10 MeV) that are poorly mapped, and affect the cross-section at lower energies; see El Eid (2005) for a discussion. Depending on the data and extrapolation method used, the cross section at temperatures relevant for helium burning (a few $\times 10^8$ K) can vary by fairly large factors. Figure 2.3 shows these cross sections for a few formulations of this reaction.

With these caveats, there are several recent tabulations of nuclear reaction rates and their dependence on temperature and density. Still, the "classic" reference remains Caughlan and Fowler (1988) and references therein, with updated rates available on the NACRE database/website (Angulo *et al.*, 1999).

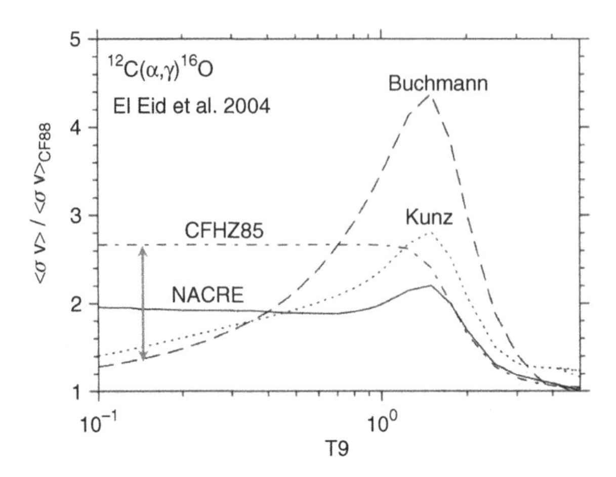

FIG. 2.3. Variation of the cross section of the $^{12}C(\alpha,\gamma)^{16}O$ reaction for various formulations, from El Eid *et al.* (2004). At typical relevant stellar energies are approximately a few $\times 10^8 K$ – indicated by the red line – the range of suggested values is significant. See El Eid *et al.* (2004) for details.

2.2.2.2 "Thermal" neutrino emission

One of the most important coolants for white dwarfs during their early phases is neutrino emission. The processes that produce neutrinos in white dwarfs differ from the neutrino production associated with nuclear fusion. In a dense plasma, neutrinos can play the role that photons play in more ordinary stellar material. When photons are produced in these interactions, they quickly thermalize with the plasma because of electromagnetic scattering. But when neutrinos are produced, they are not thermalized and leave the star, taking their share of the interaction energy away with them. Thus, neutrino emission is an energy sink, rather than a source.

The ability of neutrinos to act in this way is enhanced when the density is high – and so thermal neutrino processes can be important coolants in the cores of red giants, and in white dwarf stars. The important thermal neutrino processes include Bremmstrahlung, where neutrinos are involved in free-free scattering instead of photons. "Plasmon neutrinos" are an even more efficient energy sink in hot white dwarfs – they result from decay of plasmons within a dense plasma. Plasmons are similar to photons, but are coupled with the plasma in such a way that they can decay (into a neutrino-antineutrino pair) and still conserve momentum and energy. For an excellent and compact summary of neutrino production mechanisms in white dwarfs, see Winget *et al.* (2004).

The neutrino emission rates are, as is easy to see, impossible to measure experimentally. Even if they could be produced under experimental conditions, detection of those produced would be an exciting challenge, because thermal processes do not produce coherent neutrino beams. The rates are calculated using the Standard Model of particle physics. A current algorithm for using those calculations in the form of energy loss rates in stars is found in Itoh *et al.* (1996). Verifying those computations does fall into the realm of astrophysics – neutrinos are the dominant coolant for most white dwarfs during a significant fraction of their early cooling stages. So, measuring the cooling rate of white dwarfs can provide an experimental, if indirect, test of the theoretical thermal neutrino production rates. This measurement can be done in a statistical way by looking for features in the luminosity function of a collection of white dwarfs, but a more direct way is to measure the cooling rate of a single white dwarf. Figure 2.4 gives a preview of this, showing that the neutrio luminosity can exceed the photon luminosity during a portion of the cooling history of a white dwarf. We will discuss this further in a later section, but clearly a sensitive asteroseismic probe of white dwarf structure can enable this measurement.

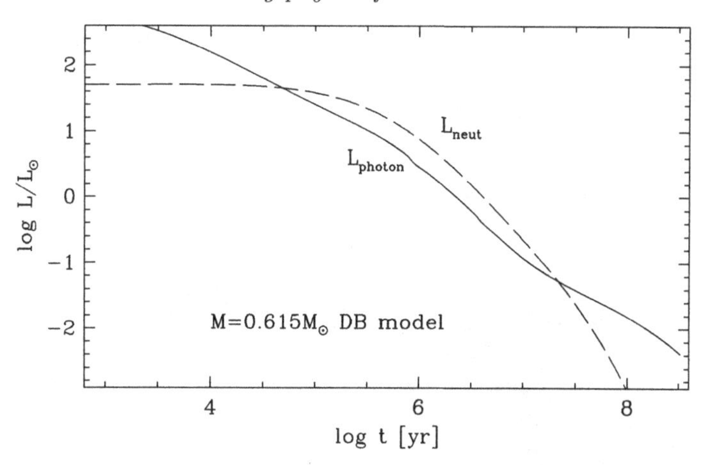

FIG. 2.4. Photon and neutrino luminosity as a function of cooling age for a $0.615M_\odot$ white dwarf with a helium-rich atmosphere. Note that the energy loss by neutrinos exceeds the photon luminosity over a significant time in this early cooling phase. The plasmon neutrino process dominates Bremmstrahlung for this model.

2.2.3 *Opacities:* $\kappa(\rho, T, X_i)$

For regions of the star that are not convectively unstable, reliable calculation of the opacity can be nontrivial, especially in the outer layers where the radiative opacity is dominated by atomic processes. The Rosseland mean opacity (Equation 2.31) hides these difficulties in the frequency-dependent opacity κ_ν. These processes include electron scattering (which is easy, as it is not frequency dependent), free–free scattering, bound–free absorption, and bound–bound absorption. At cooler temperatures, H^- opacity becomes significant for solar-type stars, and even cooler stars require accurate treatment of molecular absorption by CO, OH, H_2O, CH_4, and so on. An issue in computing the Rosseland mean opacity is that the frequency-dependent cross section is an extremely rapidly varying function at temperatures near and below complete ionization. Figure 2.5 illustrates this for iron, from Rogers and Iglesias (1992), at 2.51×10^5 K, with the relevant atomic transitions labeled.

Considering the fact that one needs to include all species that contribute to the opacities, integrals for the Rosseland mean opacity must cover dozens of elements (each in their appropriate ionization and excitation states). Indeed, advances in modeling atomic states and in computational scale resulted in significant revision of the atomic opacities used in stellar interiors calculations in the mid-1990s by Rogers and Iglesias (1992) and Seaton *et al.* (1994). It is notable that the need for updated opacities (particularly the contribution of iron) was pointed out in part as the result of seismic modeling of Cepheid variables (Simon, 1982; Iglesias and Rogers, 1991), as we will see in a later section.

For cooler stars, the contribution of molecular transitions (and, at lower temperatures, dust opacity) complicates the calculation further. Figure 2.6, from Alexander *et al.* (2003), shows a sample run of the absorption coefficient (per molecule) for CO. The relative strengths of the various low–temperature opacity sources are summarized by Ferguson *et al.* (2005).

2.2.4 *Summary*

Overall, stellar models do a very good job of describing many of the observables of stars with the current state-of-the-art of physics inputs (as long as we continue to not mention convection). When asteroseismic efforts are introduced, though, gaps in our physics understanding are quickly exposed, and this in turn can shake our confidence in the way we treat the input physics for stellar models. The current section reviewed the

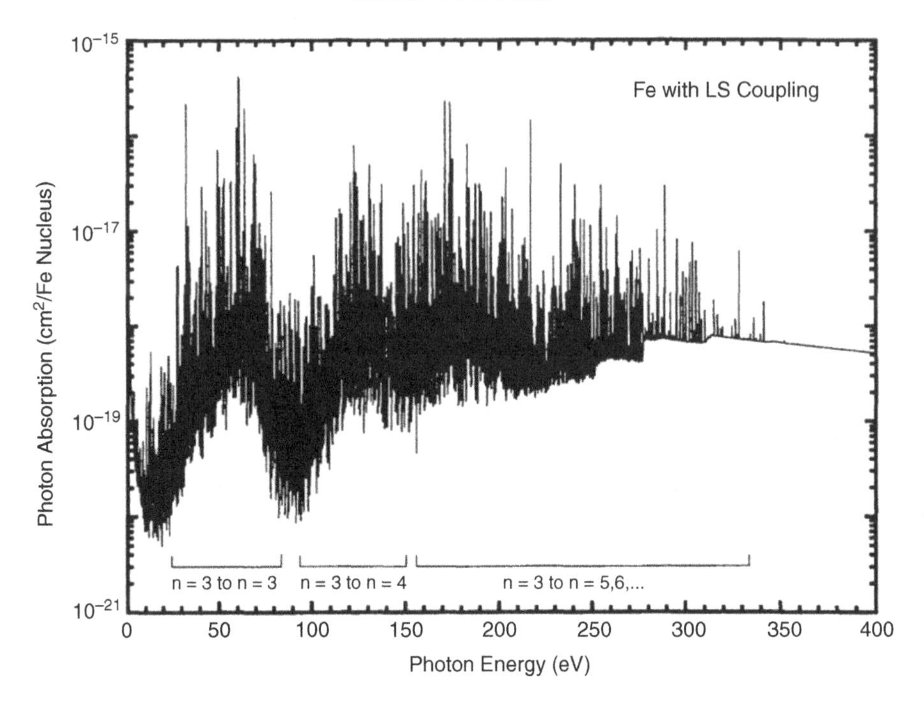

FIG. 2.5. Frequency–dependent absorption coefficient for iron at log $T = 5.4$, from Rogers and Iglesias (1992).

approach taken in modern stellar modeling – and in various places, improvements have been made based on seismic analysis.

In the following sections, we will show how observations of pulsating white dwarfs can test or constrain the physics of dense matter under conditions when crystallization might occur. Those pulsating white dwarfs also place constraints on interesting nuclear cross sections, in particular, for the $^{12}C(\alpha,\gamma)^{16}O$ reaction, and on the neutrino emission rates. Cepheid pulsation systematics demanded recalculation of radiative opacities that include

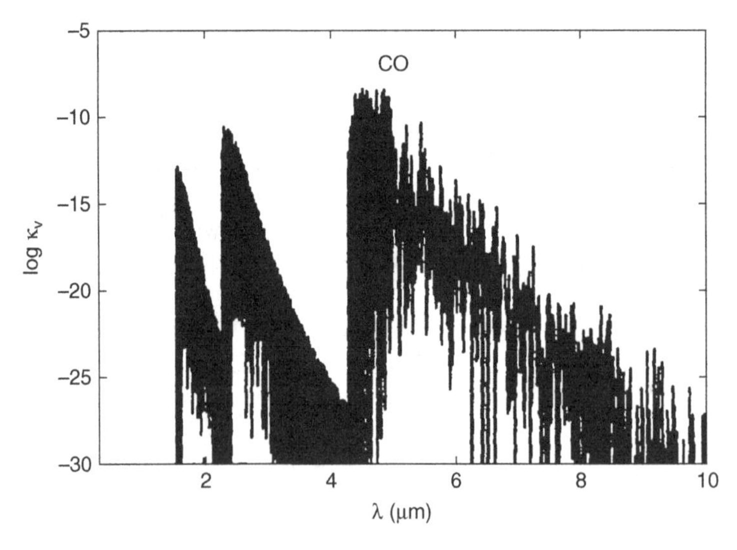

FIG. 2.6. The absorption coefficient for CO as a function of wavelength, from Alexander *et al.* (2003).

a realistic treatment of heavy elements. Observations of pulsating sdB stars probe other effects such as radiative levitation (as we will see), and solar-like oscillations of main sequence stars, as well as white dwarfs, provide new probes of convection in stars that may finally shake our reliance on the antiquated mixing length theory.

2.3 Seismology to the rescue: feedback between pulsation studies and input physics

In the previous section we reviewed some of the more difficult or problematic aspects of the physics of stellar interiors and how that physics can be compactly implemented in stellar models. Now, we can discuss how asteroseismic studies have impacted this area of stellar astrophysics (mostly for the better). The asteroseismic signal of greatest utility is the set of pulsation frequencies – using relatively simple adiabatic pulsation theory provides the tool we can apply. We'll cover some examples without trying to be exhaustive, drawing heavily on results from compact pulsating stars such as white dwarfs – since that is an area that has been around for a while and one that I'm familiar with. Althaus *et al.* (2010) provide a comprehensive review of white dwarf evolution in the context of asteroseismology that is complementary to the review by Kawaler (1995).

We start with white dwarf asteroseismology and how we can probe the crystallization of stellar interiors, and also how seismic probes of the internal composition profile can constrain important nuclear cross-sections. Next, we'll briefly discuss how timing of white dwarf pulsations over many years can provide a measurement of the cooling rate of individual stars, and therefore constrain neutrino emission rates. For variety, we'll then consider how the theory of Cepheid pulsations led to a revision of the radiative opacity calculations.

2.3.1 *White dwarf crystallization revealed by asteroseismology*

In Section 2.2.1.2 we told the story of how Salpeter (1961) concluded that under conditions found within white dwarf stars, the interior could undergo a phase transition and develop a crystalline core. What Salpeter did not anticipate was that we would find that white dwarfs undergo nonradial pulsation when the (pure hydrogen) envelopes reach temperatures where hydrogen is partially ionized at depth.

Generally, at $T_{\rm eff} \approx 11{,}000$ K to 13,000 K, the envelope partial ionization drives convection and also pulsations, through processes to be reviewed later. The historical development of white dwarf asteroseismology, which dates from the discovery of the first pulsating white dwarf by Arlo Landolt in the late 1960s (Landolt, 1968); for an excellent overview of the current state of the subject, see Winget and Kepler (2008) and Fontaine and Brassard (2008). There is a correlation between the luminosity of a white dwarf and the core temperature that follows closely the seminal work by Mestel (1952) updated by van Horn (1971):

$$\frac{L}{L_\odot} = 1.7 \times 10^{-3} \left(\frac{M}{M_\odot}\right) \frac{\mu}{\mu_e^2} T_{\rm c,7}^{3.5} \tag{2.47}$$

where the core temperature is in units of 10^7K. Recalling Section 2.2.1.2 and in particular Equation 2.46, the central temperature at crystallization for oxygen is approximately 3.4×10^6 K, which is at a luminosity of only $\approx 2 \times 10^{-5}$ $L_\odot$. This is at the lower limit of observed white dwarf luminosities, suggesting that any white dwarfs that are cool enough to crystalize will be too cool to pulsate. However, one must remember that the radius of a white dwarf is determined almost entirely by its mass (the famous mass–radius relationship for degenerate stars) – so if the mass of a white dwarf at that

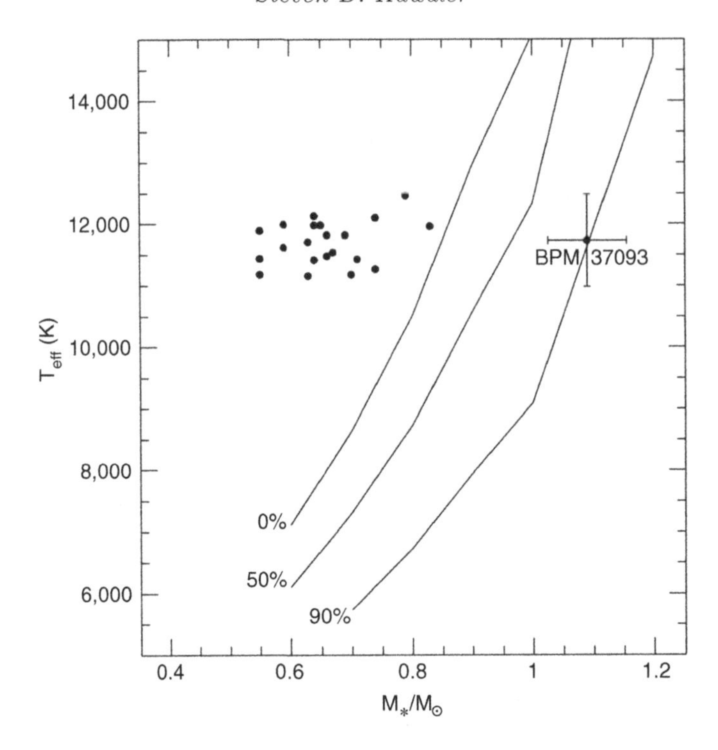

FIG. 2.7. Crystalline core fractions for white dwarf models as a function of mass and $T_{\rm eff}$ compared with some pulsating white dwarfs, from Montgomery and Winget (1999).

luminosity is large enough, it would still require a sufficiently high $T_{\rm eff}$ to be seen at this luminosity.

In fact, white dwarfs with masses greater than about 1 $M_\odot$ can crystalize at effective temperatures that place them in the pulsational instability strip. This possibility was first explored by Mike Montgomery and Don Winget (Montgomery and Winget, 1999) following the discovery of a massive pulsating white dwarf, BPM 37093, with a mass of approximately 1 $M_\odot$ (Kanaan *et al.*, 1992). Computations of evolutionary white dwarf models by Montgomery and Winget (1999) (see Fig. 2.7) compared with the then-known pulsating white dwarfs confirmed the theoretical expectation that BPM 37093 should be a pulsating white dwarf with a crystalline core. The seismic diagnostic of a crystalline core is the fact that nonradial modes propagate quite differently through solids than through gasseous material. For g modes such as we see in white dwarfs, the eigenfunctions carry very small amplitude in crystalline material when compared to noncrystalline white dwarf cores, which in turn affects the pulsation frequencies in a measurable way. Montgomery and Winget (1999) demonstrate how g mode's overtones can determine the mean period spacing, and coupled with other data can reveal whether the core is crystalline or not.

In an effort to determine the frequencies of as many modes as possible, BPM 37093 was the primary target for two Whole Earth Telescope campaigns that ultimately were able to expose several $l = 2$ g modes in the star (Kanaan *et al.*, 2000, 2005). Figure 2.8, from Kanaan *et al.* (2005), shows the measured period spacing superimposed over various models. Modeling of these pulsations concluded that there is strong evidence that the core of BPM 37093 is indeed crystalized, in accordance with theoretical expectations. With the conclusion that BPM 37093 was largely crystalline, the public became highly engaged. This asteroseismic result was announced in February 2004, making the discovery of an enormous "diamond star" a terrific Valentine's Day story!

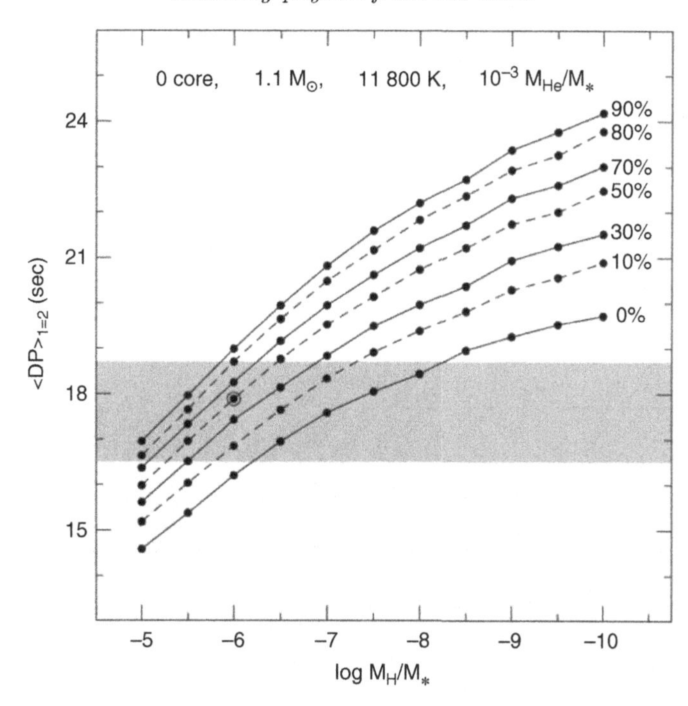

FIG. 2.8. The observed period spacing in BPM 37093 compared with relevant stellar models, from Kanaan *et al.* (2005).

2.3.2 *The* $^{12}C(\alpha,\gamma)^{16}O$ *reaction rate: constraints from white dwarf seismology*

We saw in Section 2.2.2 that the reaction that is principally responsible for the C/O ratio in the Universe has large uncertainties associated with it because of the difficulty mapping low–energy resonances. The $^{12}\mathrm{C}(\alpha,\gamma)^{16}\mathrm{O}$ reaction occurs at slightly higher temperatures than the triple–α reaction that produces $^{12}\mathrm{C}$ during core helium burning, and so the rate of production of $^{16}\mathrm{O}$ increases toward the end of helium core burning. White dwarfs therefore should have a decreasing C/O ratio as one progresses from the outer core to the inner core, reflecting this evolution. The central C/O ratio will therefore depend on the $^{12}\mathrm{C}(\alpha,\gamma)^{16}\mathrm{O}$ reaction rate.

White dwarf pulsation periods principally depend on the envelope structure (Winget and Kepler, 2008), but the core properties also influence the periods. In matching observed periods with model periods, near–surface composition gradients (accentuated by diffusion processes) produce mode trapping, which allows determination of the surface layering structure. But composition gradients in the core (caused by the increasing production of $^{16}\mathrm{O}$, for example) are also influential (Metcalfe *et al.*, 2003; Montgomery *et al.*, 2003).

With several DB (helium atmosphere) white dwarfs having successful ground-based observing campaigns that revealed a large number of modes, several groups have investigated whether seismic probes could in fact measure the core oxygen abundance and the composition gradients in the core caused by the sequential stages of helium burning. Metcalfe *et al.* (2001, 2002) showed that by varying the assumed $^{12}\mathrm{C}(\alpha,\gamma)^{16}\mathrm{O}$ reaction rate, the core oxygen mass fraction in the remnant white dwarfs could change significantly. Using data from GD 358 (Winget *et al.*, 1994), and CBS 114 (Handler *et al.*, 2002), Metcalfe (2003) concluded that the rates for this problematic reaction are in fact in line with recent determinations discussed by El Eid *et al.* (2004). The seismic results are shown in Fig. 2.9.

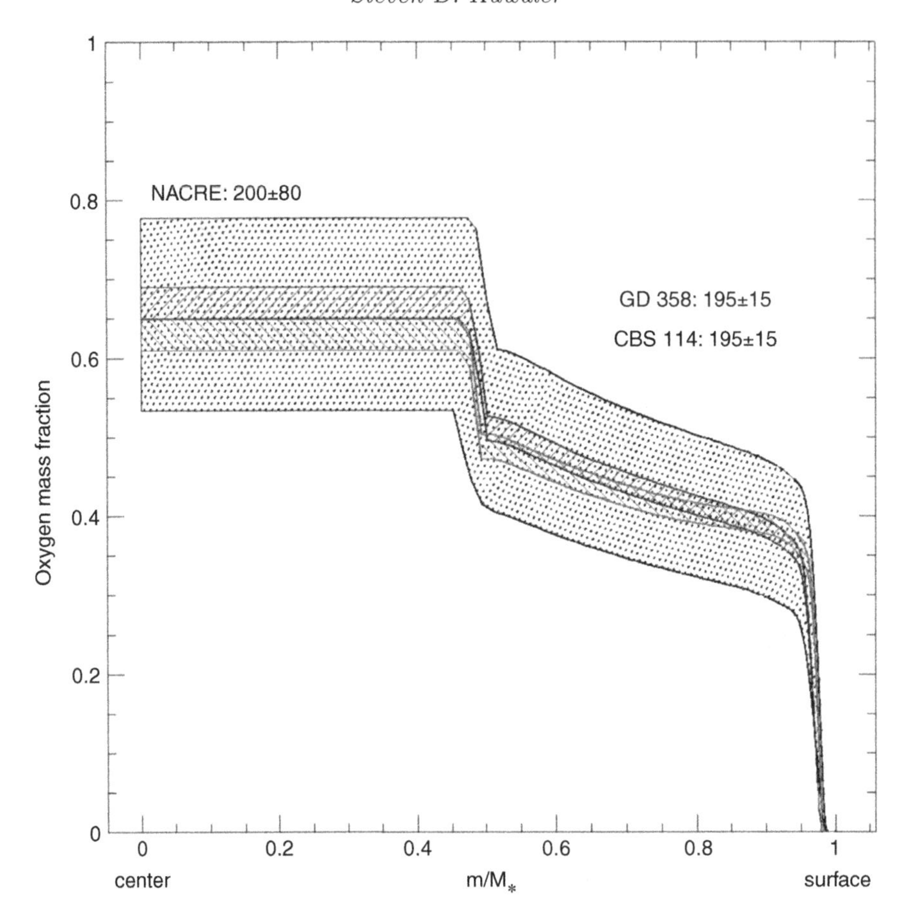

FIG. 2.9. Oxygen abundance profile in a pulsating white dwarf interior, from Metcalfe (2003). The wide (gray) shaded region is the range of profiles given the $\pm 1\sigma$ range from the NACRE compilation (Angulo *et al.*, 1999); the seismic constraints are given in the narrower band for two white dwarf pulsators.

2.3.3 *Measuring white dwarf cooling via asteroseismology*

In Section 2.2.2.2, and in particular Fig. 2.4, we saw that the cooling (and therefore overall evolution) of white dwarfs can be dominated by plasmon neutrino emission during the drop in luminosity from the planetary nebula nucleus phase into the low-luminosity phase. By measuring the cooling of an individual white dwarf whose global properties are otherwise constrained by spectroscopy and asteroseismology, we can in principle test the theoretically expected rate via observation.

2.3.3.1 Measuring cooling rates via pulsations

Consider the accuracy to which we can measure most stellar properties through spectroscopy. On a good day, we can measure, for example, $T_{\rm eff}$ to 0.1%. With knowledge of the distance to a star, we might gain a precision in luminosity of perhaps 1%. Now, if we want to measure *changes* in those quantities brought about by the slow secular evolution of the star, which occurs on time scales of millions or billions of years, we'd have to stick around for a thousand years or more before being able to detect any changes at all.

But through time-series photometry or spectroscopy, we can measure oscillation frequencies to a much higher precision, especially in those stars that show self-excited (and therefore extremely coherent) pulsations. Stars for which we can get parts-per-million accuracy in periods include the pulsating white dwarfs, δ Scuti stars, and other classical

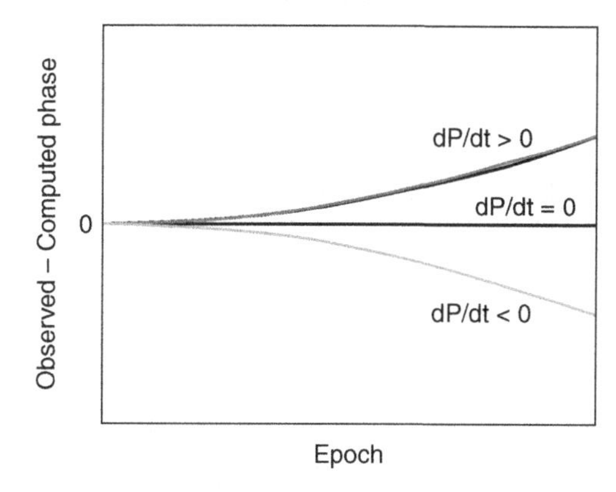

FIG. 2.10. The change in $(O - C)$ as a function of time for constant rate of increase in period (upper curve) or decrease in period (lower curve).

pulsators such as RR Lyr and Cepheid pulsators. With measurement of a global property of a star (a pulsation period) at that level of accuracy, we can indeed hope to measure a change in the pulsation period, caused by secular evolution, in a relatively short period of time (within the career of a graduate student, for example).

The direct measurement of a pulsation period change requires patience and a cooperative star that shows tractable phase or amplitude variations over many years. While one could in principle measure the pulsation period itself directly at two widely–spaced epochs, a more integrated method would be to look for continuous phase changes caused by the increasing or decreasing period. If we measure the period very precisely, and use it to predict the phase in the future under the assumption that the star's period is not changing with time, then any real changes in the period within the star will cause the phase to advance (or retreat) in a way that will reveal the period change. The traditional method for accessing these phase changes is through a so-called $(O - C)$ analysis, where $(O - C)$ is the difference between the observed phase at a given time, and that is calculated assuming that the pulsation period is constant.

The phase, as determined from the time of next maximum in the light curve at a given epoch, is given by

$$t_{\rm max,i} = t_{\rm max,0} + i \times P_0 + i^2 P \frac{dP}{dt} \tag{2.48}$$

for the ith maximum after an agreed–upon first maximum, where we have expanded the equation through a Taylor expansion assuming dP/dt is small. If the period itself does not change with time, the predicted maximum of the ith cycle will be perfectly accurate with only the first two terms in the above equation, and the value of $(O - C)_i$ will be a constant at zero. If there is an error in the period, then $(O - C)$ will increase linearly with time (if the assumed period is too short) or decrease linearly with time (if too long).

If the P_0 is not in error but the period is changing (slowly) with time, then Equation 2.48 yields

$$(O - C) = (\Delta t)^2 \frac{1}{P} \frac{dP}{dt}, \tag{2.49}$$

where Δt is the time elapsed between the chosen epoch and the measurement of the current time of maximum $t_{\rm max}$. Clearly, the departure of $(O - C)$ from zero increases as $({\rm time})^2$ goes by, enabling quadratically increasing sensitivity to dP/dt over time. A change in $(O - C)$ will have a parabolic form if dP/dt is constant, as shown in Fig. 2.10.

Since the pulsation period is some measure of the global structure of the star (such as the radius as it affects the mean density), the periods of pulsating stars change greatly as they evolve, and must change on approximately the evolutionary time scale. Assuming

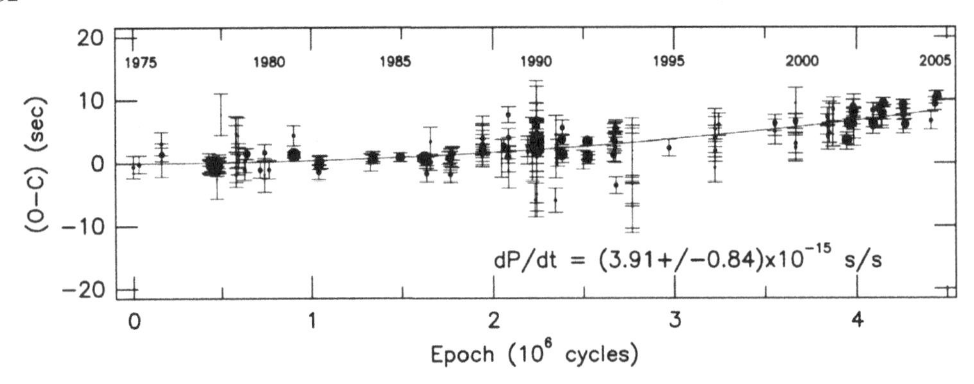

FIG. 2.11. An $(O-C)$ diagram for the 215s g mode in the pulsating DA white dwarf G117-B15A, revealing the upward parabola that results from a positive rate of period change, from Kepler *et al.* (2005). The time scale for period change is approximately 1.7×10^9 years.

a a time scale for period change (inversely proportional to $\dot{P}/P$) of 10^7 years, and a precision of measuring $t_{\max}$ of 10 seconds (both typical for young, hot white dwarfs), one finds that the time needed to detect an evolutionary change is only about 2 years for a 1σ result, and 3 years for a 3σ result.

Of course, these expectations are for ideal circumstances. For this process to work, we need to ensure that the frequencies are precisely known, and that the analysis is not hampered by cycle counting errors during the inevitable daily, monthly, and seasonal gaps in the data. In addition, not all oscillation modes are as stable in phase and/or amplitude as needed to reveal this effect. Nonlinear interactions between modes, shorter-timescale phenomena such as convection or magnetic fields, and other effects can hamper our ability to do the necessary high–precision cycle-counting required. All of these effects would only lead to larger values for $\dot{P}$, so that the estimates for the rate of period change (and therefore the evolutionary time scale) are always lower limits to the true time scales of evolution.

White dwarfs evolve at roughly constant radius, so their evolution is a cooling process, with the fading over time resulting entirely from a decreasing core (and envelope) temperature. The pulsations are g modes, which are sensitive to the Brünt-Väisäla frequency. That, in turn, is set by the temperature stratification in the star – hence the cooling of the star leads to a decrease in the Brünt-Väisäla frequency, and an increase in the pulsation period. Thus, for white dwarfs we expect to see periods increase with time, on time scales equal to the cooling rates of those stars.

An excellent demonstration of this is the pulsating DA white dwarf G117-B15A. This star has been continually monitored for well over 30 years, mostly by the Brazilian astronomer S. O. Kepler (no relation). Figure 2.11 shows the $(O-C)$ diagram for this star, and the (now) statistically significant measurement of the rate of cooling of this star (Kepler *et al.*, 2005). This star is significantly cooler than the example just cited, and has an expected cooling time scale of approximately 10^9 years.

2.3.3.2 Application to white dwarf pulsators

There are two classes of pulsating white dwarf that overlap with this neutrino cooling phase. At the higher–luminosity end are the GW Vir stars, with the prototype PG 1159-035. The luminosities of those stars are generally higher than of interest for neutrino cooling studies, but the lower–luminosity members are in the range where $L_\nu > L_\gamma$. In particular, O'Brien *et al.* (1998) and O'Brien and Kawaler (2000) show that the GW Vir star PG 0122+200 would be a good candidate for measuring the cooling rate and therefore constraining neutrino emission. Unfortunately, a recent $(O-C)$ analysis suggests that

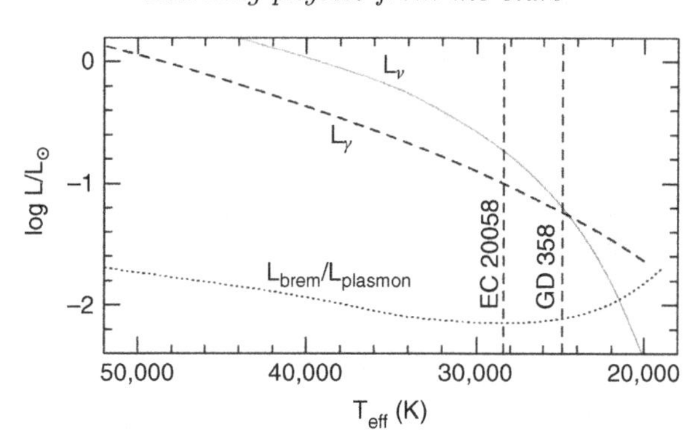

FIG. 2.12. Relative photon and neutrino luminosities in the region where we find DB pulsating white dwarfs, from Winget *et al.* (2004).

the period of PG 0122+200 is changing much faster than the models predict (Vauclair *et al.*, 2011), suggesting that other factors are influencing the pulsations.

Winget *et al.* (2004) show an example of this approach (see Fig. 2.13), and demonstrate that the rate of period change can be an effective probe, as the rate of period change depends nearly linearly on the neutrino rates used in the calculations.

More promising are stars at the lower end of the range of luminosities where neutrino emission dominates photon cooling (see Fig. 2.4). Expanding the lower luminosity side in Fig. 2.12 (from Winget *et al.*, 2004), we see that two well-studied pulsating DB white dwarf stars straddle the luminosity where neutrino emission yields to photon emission as the dominant cooling process.

In particular, EC 20058 should be experiencing neutrino cooling that exceeds photon cooling by nearly a factor of two. The influence of neutrinos on the cooling rate can be estimated by computing evolutionary models of the stars with the current neutrino rates, and with modified neutrino rates. Self-consistent calculations with and without neutrino emission, for example, both constrained to match the observed pulsation periods themselves, can then be probed to provide the expected rate of period change as a function of the neutrino emission rate.

Sullivan *et al.* (2008) report on Whole Earth Telescope observations of EC 20058, which is part of a long-term observing campaign that spans the time from its discovery as a

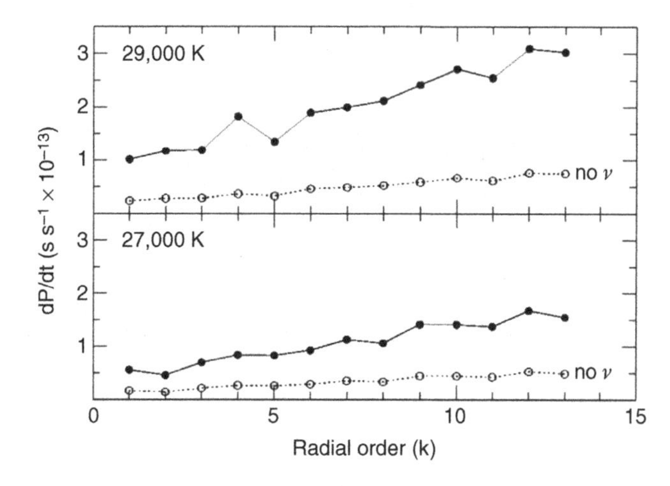

FIG. 2.13. Rate of period change in various DB white dwarf models, from Winget *et al.* (2004). The solid curve connects modes in a model with the standard neutrino emission, whereas the dotted line is for modes in a model with neutrino emission suppressed.

pulsator in the mid-1990s through the present. The star is an extremely stable pulsator (in terms of phase and amplitude changes) and should, in the near future, provide the first strong constraints on the rate of plasmon neutrino emission from a dense plasma. As Denis Sullivan would say, "Stay tuned!"

2.3.4 *Cepheid masses as opacity probes*

Here, I will briefly recount a saga from the annals of stellar pulsation theory that was an early demonstration of the power of asteroseismology (though it was yet to be named that) to expose and fix problems with stellar interior physics. The centerpieces of this story are multimode pulsators: the so–called "beat Cephieds" and "bump Cepheids." The beat Cepheids (also sometimes called double-mode Cepheids) pulsate in the (radial) fundamental and first–overtone modes. The period range of the beat Cepheids is in the 2–7 day range, and the ratio of the period of the first overtone to the fundamental (P_{10}) ranges from about 0.695 (at the long period end) to 0.715 (at the short period end). The "bump Cepheids" display a secondary maximum that progresses slowly in phase with respect to the fundamental pulsation mode; the bump is the manifestation of a near–resonance between the pulsation of the second overtone P_2 and the fundamental P_0. The bump Cepheids are a more homogeneous class with periods close to 10 days, and P_{20} very nearly 0.5 – that is, close to a 1:2 resonance.

As summarized very nicely in Moskalik *et al.* (1992) and Simon (1987), evolutionary models of beat Cepheids with the proper fundamental period P_0 had period ratios P_{10} that were significantly larger than the observed value. To reach the observed P_{10} models using the then-standard opacities (i.e., in the early 1980s) required unrealistically low masses; even so, the models had a steeper dependence of P_{10} on P_0 than what is observed. For bump Cepheids, the period of 10 days corresponded to a mass of 6.3 $M_\odot$ – but then the theoretical value of P_{20} was too high to explain the observed resonance. Thus, as of 1981, the evolutionary model masses for Cepheids did not match the pulsation masses, with the discrepancy being large enough (a factors of two more) to make this a "famous problem."

In 1981 and 1982, Norman Simon made a bold statement that these problems with Cepheid masses could be solved if the opacity caused by heavy elements was being underestimated by a factor of 2–3 in the then-current generation of opacity calculations (Simon, 1981, 1982). He found that by making such an increase, the resulting models were able to explain the observed periods and period ratios with the same mass as required by evolutionary calculations. This same augmentation also would provide a driving mechanism for the pulsating B stars known as β Cepheids, another "famous problem" of the era. His 1982 paper, "*A Plea for Reexamining Heavy Element Opacities in Stars*," states:

> It is thus quite possible that, by the single stroke of augmenting the heavy element opacities by factors of 2–3, we can bring into line with the theory of stellar structure and evolution not only the double-mode and bump Cepheids, but the β Cepheid pulsators as well.

Simon's augmented opacities from 1982 are shown in Fig. 2.14, from Simon (1982). The enhancements at temperatures from $10^{4.8}$ and above, he postulated, should result from a more careful treatment of b-f and b-b transitions in heavy elements than were implemented in the standard opacity tables available at that time. The augmented opacities in Fig. 2.14 produced models that fit the beat and bump Cepheid phenomena, and solved other problems in stellar pulsation theory.

This suggestion, in part, motivated Carlos Iglesias and Forrest Rogers, of Lawrence Livermore National Laboratories, to undertake a new computation of astrophysical opacities. They eventually turned the computational machinery of their laboratory onto the problem, and found that indeed, just as predicted by Simon nearly a decade earlier,

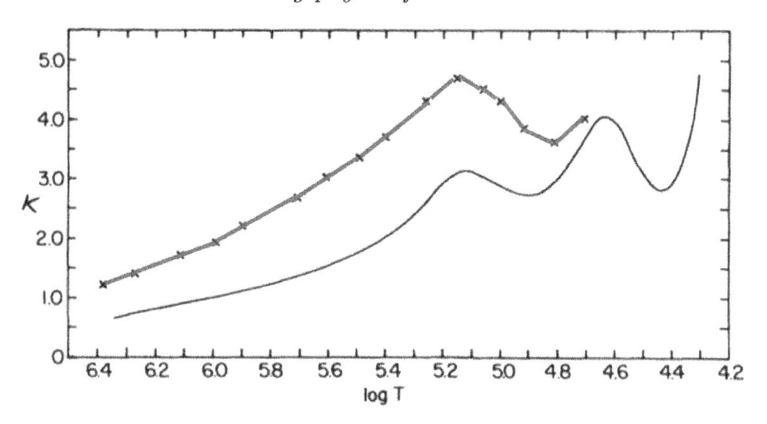

FIG. 2.14. Interior opacity in a Cepheid model from the Los Alamos compilation (Cox and Tabor, 1976), along with the suggested enhancement needed to reconcile evolutionary and pulsation models of Cepheids. The thicker line is for augmented metal opacities; the thinner, for "normal" opacities. From Simon (1982).

inclusion of an improved treatment of the physics of bound states in heavy elements led to an enhanced opacity. Figure 2.15 shows that, as Simon (1982) predicted, the opacity enhancement was a factor of 2–3 at the temperatures in that earlier work.

Using the (then) new OPAL opacities, Moskalik *et al.* (1992) revisited pulsation and evolution models of Cepheids (and β Cepheids) to verify that the OPAL opacities did indeed solve the Cepheid mass problem. Their results provided a dramatic closure to the problem, confirming that the newer input physics resulted in a much closer match

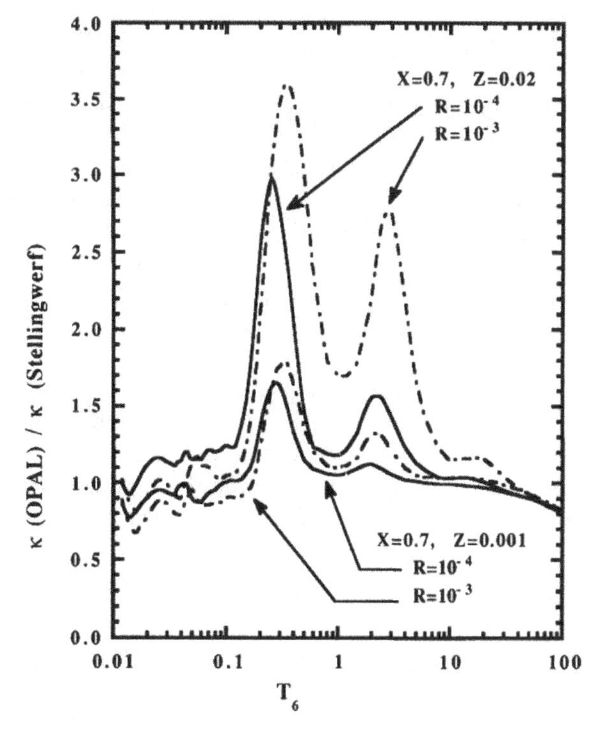

FIG. 2.15. Early OPAL opacities relevant to Cepheid interiors, compared with analytic fits by Stellingwerf (1975a,b) to the Los Alamos compilation (Cox and Tabor, 1976), from Iglesias and Rogers (1991).

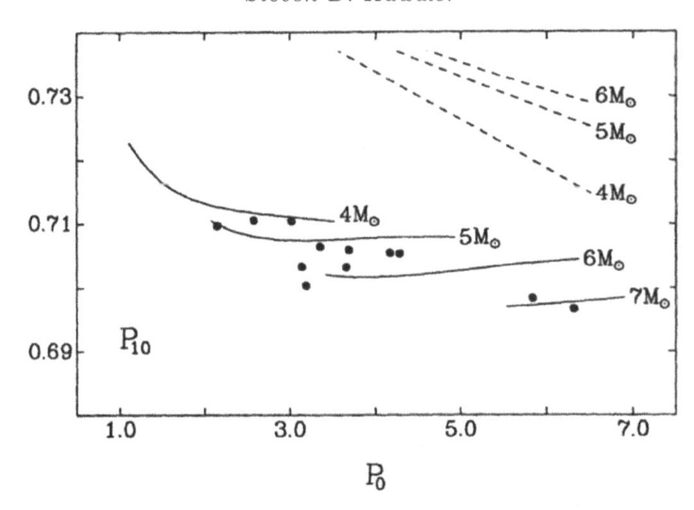

FIG. 2.16. Period ratios for beat Cepheid models compared with observations, from Moskalik *et al.* (1992). Dashed lines correspond to models with older opacities, solid lines are tracks for models with OPAL opacities.

between the masses of Cepheids from evolutionary consideration and the masses determined through pulsation period and period ratio matching. Figure 2.16, for example, shows tracks for Cepheid models with the Los Alamos opacities (dashed lines) falling well above the observed beat Cepheid period ratios in the P_{10}-P_0 plane, although models with the OPAL opacities fit the data quite closely and at reasonable masses. Moskalik *et al.* (1992) also show concordance for the bump Cepheids.

2.4 Summary

The previous section gave only a few illustrations of the ways that stellar pulsation and asteroseismology have been able to teach us some basic physics. Many other examples are in the current literature and form the core of active research in asteroseismology. We have not even mentioned the impact that helioseismology has had on the discovery and characterization of neutrino oscillations (Bahcall and Ulrich, 1988; Bahcall *et al.*, 2002) or on probes of the details of convection in the outer layers of the Sun such as in Christensen-Dalsgaard *et al.* (1991). Details of convective efficiency in white dwarfs, too, can be exposed through analysis of the shape of the pulsations (Montgomery, 2007; Montgomery *et al.*, 2010). Even more fun is the possibility that measuring the evolution rate of white dwarfs could reveal the presence of axions or other exotic particles, as suggested by Bischoff-Kim *et al.* (2008) and others.

These results are meaningful for our selfish purposes of improving stellar models for application to other areas of astrophysics, but also have value well outside the core astrophysics disciplines – atomic physics, nuclear physics, and condensed matter physics in particular. All this from a handful of coefficients in four differential equations!

Acknowledgements

It was an absolute delight to have the opportunity to participate as an instructor at the Canary Islands Winter School in Astrophysics. The IAC staff did an exemplary job in making things run smoothly, and the students who attended provided stimulating questions and discussions. Pere Pallé in particular put together a great program and kept all of us instructors on task and on time. He also provided excellent advice for all aspects of our time in La Laguna, Casa Peter, most notably.

REFERENCES

Alexander, D. R., Ferguson, J. W., Tamanai, A., Bodnarik, J., Allard, F., and Hauschildt, P. H. 2003 (Jan.). Opacities of molecules and dust. Pages 289–302 of: I. Hubeny, D. Mihalas, and K. Werner (ed), *Stellar Atmosphere Modeling*. Astronomical Society of the Pacific Conference Series, vol. 288.

Althaus, L. G., Córsico, A. H., Isern, J., and García-Berro, E. 2010. Evolutionary and pulsational properties of white dwarf stars. *A&A Rev.*, **18**(Oct.), 471–566.

Angulo, C., Arnould, M., Rayet, M., Descouvemont, P., Baye, D., Leclercq-Willain, C., Coc, A., Barhoumi, S., Aguer, P., Rolfs, C., Kunz, R., Hammer, J. W., Mayer, A., Paradellis, T., Kossionides, S., Chronidou, C., Spyrou, K., degl'Innocenti, S., Fiorentini, G., Ricci, B., Zavatarelli, S., Providencia, C., Wolters, H., Soares, J., Grama, C., Rahighi, J., Shotter, A., and Lamehi Rachti, M. 1999. A compilation of charged-particle induced thermonuclear reaction rates. *Nuclear Physics A*, **656**(Aug.), 3–183.

Bahcall, J. N., and Ulrich, R. K. 1988. Solar models, neutrino experiments, and helioseismology. *Reviews of Modern Physics*, **60**(Apr.), 297–372.

Bahcall, J. N., Gonzalez-Garcia, C. M., and Pena-Garay, C. 2002. Before and after: how has the SNO NC measurement changed things? *Journal of High Energy Physics*, **7**(July), 54–86.

Bischoff-Kim, A., Montgomery, M. H., and Winget, D. E. 2008. Strong limits on the DFSZ axion mass with G117-B15A. *ApJ*, **675**(Mar.), 1512–17.

Böhm-Vitense, E. 1958. Über die Wasserstoffkonvektionszone in Sternen verschiedener Effektivtemperaturen und Leuchtkräfte. Mit 5 Textabbildungen. *ZAp*, **46**, 108.

Caughlan, G. R., and Fowler, W. A. 1988. Thermonuclear Reaction Rates V. *Atomic Data and Nuclear Data Tables*, **40**, 283.

Christensen-Dalsgaard, J., Gough, D. O., and Thompson, M. J. 1991. The depth of the solar convection zone. *ApJ*, **378**(Sept.), 413–37.

Cox, A. N., and Tabor, J. E. 1976. Radiative opacity tables for 40 stellar mixtures. *ApJS*, **31**(June), 271–312.

El Eid, M. 2005. Astrophysics: the process of carbon creation. *Nature*, **433**(Jan.), 117–19.

El Eid, M. F., Meyer, B. S., and The, L.-S. 2004. Evolution of massive stars up to the end of central oxygen burning. *ApJ*, **611**(Aug.), 452–65.

Ferguson, J. W., Alexander, D. R., Allard, F., Barman, T., Bodnarik, J. G., Hauschildt, P. H., Heffner-Wong, A., and Tamanai, A. 2005. Low-temperature opacities. *ApJ*, **623**(Apr.), 585–96.

Fontaine, G., and Brassard, P. 2008. The pulsating white dwarf stars. *PASP*, **120**(Oct.), 1043–96.

Handler, G., Metcalfe, T. S., and Wood, M. A. 2002. The asteroseismological potential of the pulsating DB white dwarf stars CBS 114 and PG 1456+103. *MNRAS*, **335**(Sept.), 698–706.

Hansen, C. J., Kawaler, S. D., and Trimble, V. 2004. *Stellar interiors: physical principles, structure, and evolution.*

Iglesias, C. A., and Rogers, F. J. 1991. Opacity tables for Cepheid variables. *ApJ*, **371**(Apr.), L73–L75.

Itoh, N., Hayashi, H., Nishikawa, A., and Kohyama, Y. 1996. Neutrino energy loss in stellar interiors. VII. pair, photo-, plasma, bremsstrahlung, and recombination neutrino processes. *ApJS*, **102**(Feb.), 411.

Kanaan, A., Kepler, S. O., Giovannini, O., and Diaz, M. 1992. The discovery of a new DAV star using IUE temperature determination. *ApJ*, **390**(May), L89–L91.

Kanaan, A., Nitta-Kleinman, A., Winget, D. E., Kepler, S. O., Montgomery, M., and WET team. 2000. BPM 37093: preliminary results from XCOV 16 and XCOV 17. *Baltic Astronomy*, **9**, 87–96.

Kanaan, A., Nitta, A., Winget, D. E., Kepler, S. O., Montgomery, M. H., Metcalfe, T. S., Oliveira, H., Fraga, L., da Costa, A. F. M., Costa, J. E. S., Castanheira, B. G., Giovannini, O., Nather, R. E., Mukadam, A., Kawaler, S. D., O'Brien, M. S., Reed, M. D., Kleinman, S. J., Provencal, J. L., Watson, T. K., Kilkenny, D., Sullivan, D. J., Sullivan, T., Shobbrook, B., Jiang, X. J., Ashoka, B. N., Seetha, S., Leibowitz, E., Ibbetson, P., Mendelson, H., Meištas, E. G., Kalytis, R., Ališauskas, D., O'Donoghue, D., Buckley, D., Martinez, P., van Wyk, F., Stobie, R., Marang, F., van Zyl, L., Ogloza, W., Krzesinski, J., Zola, S.,

Moskalik, P., Breger, M., Stankov, A., Silvotti, R., Piccioni, A., Vauclair, G., Dolez, N., Chevreton, M., Deetjen, J., Dreizler, S., Schuh, S., Gonzalez Perez, J. M., Østensen, R., Ulla, A., Manteiga, M., Suarez, O., Burleigh, M. R., and Barstow, M. A. 2005. Whole Earth Telescope observations of BPM 37093: a seismological test of crystallization theory in white dwarfs. *A&A*, **432**(Mar.), 219–24.

Kawaler, S. D. 1995. White dwarf stars (with 37 figures). A. O. Benz and T. J.-L. Courvoisier (ed.), *Saas-Fee Advanced Course 25: Stellar Remnants.*

Kepler, S. O., Costa, J. E. S., Castanheira, B. G., Winget, D. E., Mullally, F., Nather, R. E., Kilic, M., von Hippel, T., Mukadam, A. S., and Sullivan, D. J. 2005. Measuring the evolution of the most stable optical clock G 117-B15A. *ApJ*, **634**(Dec.), 1311–18.

Landolt, A. U. 1968. A new short-period blue variable. *ApJ*, **153**(July), 151.

Mestel, L. 1952. On the theory of white dwarf stars. I. The energy sources of white dwarfs. *MNRAS*, **112**, 583.

Metcalfe, T. S. 2003. White dwarf asteroseismology and the $^{12}C(\alpha,\gamma)^{16}O$ rate. *ApJ*, **587**(Apr.), L43–L46.

Metcalfe, T. S., Winget, D. E., and Charbonneau, P. 2001. Preliminary constraints on $^{12}C(\alpha,\gamma)^{16}O$ from white dwarf seismology. *ApJ*, **557**(Aug.), 1021–27.

Metcalfe, T. S., Salaris, M., and Winget, D. E. 2002. Measuring $^{12}C(\alpha, \gamma)^{16}O$ from white dwarf asteroseismology. *ApJ*, **573**(July), 803–11.

Metcalfe, T. S., Montgomery, M. H., and Kawaler, S. D. 2003. Probing the core and envelope structure of DBV white dwarfs. *MNRAS*, **344**(Oct.), L88–L92.

Montgomery, M. H. 2007. Using non-sinusoidal light curves of multi-periodic pulsators to constrain convection. R. Napiwotzki and M. R. Burleigh (ed.), *15th European Workshop on White Dwarfs*. Astronomical Society of the Pacific Conference Series, vol. 372.

Montgomery, M. H., and Winget, D. E. 1999. The effect of crystallization on the pulsations of white dwarf stars. *ApJ*, **526**(Dec.), 976–90.

Montgomery, M. H., Metcalfe, T. S., and Winget, D. E. 2003. The core/envelope symmetry in pulsating stars. *MNRAS*, **344**(Sept.), 657–64.

Montgomery, M. H., Provencal, J. L., Kanaan, A., Mukadam, A. S., Thompson, S. E., Dalessio, J., Shipman, H. L., Winget, D. E., Kepler, S. O., and Koester, D. 2010. Evidence for temperature change and oblique pulsation from light curve fits of the pulsating white dwarf GD 358. *ApJ*, **716**(June), 84–96.

Moskalik, P., Buchler, J. R., and Marom, A. 1992. Toward a resolution of the bump and beat Cepheid mass discrepancies. *ApJ*, **385**(Feb.), 685–93.

O'Brien, M. S., and Kawaler, S. D. 2000. The predicted signature of neutrino emission in observations of pulsating pre-white dwarf stars. *ApJ*, **539**(Aug.), 372–8.

O'Brien, M. S., Vauclair, G., Kawaler, S. D., Watson, T. K., Winget, D. E., Nather, R. E., Montgomery, M., Nitta, A., Kleinman, S. J., Sullivan, D. J., Jiang, X. J., Marar, T. M. K., Seetha, S., Ashoka, B. N., Bhattacharya, J., Leibowitz, E. M., Hemar, S., Ibbetson, P., Warner, B., van Zyl, L., Moskalik, P., Zola, S., Pajdosz, G., Krzesinski, J., Dolez, N., Chevreton, M., Solheim, J.-E., Thomassen, T., Kepler, S. O., Giovannini, O., Provencal, J. L., Wood, M. A., and Clemens, J. C. 1998. Asteroseismology of a star cooled by neutrino emission: the pulsating pre–white dwarf PG 0122+200. *ApJ*, **495**(Mar.), 458.

Paxton, B., Bildsten, L., Dotter, A., Herwig, F., Lesaffre, P., and Timmes, F. 2011. Modules for experiments in stellar astrophysics (MESA). *ApJS*, **192**(Jan.), 3.

Potekhin, A. Y., and Chabrier, G. 2010. Thermodynamic functions of dense plasmas: analytic approximations for astrophysical applications. *Contributions to Plasma Physics*, **50**(Jan.), 82–7.

Rogers, F. J., and Iglesias, C. A. 1992. Radiative atomic Rosseland mean opacity tables. *ApJS*, **79**(Apr.), 507–68.

Rogers, F. J., Swenson, F. J., and Iglesias, C. A. 1996. OPAL equation-of-state tables for astrophysical applications. *ApJ*, **456**(Jan.), 902.

Salpeter, E. E. 1961. Energy and pressure of a zero-temperature plasma. *ApJ*, **134**(Nov.), 669.

Saumon, D., Chabrier, G., and van Horn, H. M. 1995. An equation of state for low-mass stars and giant planets. *ApJS*, **99**(Aug.), 713.

Seaton, M. J., Yan, Y., Mihalas, D., and Pradhan, A. K. 1994. Opacities for stellar envelopes. *MNRAS*, **266**(Feb.), 805.

Simon, N. R. 1981 (Mar.). Experimental envelope models for cepheids. *Bulletin of the American Astronomical Society*. Bulletin of the American Astronomical Society, vol. 13.

Simon, N. R. 1982. A plea for reexamining heavy element opacities in stars. *ApJ*, **260**(Sept.), L87–L90.

Simon, N. R. 1987. Cepheids: problems and possibilities. Pages 148–158 of: A. N. Cox, W. M. Sparks, and S. G. Starrfield (ed.), *Stellar Pulsation*. Lecture Notes in Physics, Berlin Springer Verlag, vol. 274.

Stellingwerf, R. F. 1975a. Modal stability of RR Lyrae stars. *ApJ*, **195**(Jan.), 441–66.

Stellingwerf, R. F. 1975b. Nonlinear effects in double-mode Cepheids. *ApJ*, **199**(Aug.), 705–709.

Sullivan, D. J., Metcalfe, T. S., O'Donoghue, D., Winget, D. E., Kilkenny, D., van Wyk, F., Kanaan, A., Kepler, S. O., Nitta, A., Kawaler, S. D., Montgomery, M. H., Nather, R. E., O'Brien, M. S., Bischoff-Kim, A., Wood, M., Jiang, X. J., Leibowitz, E. M., Ibbetson, P., Zola, S., Krzesinski, J., Pajdosz, G., Vauclair, G., Dolez, N., and Chevreton, M. 2008. Whole Earth Telescope observations of the hot helium atmosphere pulsating white dwarf EC20058-5234. *MNRAS*, **387**(June), 137–52.

van Horn, H. M. 1971. Cooling of white dwarfs. W. J. Luyten (ed.), *White Dwarfs*. IAU Symposium, vol. 42.

Vauclair, G., Fu, J.-N., Solheim, J.-E., Kim, S.-L., Dolez, N., Chevreton, M., Chen, L., Wood, M. A., Silver, I. M., Bognár, Z., Paparó, M., and Córsico, A. H. 2011. The period and amplitude changes in the coolest GW Virginis variable star (PG 1159-type) PG 0122+200. *A&A*, **528**(Apr.), A5.

Winget, D. E., and Kepler, S. O. 2008. Pulsating white dwarf stars and precision asteroseismology. *ARA&A*, **46**(Sept.), 157–99.

Winget, D. E., Nather, R. E., Clemens, J. C., Provencal, J. L., Kleinman, S. J., Bradley, P. A., Claver, C. F., Dixson, J. S., Montgomery, M. H., Hansen, C. J., Hine, B. P., Birch, P., Candy, M., Marar, T. M. K., Seetha, S., Ashoka, B. N., Leibowitz, E. M., O'Donoghue, D., Warner, B., Buckley, D. A. H., Tripe, P., Vauclair, G., Dolez, N., Chevreton, M., Serre, T., Garrido, R., Kepler, S. O., Kanaan, A., Augusteijn, T., Wood, M. A., Bergeron, P., and Grauer, A. D. 1994. Whole Earth Telescope observations of the DBV white dwarf GD 358. *ApJ*, **430**(Aug.), 839–49.

Winget, D. E., Sullivan, D. J., Metcalfe, T. S., Kawaler, S. D., and Montgomery, M. H. 2004. A strong test of electroweak theory using pulsating DB white dwarf stars as plasmon neutrino detectors. *ApJ*, **602**(Feb.), L109–L112.

3. Solar-like oscillations: An observational perspective

TIMOTHY R. BEDDING

3.1 What are solar-like oscillations?

Oscillations in the Sun are excited stochastically by convection. We use the term "solar-like" to refer to oscillations in other stars that are excited by the same mechanism, even though some of these stars may be very different from the Sun. The stochastic nature of the excitation produces oscillations over a broad range of frequencies, which in the Sun is about 1 to 4 mHz (the well-known 5-minute oscillations). Stellar oscillations can also be excited via opacity variations (the heat-engine mechanism, also called the κ mechanism), as seen in various types of classical pulsating stars (Cepheids, RR Lyraes, Miras, white dwarfs, δ Scuti stars, etc.).

For a star to show solar-like oscillations, it must be cool enough to have a surface convective zone. In practice, this means being cooler than the red edge of the classical instability strip, which includes the lower main sequence, as well as cool subgiants and even red giants. Indeed, solar-like oscillations with periods of hours (and longer) have now been observed in thousands of G and K giants (see Section 3.9). There is also good evidence that the pulsations in semiregular variables (M giants) are solar-like (Christensen-Dalsgaard *et al.*, 2001; Bedding, 2003; Tabur *et al.*, 2010), as perhaps are those in M supergiants such as Betelgeuse (Kiss *et al.*, 2006).

What about hotter stars? By definition, solar-like oscillations are excited stochastically in the outer convection zone. Given that surface convection is thought to inhibit operation of the κ mechanism (e.g., Gastine and Dintrans, 2011), it seems that solar-like and heat-engine oscillations would be mutually exclusive. In fact, both may operate in a star provided the surface convection layer is thin (Samadi *et al.*, 2002; Kallinger and Matthews, 2010) and there is evidence from Kepler data for solar-like oscillations in at least one δ Scuti star (Antoci *et al.*, 2011). There has even been a suggestion that subsurface convection in B-type stars could excite solar-like oscillations (Cantiello *et al.*, 2009; Belkacem *et al.*, 2010). There is some evidence from CoRoT observations that the β Cephei star V1449 Aql shows solar-like oscillations (Belkacem *et al.*, 2009), although other interpretations are possible (Degroote *et al.*, 2009) and Kepler observations of similar stars have so far failed to confirm stochastically excited oscillations (Balona *et al.*, 2011). Meanwhile, Degroote *et al.* (2010) suggested that the O-type star HD 46149 shows stochastically excited oscillations, based on CoRoT photometry.

3.2 Properties of oscillations

We now summarize the main properties of solar-like oscillations. For more extensive discussion, see, for example, Brown and Gilliland (1994), Bedding and Kjeldsen (2003), Christensen-Dalsgaard (2004), Aerts *et al.* (2010), and Ballot (2010).

3.2.1 *Types of oscillation modes*

Broadly speaking, there are two types of stellar oscillations. Pressure modes (*p modes*) are acoustic waves, for which the restoring force arises from the pressure gradient. These are seen in the Sun and Sun-like stars, and also in most of the classical pulsators. For gravity modes (*g modes*), the restoring force is buoyancy. The best-studied examples occur in white dwarfs (see Althaus *et al.*, 2010, for a recent review), but g modes are also seen in other classical pulsators such as γ Dor stars, slowly pulsating B stars, and some sdB stars. Some stars have been found to have both p modes and g modes, including hybrid δ Scuti/γ Dor stars (Grigahcène *et al.*, 2010) and possibly the Sun (García *et al.*, 2008). Finally, some stars oscillate in *mixed modes*, which have p-mode character in the envelope and g-mode character in the core. These are discussed in Sections 3.8 and 3.9. First, however, we consider p modes in some detail because they are most relevant to the topic of solar-like oscillations.

Mathematically, each p mode can be described by three integers. The radial order (n) specifies the number of nodal shells of the standing wave. In the Sun, the modes with the highest amplitude have n in the range 19–22. The angular degree ($l = 0, 1, 2, \ldots$) specifies the number of nodal lines at the surface. Modes with $l = 0$ are called *radial modes* and those with $l \geq 1$ are *non-radial modes*. In addition, modes with $l = 0$ are sometimes called *monopole modes*, while those with $l = 1$ are *dipole modes*, those with $l = 2$ are *quadrupole modes*, and those with $l = 3$ are *octupole modes*. Because the surface of a distant star is generally not resolved, cancellation effects mean that only modes with $l \leq 3$ (or perhaps 4) are observable.

The third integer that specifies a mode is the *azimuthal order* (m), which takes on values from $-l$ to $+l$. However, the oscillation frequencies do not depend on m unless the star is rotating (or spherical symmetry is broken in some other way, such as by the presence of a magnetic field). For a rotating star, the modes with a given n and l are split in frequency into a multiplet with $2l + 1$ components (one for each value of m). The frequency separation between these components is proportional to the rotational frequency of the star and their relative amplitudes depend on the inclination angle between the stellar rotation axis and the line of sight. The study of rotational splitting in solar-like oscillations, while showing great promise, is still in its infancy and will not be discussed further in this chapter.

The comments in the preceding two paragraphs also apply to g modes, but note that (1) radial g modes do not exist (and, therefore, neither do radial mixed modes), (2) n is generally taken to be negative for g modes (see Section 3.5.2 of Aerts *et al.*, 2010, for a full discussion), and (3) rotational splitting of g modes depends on l (Section 3.8.4 of Aerts *et al.*, 2010).

3.2.2 *The frequency spectrum of oscillations*

Figure 3.1a shows the frequency spectrum of p modes in the Sun. The envelope of the peak heights defines the frequency of maximum power, $\nu_{\rm max}$, which has a value of about 3,100 μHz in the Sun. The mode frequencies show a very regular pattern that is characteristic of an oscillating sphere. Figure 3.1b shows a close-up labeled with the (n, l) values of the modes, which were determined by comparing with theoretical models. Note the alternating pattern of large and small separations as l cycles repeatedly through the values $2, 0, 3, 1, 2, 0, \ldots$ The quantity $\Delta\nu$ is the spacing between consecutive radial overtones (i.e., modes with a given l whose n values differ by 1). It is called the *large frequency separation* and has a value of 135 μHz in the Sun. To a good approximation, $\Delta\nu$ is proportional to the square root of the mean stellar density (see Section 3.6).

In Fig. 3.1b we see two so-called small frequency separations: $\delta\nu_{02}$ is the spacing between close pairs with $l = 0$ and $l = 2$ (for which n differs by 1) and $\delta\nu_{13}$ is the spacing between adjacent modes with $l = 1$ and $l = 3$ (ditto). Another small separation can be defined by noting that the $l = 1$ modes do not fall exactly halfway between $l = 0$ modes. We define $\delta\nu_{01}$ to be the amount by which $l = 1$ modes are offset from the midpoint of the $l = 0$ modes on either side.[1] To be explicit, for a given radial order, n,

[1] One can also define an equivalent quantity, $\delta\nu_{10}$, as the offset of $l = 0$ modes from the midpoint between the surrounding $l = 1$ modes, so that $\delta\nu_{10} = \nu_{n,0} - \frac{1}{2}(\nu_{n-1,1} + \nu_{n,1})$.

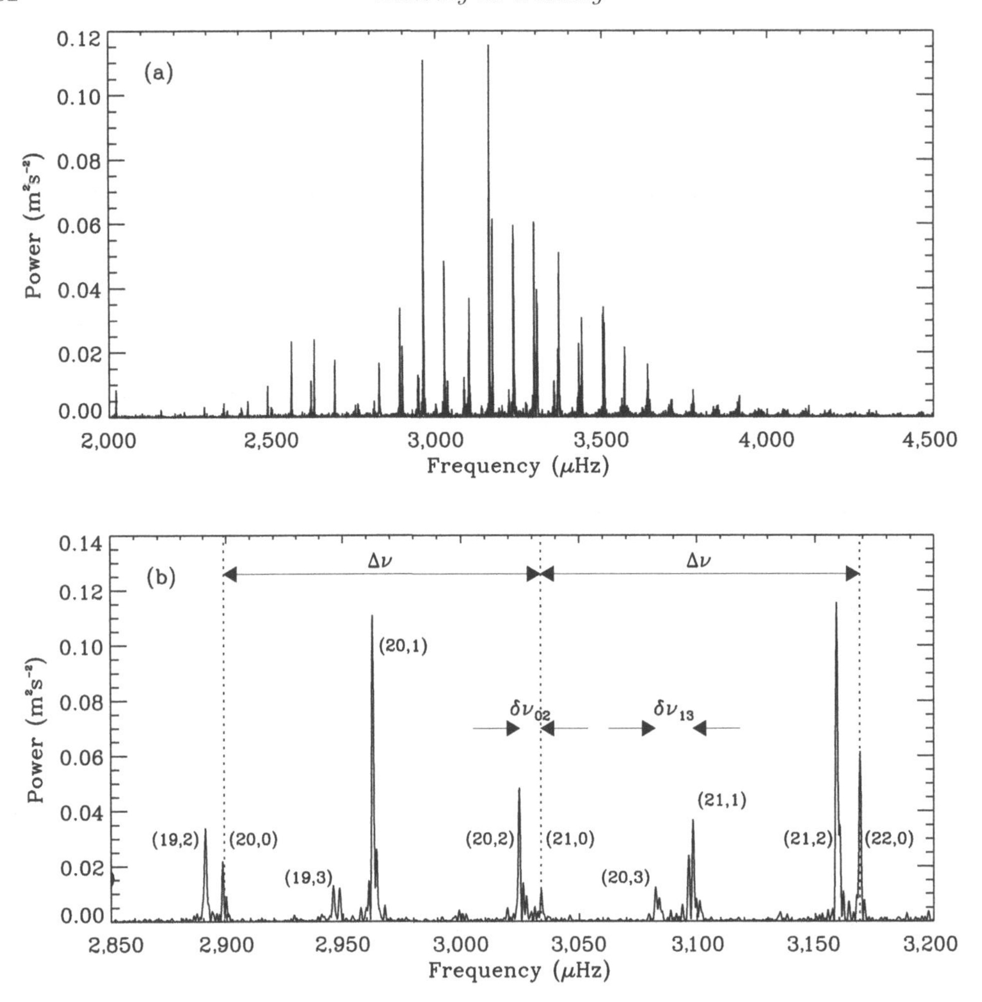

FIG. 3.1. The solar power spectrum from ten days of disk-integrated velocity measurements with the BiSON instrument (e.g., Chaplin *et al.*, 1997). Panel (b) shows a close-up labeled with (n, l) values for each mode. Dotted lines show the radial modes, and the large and small separations are indicated.

these separations are defined as follows:

$$\delta\nu_{02} = \nu_{n,0} - \nu_{n-1,2} \tag{3.1}$$

$$\delta\nu_{01} = \tfrac{1}{2}(\nu_{n,0} + \nu_{n+1,0}) - \nu_{n,1} \tag{3.2}$$

$$\delta\nu_{13} = \nu_{n,1} - \nu_{n-1,3}. \tag{3.3}$$

In practice, the oscillation spectrum is not precisely regular and so all of these spacings vary slightly with frequency.

The $l = 4$ modes are weakly visible in Fig. 3.1b (at 3,000 and 3,136 μHz). By analogy, with $l = 2$, we can define the small separation for these modes as

$$\delta\nu_{04} = \nu_{n,0} - \nu_{n-2,4}. \tag{3.4}$$

Finally, we note that each peak in the power spectrum in Fig. 3.1b is slightly broadened, which reflects the finite lifetimes of the modes. If we define the lifetime τ of a damped oscillation mode to be the time for the amplitude to decay by a factor of e then this mode will produce a Lorentzian shape in the power spectrum with a linewidth (FWHM)

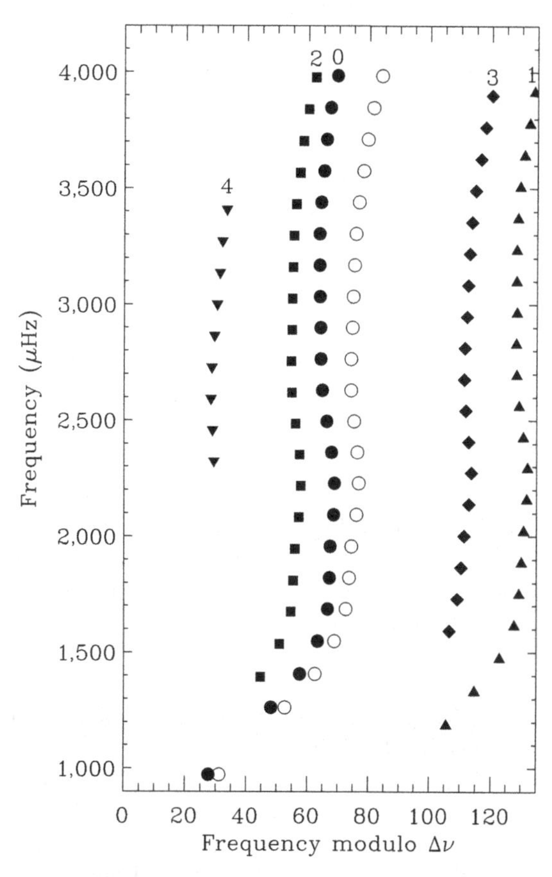

FIG. 3.2. Échelle diagram of the observed frequencies in the Sun (Broomhall *et al.*, 2009). The filled symbols show the frequencies of modes with $l = 0$, 1, 2, 3, and 4, using a large separation of $\Delta\nu = 135.0\,\mu$Hz. The open symbols show the result of plotting the $l = 0$ frequencies modulo: a slightly smaller large separation ($\Delta\nu = 134.5\,\mu$Hz).

of $\Gamma = (\pi\tau)^{-1}$. The dominant modes in the Sun have linewidths of 1–2 μHz and hence lifetimes of 2–4 days (e.g., Chaplin *et al.*, 1997).

3.2.3 *The échelle diagram*

The échelle diagram, first introduced by Grec *et al.* (1983) for global helioseismology, is used extensively in asteroseismology to display oscillation frequencies. It involves dividing the spectrum into equal segments of length $\Delta\nu$ and stacking them one above the other so that modes with a given degree align vertically in ridges. Any departures from regularity are clearly visible as curvature in the échelle diagram. For example, variations in the small separations appear as a convergence or divergence of the corresponding ridges.

Figure 3.2 shows the observed frequencies in the Sun in échelle format for modes with $l \leq 4$ (filled symbols), using a large separation of $\Delta\nu = 135.0\,\mu$Hz. Using a slightly smaller value of $\Delta\nu$ causes the ridges to tilt to the right, as shown by the open symbols for $l = 0$ only. It also shifts the ridges sideways, which is important if we are trying to measure the quantity ϵ (see Section 3.2.4).

Sometimes, the ridges wrap around, which has almost happened to the $l = 1$ modes at the top of Fig. 3.2 (filled triangles). Of course, the diagram can be made wider by plotting the ridges more than once (an example is shown in the right panel of Fig. 3.10). Also note that one can shift the ridges sideways by subtracting a fixed reference frequency, which must be kept in mind when reading ϵ from the diagram (again, see Fig. 3.10).

When making an échelle diagram, it is usual to plot ν versus (ν mod $\Delta\nu$), in which case each order slopes upwards slightly (see Fig. 3.2). However, for grayscale images it can be preferable to keep the orders horizontal (an example is shown Fig. 3.4).

3.2.4 *The asymptotic relation*

The approximate regularity of the p-mode spectrum means we can express the mode frequencies in terms of the large and small separations, as follows:

$$\nu_{n,l} \approx \Delta\nu\left(n + \tfrac{1}{2}l + \epsilon\right) - \delta\nu_{0l}. \tag{3.5}$$

Here, $\nu_{n,l}$ is the frequency of the p mode with radial order n and angular degree l (and assuming no dependence on m, which means neglecting any rotational splitting). By definition, the small separation $\delta\nu_{0l}$ is zero for $l = 0$. Its values for $l = 1$, 2 and 4 have already been discussed (Equations 3.1, 3.2, and 3.4). For $l = 3$, we have $\delta\nu_{03} = \delta\nu_{01} + \delta\nu_{13}$, and we note that $\delta\nu_{03}$ is the amount by which the $l = 3$ mode is offset from the midpoint of the adjacent $l = 0$ modes. This is arguably a more sensible measurement of the position of the $l = 3$ ridge than $\delta\nu_{13}$ because it is made with reference to the radial modes (see Bedding *et al.*, 2010b).

Equation 3.5 describes the oscillation frequencies from an observational perspective. A theoretical asymptotic expression (Tassoul, 1980; Gough, 1986, 2003) gives physical significance to $\Delta\nu$, $\delta\nu_{0l}$ and ϵ as integrals of the sound speed. It turns out that the large separation is approximately proportional to the square root of the mean stellar density, as discussed further in Section 3.6. The small separations in main-sequence stars are sensitive to the gradient of the sound speed near the core, and hence to the age of the star (see Section 3.7). The asymptotic analysis (see previous references) indicates that $\delta\nu_{0l}$ is proportional to $l(l+1)$. Finally, ϵ is sensitive to upper turning point of the modes (e.g., Gough, 1986; Pérez Hernández and Christensen-Dalsgaard, 1998; Roxburgh, 2010). In the échelle diagram, ϵ can be measured from the horizontal position of the $l = 0$ ridge. In practice there is an ambiguity in ϵ if n is not known (see Equation 3.5). In the Sun, we know from theoretical models that $\epsilon \approx 1.5$ (rather than 0.5), which allows us to find the correct value of n for each mode. But it is important to remember that n is not directly observable.

3.2.5 *Departures from the asymptotic relation*

The asymptotic relation in Equation 3.5 is only an approximation. Looking at Fig. 3.2, we see that the ridges all curve in an "S" shape that is particularly pronounced at low frequencies. We also see that the $l = 0$ and $l = 2$ ridges converge toward the top, as do the $l = 1$ and $l = 3$ ridges. How should we best describe these departures? One approach is to say that $\Delta\nu$ is a function of both l and frequency. In fact, doing this is sufficient to describe all of the features just mentioned. However, it is physically more sensible to say instead that the convergence of the ridges is due to changes of the various *small* separations with frequency. Indeed, a linear dependence of $\delta\nu_{02}$ on frequency fits the solar data quite well (Elsworth *et al.*, 1990), and has been adopted for other stars (White *et al.*, 2011). If we allow the small separations to vary with frequency then there is no longer any need to allow $\Delta\nu$ to be a function of l.

What about the overall S-shaped curvature? This can be described as a change of $\Delta\nu$ with frequency, and this is often done. However, it seems physically more sensible to ascribe the curvature to variations with frequency of ϵ (e.g., Gough, 1986; Pérez Hernández and Christensen-Dalsgaard, 1998; Roxburgh, 2010). This allows $\Delta\nu$ to have a single value that is independent of both l and frequency. However, in practice, if we are given a set of frequencies (either observed or calculated) then $\Delta\nu$ and ϵ must be measured together, preferably in the region centred on $\nu_{\max}$. As we see from the open

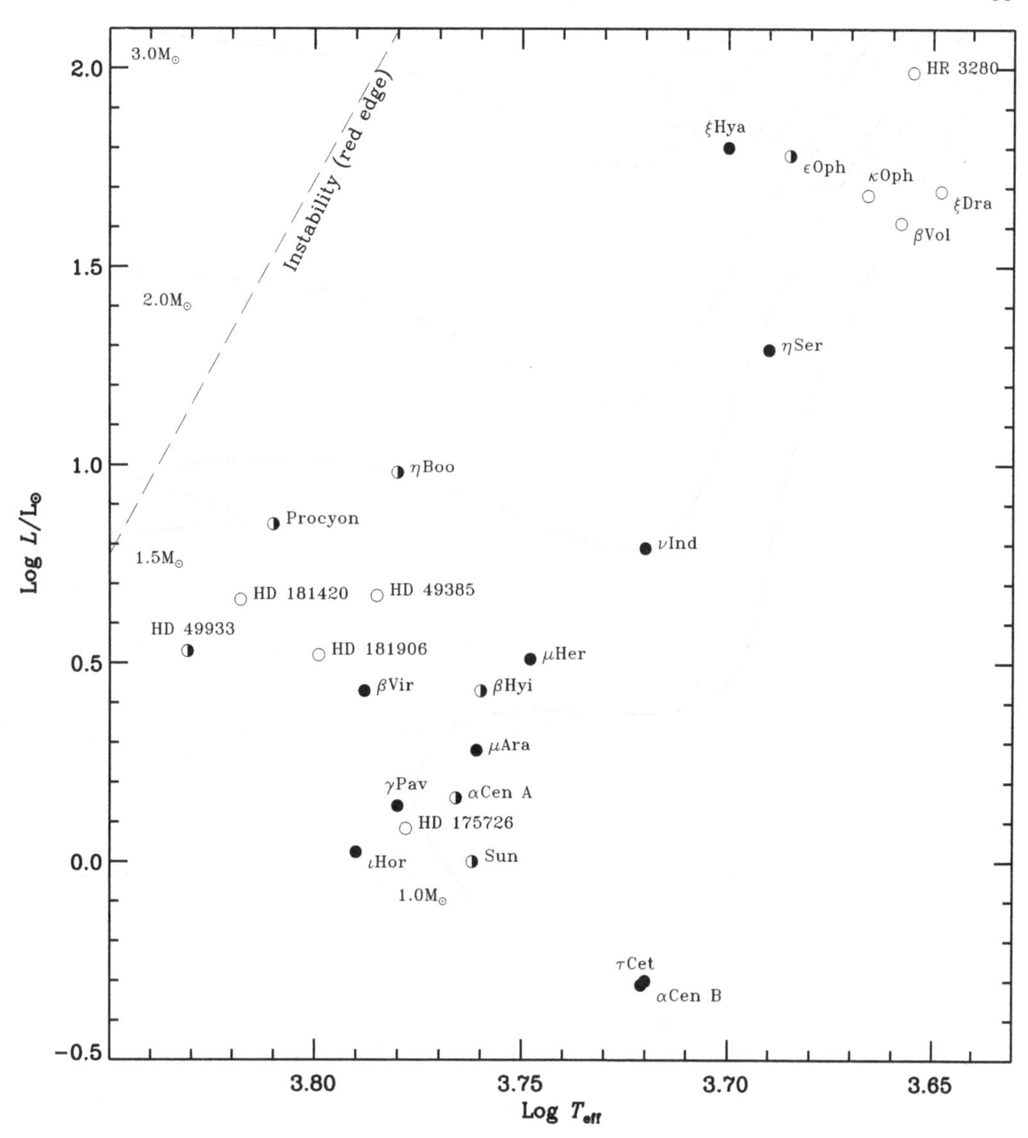

FIG. 3.3. Hertzsprung-Russell diagram showing stars with solar-like oscillations for which $\Delta\nu$ has been measured. Filled symbols indicate observations from ground-based spectroscopy and open symbols indicate space-based photometry (some stars were observed using both methods). Note that stars observed by Kepler are not included. Figure courtesy of Dennis Stello.

symbols in Fig. 3.2, a small change in $\Delta\nu$ translates to quite a large change in ϵ. These issues are discussed in some detail by White *et al.* (2011).

3.3 Ground-based observations

We now review some of the observations of solar-like oscillations. More details and additional references can be found in various review articles (e.g., Brown and Gilliland, 1994; Heasley *et al.*, 1996; Bedding and Kjeldsen, 2003; Bedding and Kjeldsen, 2008; Christensen-Dalsgaard, 2004; Cunha *et al.*, 2007; Bazot *et al.*, 2008; Aerts *et al.*, 2008). Figure 3.3 is an HR diagram showing most of the stars that will be mentioned here.

When global oscillations were discovered in the Sun, it was quickly realised that measuring similar oscillations in Sun-like stars would be tremendously valuable. Several

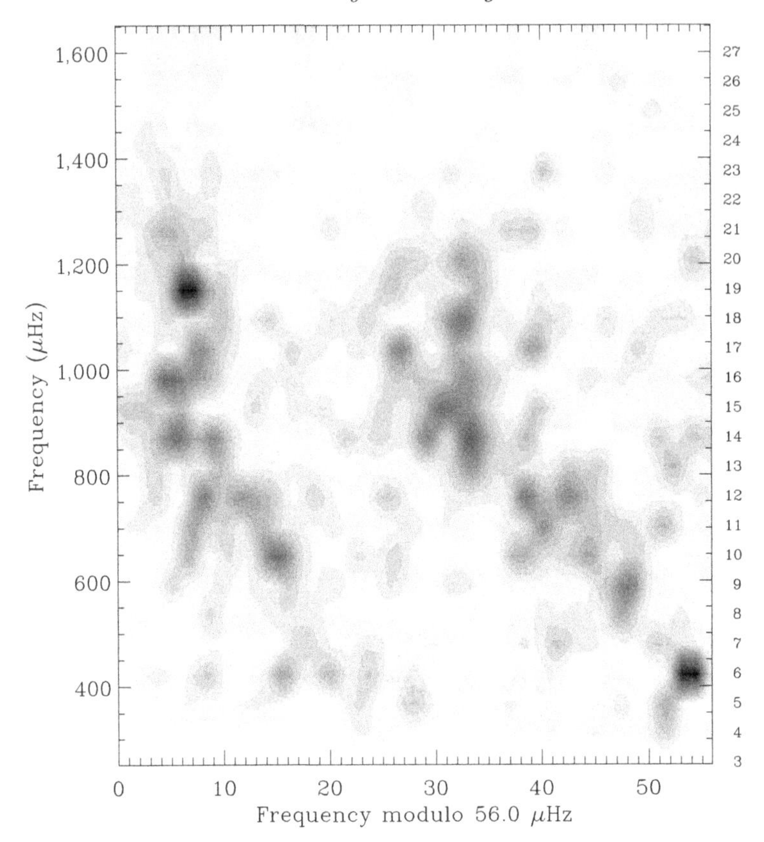

FIG. 3.4. Échelle diagram of Procyon from the multisite campaign, showing broadening due to short mode lifetimes (Bedding *et al.*, 2010a). Which ridge is $l = 1$ and which is $l = 0$ and 2? Note that the power spectrum has been smoothed to a resolution of $2\,\mu$Hz.

unsuccessful attempts were made (e.g., Traub *et al.*, 1978; Baliunas *et al.*, 1981; Smith, 1982) and some detections were claimed over the next decade but a review by Kjeldsen and Bedding (1995) concluded that none were convincing. However, it now seems clear that the excess power in the F5 star Procyon reported by Brown *et al.* (1991) was indeed due to oscillations and should be recognized as the first such detection, although it took some time and more observations for the large separation (about $55\,\mu$Hz) to be established (Mosser *et al.*, 1998; Barban *et al.*, 1999; Martić *et al.*, 1999). Furthermore, even a multisite campaign on Procyon using eleven telescopes at eight observatories (Arentoft *et al.*, 2008) has failed to give an unambiguous mode identification (Bedding *et al.*, 2010a), due to the short lifetimes of the oscillation modes (see Fig. 3.4). We refer to this as the F star problem (see also Section 3.4).

Most early oscillation searches measured Doppler shifts in radial velocities. However, a major effort to measure oscillations in intensity was made in a multi-site photometric campaign on the open cluster M67 by Gilliland *et al.* (1993). Photometric measurements have the big advantage of allowing simultaneous measurements of many stars. This has now been done from space by the CoRoT and Kepler missions, but ground-based photometry is severely hampered by atmospheric scintillation. The Gilliland *et al.* (1993) M67 campaign did not produce definite detections, although it did indicate that oscillation amplitudes of F stars are lower than expected from theoretical calculations (Kjeldsen and Bedding, 1995). It now seems clear that low-oscillation amplitudes in F stars are a direct result of the short mode lifetimes that were mentioned for Procyon.

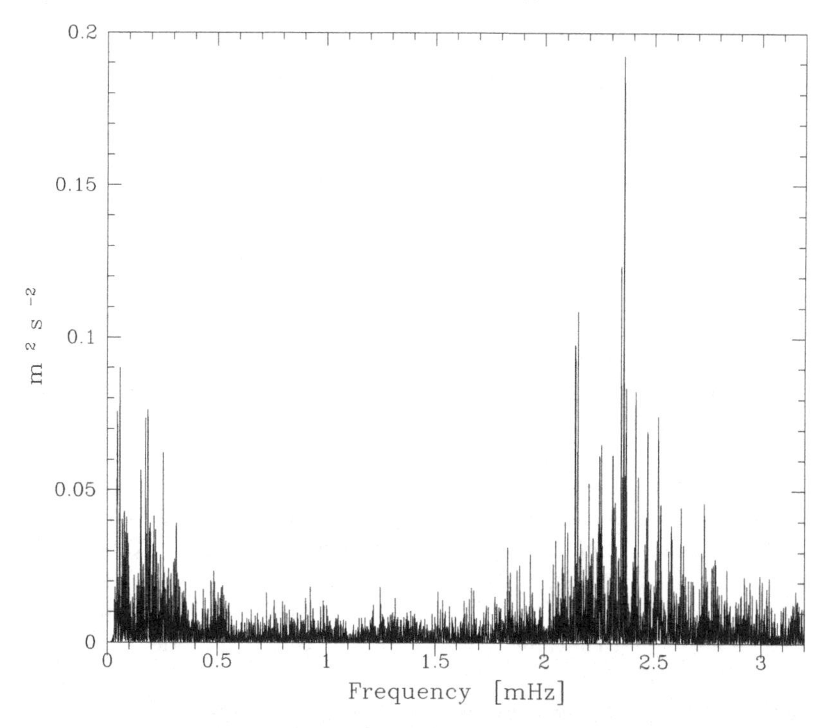

FIG. 3.5. Power spectrum of α Cen A from velocity observations over 13 nights with the CORALIE spectrograph by Bouchy and Carrier (2002).

A third method for detecting oscillations was applied by Kjeldsen *et al.* (1995) to the G0 subgiant η Boo. This method involved measuring fluctuations in the equivalent widths of the temperature-sensitive Balmer lines using low-resolution spectroscopy (see also Bedding *et al.*, 1996). It produced good evidence for multiple oscillation modes in η Boo with a large spacing of 40 μHz. Subsequent observations confirmed this result (Kjeldsen *et al.*, 2003; Carrier *et al.*, 2005), establishing η Boo as the first star apart from the Sun with a clear detection of solar-like oscillations. A very interesting result for η Boo, already apparent from the data obtained by Kjeldsen *et al.* (1995), was the departure of some of the dipole ($l = 1$) modes from their regular spacing. This was interpreted by Christensen-Dalsgaard *et al.* (1995) as arising from bumping of mixed modes, as discussed in more detail in Section 3.8. Meanwhile, the equivalent-width method for observing oscillations turned out to be more sensitive to instrumental drifts than we hoped, although observations of α Cen A did produce tentative evidence for p modes (Kjeldsen *et al.*, 1999).

Major breakthroughs came from improvements in Doppler precision that were driven by the push to find extra-solar planets. Observations of the G2 subgiant star β Hyi, made with UCLES on the 3.9-m Anglo-Australian Telescope (Bedding *et al.*, 2001) and confirmed with CORALIE on the 1.2-m Swiss Telescope (Carrier *et al.*, 2001), showed a clear power excess with a regular spacing ($\Delta\nu = 56\,\mu$Hz). This result was hailed by Gough (2001) as the birth of asteroseismology, but that honor should probably go to the clear detection using CORALIE of the p-mode oscillation spectrum in α Cen A by Bouchy and Carrier (2001, 2002) (Fig. 3.5).

Following on from these successes, a series of observations were made using high-resolution spectrographs on medium-to-large telescopes. Some spectrographs were equipped with an iodine cell, such as UCLES at the AAT (see earlier), UVES on the 8.2-m Very Large Telescope (e.g., Butler *et al.*, 2004) and SARG on the 3.6-m Telescopio

Nazionale Galileo (e.g., Bonanno *et al.*, 2008). Other spectrographs used a thorium-argon lamp, including CORALIE (see earlier), as well as HARPS on the European Southern Observatory's 3.6-m telescope. HARPS has produced some particularly impressive results, including the first application of asteroseismology to a planet-hosting star (μ Ara; see Bouchy *et al.*, 2005; Bazot *et al.*, 2005), an age for ι Hor that implies it is an evaporated member of the primordial Hyades cluster (Vauclair *et al.*, 2008), and the detection of oscillations in the remarkable solar twin 18 Sco (Bazot *et al.*, 2011).

Another important development was the clear detection of oscillations in the red giant star ξ Hya based on one month of observations with CORALIE (Frandsen *et al.*, 2002). This confirmed that red giants do indeed show solar-like oscillations, as already suspected from earlier observations of Arcturus and other stars. However, the oscillation periods of several hours mean that extremely long time series are needed, and the development of red-giant asteroseismology had to wait until dedicated space missions, as discussed in Section 3.4.

Single-site observations suffer from aliases, which complicate the interpretation of the frequency spectrum (see Section 3.5). Some stars have been observed from two sites, including ζ Her (Bertaux *et al.*, 2003), Procyon (Martić *et al.*, 2004; Kambe *et al.*, 2008), α Cen A (Bedding *et al.*, 2004), α Cen B (Kjeldsen *et al.*, 2005), ν Ind (a metal-poor subgiant; Bedding *et al.*, 2006; Carrier *et al.*, 2007), and β Hyi (Bedding *et al.*, 2007). The star β Hyi is a subgiant that, like η Boo, shows mixed $l = 1$ modes that are irregularly spaced (see Section 3.8).

Organizing multi-site spectroscopic campaigns is difficult and the only example so far was carried out on Procyon (see earlier). The next step is SONG, a dedicated network of 1-m telescopes equipped with iodine-stabilized spectrographs (Grundahl *et al.*, 2006). The first node is currently under construction on Tenerife. Even in the era of space asteroseismology, ground-based velocity measurements retain some advantages. The background noise from stellar granulation is lower than for intensity observations, allowing better signal-to-noise ratios at low frequencies. There is also the ability to cover the whole sky and to target nearby stars whose parallaxes and other parameters are accurately known, such as the stars mentioned in this section.

3.4 Space-based observations

Making photometric measurements from space avoids the scintillation caused by the Earth's atmosphere and allows very high precision, even with a small telescope. Depending on the spacecraft orbit, it is possible to obtain long time series with few interruptions. The first space photometer dedicated to asteroseismology was the EVRIS experiment aboard the Russian Mars 96 spacecraft. EVRIS was intended to carry out asteroseismology of bright stars during the cruise phase of the mission (Baglin, 1991; Buey *et al.*, 1997) but unfortunately the launch failed and the mission was lost.

Other asteroseismology missions were proposed, including PRISMA (Appourchaux *et al.*, 1993), STARS (Fridlund *et al.*, 1994), MONS (Kjeldsen *et al.*, 2000), and Eddington (Roxburgh and Favata, 2003). None of these made it though to construction and the first success came from an unexpected direction. After the failure of NASA's Wide-Field Infrared Explorer satellite (WIRE) immediately after launch due to the loss of coolant, its 50-mm star camera was used for asteroseismology by Buzasi (2002). WIRE was able to detect solar-like oscillations in α Cen A (Schou and Buzasi, 2001; Fletcher *et al.*, 2006) and also in several red giants (Buzasi *et al.*, 2000; Retter *et al.*, 2003; Stello *et al.*, 2008). The SMEI experiment (Solar Mass Ejection Imager) on board the Coriolis satellite has also been used to measure oscillations in bright red giants (Tarrant *et al.*, 2007, 2008). Finally, we note that the Hubble Space Telescope (HST) has been used on several occasions to observe solar-like oscillations, both via direct imaging (Edmonds and Gilliland, 1996; Gilliland, 2008; Stello and Gilliland, 2009) and also using its Fine Guidance Sensors (Zwintz *et al.*, 1999; Kallinger *et al.*, 2005; Gilliland *et al.*, 2011).

The first dedicated asteroseismology mission to be launched successfully was MOST (Microvariability and Oscillations of Stars), which is a 15-cm telescope in low-Earth orbit (Walker *et al.*, 2003). It has achieved great success with observations of classical (heat-engine) pulsators. In the realm of solar-like oscillations, controversy was generated when MOST failed to find evidence for oscillations in Procyon (Matthews *et al.*, 2004). However, Bedding *et al.* (2005) argued that the MOST nondetection was consistent with the ground-based data. Using photometry with WIRE, Bruntt *et al.* (2005) extracted parameters for the stellar granulation and found evidence for an excess due to oscillations. The discrepancy was finally laid to rest when Huber *et al.* (2011a) found an oscillation signal in a new and more accurate set of Procyon MOST data that agreed well with results from the ground-based velocity campaign (see Section 3.3).

The French-led CoRoT mission, launched in December 2006, is a 27-cm telescope in low-Earth orbit. It has contributed substantially to the list of main-sequence and subgiant stars with solar-like oscillations (e.g., Appourchaux *et al.*, 2008; Michel *et al.*, 2008; Barban *et al.*, 2009; García *et al.*, 2009; Mosser *et al.*, 2009; Deheuvels *et al.*, 2010; Mathur *et al.*, 2010). In particular, it has highlighted what might be called "the F star problem" in stars such as HD 49933 (e.g., Appourchaux *et al.*, 2008; Benomar *et al.*, 2009), which is also a problem in Procyon (Bedding *et al.*, 2010a; Bedding and Kjeldsen, 2010). This refers to stars in which the mode lifetimes are so short that the $l = 0$ and $l = 2$ modes are blended, making them indistinguishable from $l = 1$ modes. This makes it difficult to decide which ridge in the échelle diagram belongs to $l = 1$ and which belongs to $l = 0$ and 2, although measurement of ϵ may be able to resolve this ambiguity (see Section 3.7).

A major achievement by CoRoT came from observations of hundreds of G- and K-type red giants that showed clear oscillation spectra that were remarkably solar-like, with both radial and non-radial modes (De Ridder *et al.*, 2009; Hekker *et al.*, 2009; Carrier *et al.*, 2010). CoRoT continues to produce excellent results on red giants (e.g., Miglio *et al.*, 2009; Kallinger *et al.*, 2010b; Mosser *et al.*, 2010, 2011b), and also on main-sequence and subgiant stars (e.g., Mathur *et al.*, 2010; Ballot *et al.*, 2011).

Finally, we come to Kepler, which began science observations in May 2009. It has a 95-cm aperture and operates in an Earth-trailing heliocentric orbit, allowing continuous observations without interference from reflected light from the Earth. Although its primary goal is to search for extrasolar planets (Borucki *et al.*, 2010), it is no exaggeration to say that Kepler is revolutionizing asteroseismology (Gilliland *et al.*, 2010b). Most stars are observed in long-cadence mode (sampling 29.4 min; see Jenkins *et al.*, 2010), which is adequate to study oscillations in most red giants but is not fast enough for Sun-like stars. Fortunately, about 500 stars at a time can be observed in short-cadence mode (sampling 58.8 sec; see Gilliland *et al.*, 2010a).

For the first seven months of science operations, a survey was carried out with Kepler's short-cadence mode that targeted more than 2,000 main-sequence and subgiant stars for one month each. The survey detected solar-like oscillations, including a clear measurement of $\Delta\nu$, for about 500 stars (Chaplin *et al.*, 2011), representing an increase by a factor of $\sim$20 over the situation prior to Kepler.

Some of these stars have been studied individually. The first three to be published (Chaplin *et al.*, 2010) were "Java" (KIC 3656476), "Saxo" (KIC 6603624) and "Gemma" (KIC 11026764).[2] Java and Saxo are main-sequence stars, while Gemma is a subgiant showing bumped $l = 1$ modes (Metcalfe *et al.*, 2010, see Section 3.8 for a discussion of mode bumping). A small fraction of the survey stars were observed continuously

[2] Within the Kepler working groups, some stars are known by nicknames for convenience.

since the start of the mission.[3] Of these, results have so far been published on "Tigger" (KIC 11234888) and "Boogie" (KIC 11395018) by Mathur *et al.* (2011), and on "Mulder" (KIC 10273246), and "Scully" (KIC 10920273) by Campante *et al.* (2011).

As mentioned, the long-cadence mode of Kepler is perfectly adequate for red giants, and solar-like oscillations have been detected in thousands of them (e.g., Bedding *et al.*, 2010b; Huber *et al.*, 2010; Kallinger *et al.*, 2010a; Hekker *et al.*, 2011b), including some in open clusters (Stello *et al.*, 2010; Basu *et al.*, 2011; Hekker *et al.*, 2011a). Some of these results are discussed in more detail in Section 3.9 (see also reviews by Hekker, 2010, and Christensen-Dalsgaard, Chapter 7 of this book).

3.5 A few comments about Fourier analysis

Fourier analysis is an indispensable tool for asteroseismology. This section introduces some of the most important aspects. For more details including full references, see, for example, Pijpers (2006), Aerts *et al.* (2010), and Appourchaux, Chapter 5 of this book.

The *amplitude spectrum* of a time series tells us which frequencies are present, and with which amplitudes. Note that the *power spectrum* is simply the square of the amplitude spectrum. A useful way of thinking about the amplitude spectrum is as follows. We choose a frequency ν and perform a least-squares fit of a sine wave to the time series. That is, we fix the frequency of the sine wave to be ν and we vary the amplitude and phase until we get the best fit. If the quality of the time series is not uniform, as always happens with ground-based observations, it is important to weight the data using the measurement uncertainties (e.g., Frandsen *et al.*, 1995). The amplitude of this best-fitting sine wave is our result (for solar-like oscillations, which are stochastically excited, we are usually not interested in the phase). Repeating this calculation for a range of different frequencies generates the amplitude spectrum.

This way of describing the amplitude spectrum allows us to understand several important properties. First, it is clear that the observations do not need to be equally spaced in time (and ground-based observations generally are not). Now, suppose our time series consists of a single sinusoidal signal. If we choose the correct frequency, then the best-fitting amplitude will match the amplitude of the signal. If we choose the wrong frequency, the best-fitting sine wave will have a very small amplitude. But if we choose a frequency that is only slightly different from that of the signal, the best-fitting amplitude will be slightly less than the signal amplitude. Repeating this in small frequency steps will trace a sinc function ($\sin \nu/\nu$). Increasing the length of the time series will decrease the width of this sinc function.

If there are regular gaps in the time series, as occurs for ground-based single-site observations, then choosing the wrong frequency will still give a significant response in some cases. For example, suppose the sine wave we are fitting has a higher frequency than the signal, such that exactly one extra cycle fits into the gap. This sine wave will still give a fairly good fit to the data and we will get an extra peak in the amplitude spectrum, called a *sidelobe*. The same thing occurs if the chosen frequency is exactly one cycle-per-gap *lower* than the signal frequency. We therefore expect two sidelobes, one on either side of the true peak. Sidelobes are an example of *aliasing*, which occurs when two or more different sine waves give good fits to the same data. The heights of the sidelobes relative to the true peak will decrease as the gaps get smaller (and the duty cycle improves). If the period of the gaps in the time series is T_{gap}, then the sidelobes of a signal having

[3] Note that one of the 22 CCD modules on Kepler failed after 7 months of operation. Since the spacecraft rotates by 90 degrees every three months to maintain the orientation of the solar panels, about 20% of stars will fall on this failed module at some time during the year, creating annual 3-month gaps in the data for those stars.

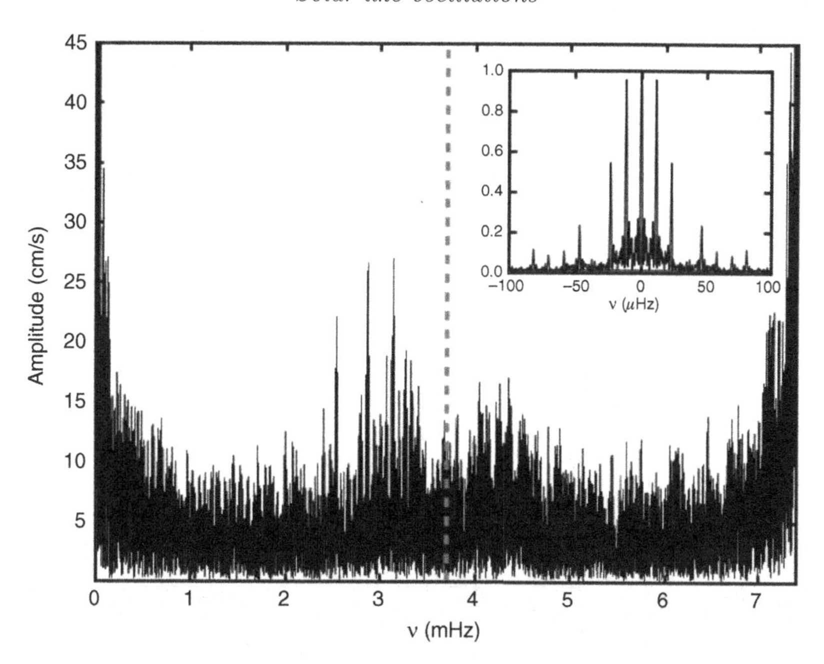

FIG. 3.6. Amplitude spectrum of the solar twin 18 Sco, based on observations taken with HARPS over 12 nights. The inset shows the spectral window in amplitude, and the vertical dashed line shows the effective Nyquist frequency. Figure from Bazot *et al.* (2011).

frequency ν will occur at frequencies of $\nu \pm 1/T_{\rm gap}$. There will also be a smaller pair of peaks separated from the true peak by $\nu \pm 2/T_{\rm gap}$, another pair at $\nu \pm 3/T_{\rm gap}$, and so on.

This leads to the concept of the *spectral window*, which is the power spectrum of a pure sine wave. It is sometimes incorrectly called the *window function*, but that term actually refers to the sampling in the time domain. In practice, the spectral window should be calculated by averaging the power spectra of a sine and a cosine wave, to eliminate end effects. Note that this averaging should be done in power and not in amplitude. The inset of Fig. 3.6 shows an example.

The another example of aliasing occurs if the observations are regularly sampled. A single sine wave with frequency ν that is sampled regularly at intervals separated by δt can be fitted equally well by a much higher-frequency sine wave having frequency $2\nu_{\rm Nyq} - \nu$, where $\nu_{\rm Nyq} = 1/(2\delta t)$ is called the Nyquist frequency. This aliasing still occurs, although with reduced amplitude, if the sampling is not exactly regular. An example is shown in Fig. 3.6 for ground-based data. In this case, one can define an effective Nyquist frequency based on the median sampling.

It is sometimes assumed that the power spectrum does not contain any useful information above the Nyquist frequency, and that any signal whose frequency is higher than $\nu_{\rm Nyq}$ cannot be measured. This is not true. For example, observations taken with Kepler in its long-cadence mode (sampling 29.4 min, $\nu_{\rm Nyq} = 283\,\mu$Hz) still contains useful information about oscillations at higher frequencies, provided the power spectrum is interpreted with care.

3.6 Scaling relations for $\Delta\nu$ and $\nu_{\rm max}$

As mentioned in Section 3.2.4, the large separation scales approximately with the square root of the mean stellar density:

$$\frac{\Delta\nu}{\Delta\nu_\odot} \approx \sqrt{\frac{\rho}{\rho_\odot}} = \left(\frac{M}{M_\odot}\right)^{0.5} \left(\frac{T_{\rm eff}}{T_{\rm eff\odot}}\right)^{3} \left(\frac{L}{L_\odot}\right)^{-0.75} . \tag{3.6}$$

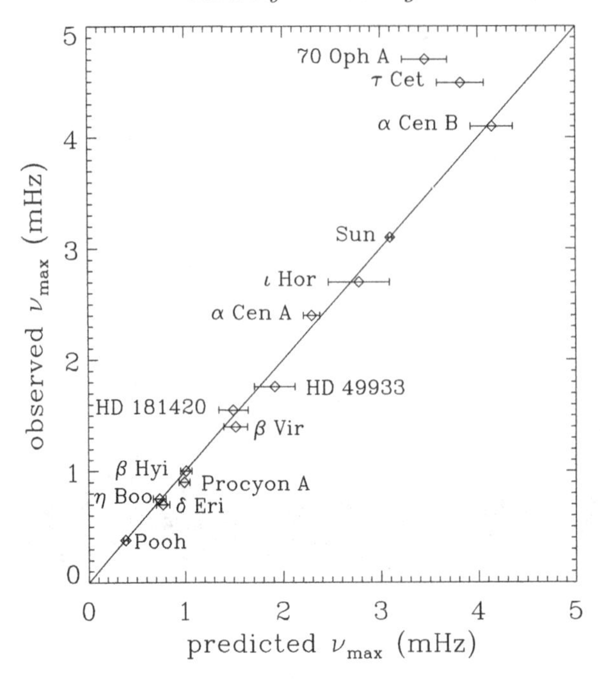

FIG. 3.7. Testing the scaling relation for $\nu_{\rm max}$. Values for the stellar parameters were taken from the compilations by Bruntt (2009) and Bruntt *et al.* (2010), with the addition of "Pooh" (KIC 4351319), a low-luminosity red giant observed by Kepler (Di Mauro *et al.*, 2011).

This scaling relation is very widely used, but how accurate is it? It was recently tested using theoretical models by White *et al.* (2011). Of course, given the curvature in the échelle diagram, the measurement of $\Delta\nu$ depends on the frequency at which it is measured (see Section 3.2.4). The approach adopted by White *et al.* (2011) was to fit to the radial modes near $\nu_{\rm max}$ (see the paper for details). They confirmed generally good agreement with Equation 3.6 but found departures reaching several percent. Interestingly, they found that these departures are predominantly a function of effective temperature and suggested a revised scaling relation that takes this into account.

Even if Equation 3.6 (or a modified version) agrees with theoretical models, this does not necessarily mean it is accurate for real stars. There are well-known discrepancies between the observed and calculated oscillation frequencies in the Sun, which are due to incorrect modeling of the surface layers (Christensen-Dalsgaard *et al.*, 1988; Dziembowski *et al.*, 1988; Rosenthal *et al.*, 1999; Li *et al.*, 2002). These discrepancies increase with frequency, so there must also be a discrepancy in $\Delta\nu$. Indeed, the value of $\Delta\nu$ of the best-fitting solar model is about 1% greater than the observed value (Kjeldsen *et al.*, 2008). We can hardly expect the situation to be better in other stars, and a solution to the problem of modeling near-surface layers is urgently needed.

Another important and widely used scaling relation expresses $\nu_{\rm max}$ in terms of stellar parameters. It is based on the suggestion by Brown *et al.* (1991) that $\nu_{\rm max}$ might be expected to be a fixed fraction of the acoustic cutoff frequency, which in turn scales as $T_{\rm eff}/\sqrt{g}$. Hence we have:

$$\frac{\nu_{\rm max}}{\nu_{\rm max,\odot}} \approx \frac{M}{M_\odot}\left(\frac{T_{\rm eff}}{T_{\rm eff\odot}}\right)^{3.5}\left(\frac{L}{L_\odot}\right)^{-1}. \tag{3.7}$$

This relation has been shown to agree quite well with observations (Bedding and Kjeldsen, 2003) and also with model calculations (Chaplin *et al.*, 2008b), and has recently been given some theoretical justification by Belkacem *et al.* (2011). Figure 3.7 shows the expected and observed values of $\nu_{\rm max}$ for some well-studied stars with accurate parallaxes.

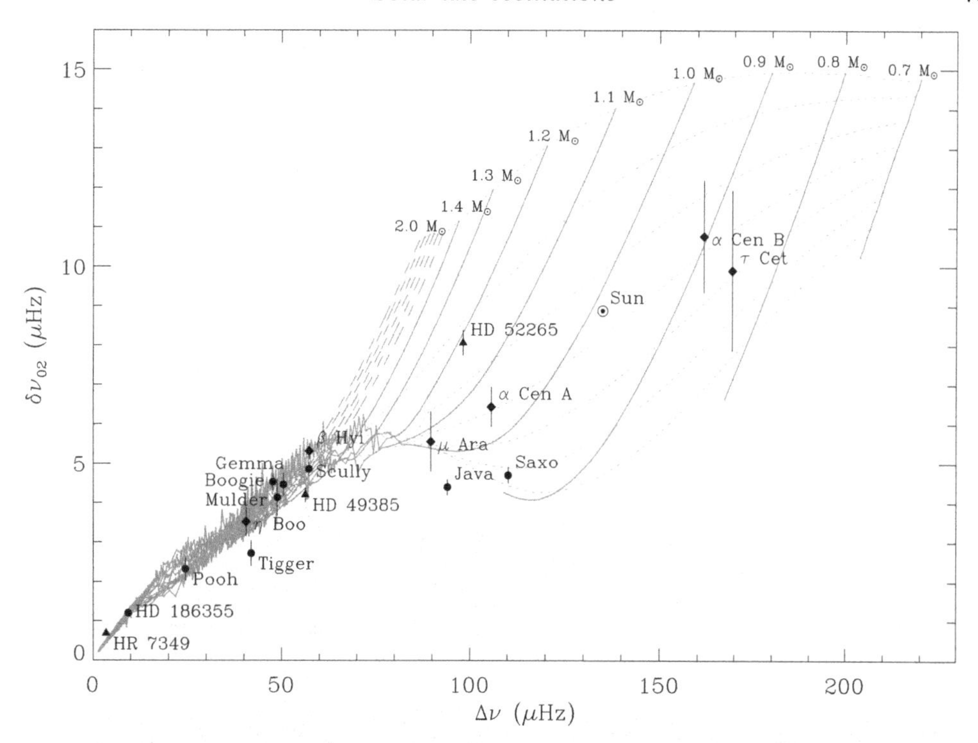

FIG. 3.8. The C-D diagram, which plots the large and small frequency separations. The solid lines show model calculations for solar metallicity using ASTEC and ADIPLS (Christensen-Dalsgaard, 2008b,a), with the dashed lines being isochrones from zero to 12 Gyr (top to bottom) in steps of 2 Gyr. The points show measurements of stars showing solar-like oscillations. Figure adapted from White *et al.* (2011).

There is generally very good agreement except for the two low-mass stars τ Cet and 70 Oph A. If confirmed, this may indicate a breakdown of the assumption that $\nu_{\rm max}$ is a fixed fraction of the acoustic cutoff frequency.

3.7 Ensemble asteroseismology

The wealth of data from CoRoT and Kepler allow us to carry out what might be called *ensemble asteroseismology.* This involves studying similarities and differences in groups of stars, and is complementary to the detailed study of a few individuals. Several groups have developed pipelines to process the large amounts of data and extract the global parameters such as $\nu_{\rm max}$, $\Delta\nu$, and amplitude. These various pipelines have been tested and compared using both simulations (Chaplin *et al.*, 2008a) and Kepler data (Hekker *et al.*, 2011c; Verner *et al.*, 2011).

Ensemble asteroseismology can be carried out using asteroseismic diagrams, in which two properties of the oscillation spectra are plotted against one another. Here, we briefly discuss some of these diagrams.

The C-D diagram. The plot of small versus large frequency separations, first introduced by Christensen-Dalsgaard (1984), has come to be known as the C-D diagram. A modern version is shown in Fig. 3.8. In main-sequence stars, the tracks for different masses and ages are well spread, allowing both to be read off the diagram. However, note that the positions of the tracks depend on metallicity, and so this must be taken into account (for details and references, see White *et al.*, 2011). For more evolved stars, the tracks converge and the small separation becomes much less sensitive as a diagnostic. C-D diagrams for

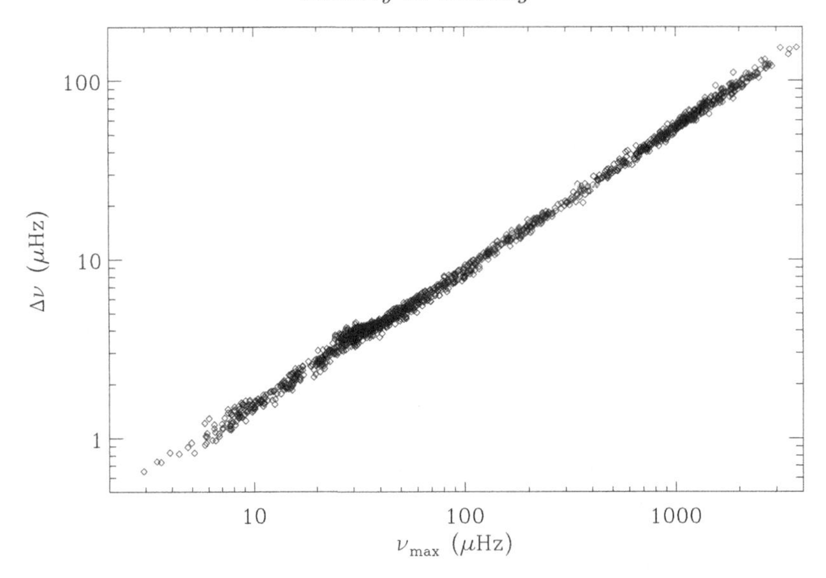

FIG. 3.9. The strong correlation between $\nu_{\rm max}$ and $\Delta\nu$, based on Kepler observations from the main sequence (upper right) to the red giants (lower left). Figure adapted from Huber *et al.* (2011b).

red giants observed with Kepler have been presented by Bedding *et al.* (2010b) and Huber *et al.* (2010).

The $\nu_{\rm max}$-$\Delta\nu$ *diagram.* The strong correlation between $\nu_{\rm max}$ and $\Delta\nu$ was first plotted for main-sequence and subgiant stars by Stello *et al.* (2009) and for red giants by Hekker *et al.* (2009). The tightness of this relationship follows from the scaling relations (Equations 3.6 and 3.7), as discussed in some detail by Stello *et al.* (2009). The relationship extends over several orders of magnitude (see Fig. 3.9), making it a useful predictor of $\Delta\nu$.

The ϵ *diagram.* As discussed in Section 3.2.4, the parameter ϵ is sensitive to the surface layers of the star but needs to be measured carefully. A diagram showing ϵ versus $\Delta\nu$ was calculated by Christensen-Dalsgaard (1984), but its diagnostic potential was not realized until recently. Bedding and Kjeldsen (2010) suggested that ϵ could be used as an aid to mode identification (the "F star problem"; see Sections 3.3 and 3.4). The first observationl ϵ diagram was made for red giants observed with Kepler by Huber *et al.* (2010), who used it to suggest mode identifications for four CoRoT red giants discussed by Hekker *et al.* (2010). Meanwhile, White *et al.* (2011) have shown that ϵ is mostly determined by the stellar effective temperature and demonstrated that it is indeed useful for addressing the F star problem.

3.8 Mode bumping and avoided crossings

As mentioned in Section 3.2.1, mixed modes have p-mode character in the envelope and g-mode character in the core. They occur in evolved stars (subgiants and red giants), in which the large density gradient outside the core effectively divides the star into two coupled cavities. This leads to the phenomenon of *mode bumping*, in which mode frequencies are shifted from their regular spacing. Here, we summarize the main features of mixed modes from an observational perspective, which necessarily includes a basic introduction to the theory. For more details on the theoretical aspects, see, for example, Scuflaire (1974); Osaki (1975); Aizenman *et al.* (1977); Dziembowski and Pamyatnykh

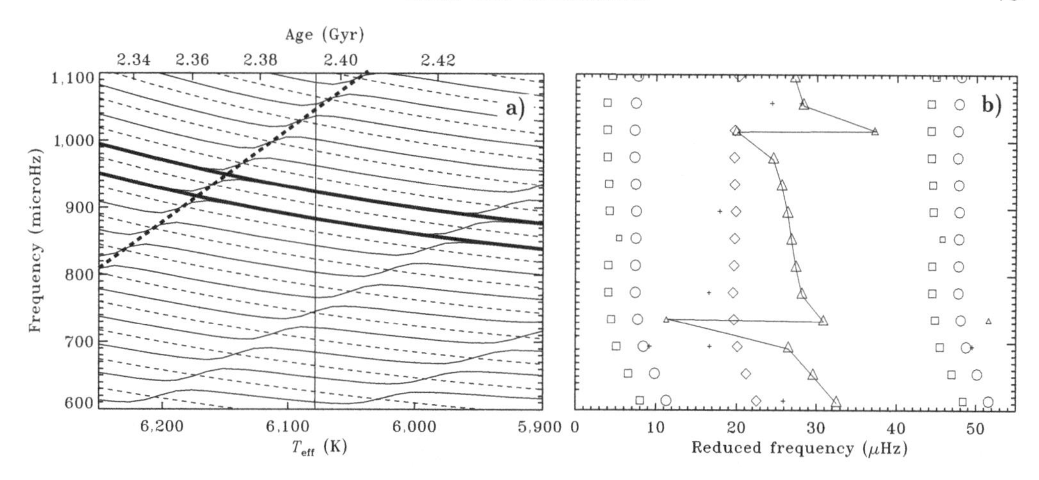

FIG. 3.10. *a)* Evolution of oscillation frequencies in models of a subgiant star of mass 1.60 $M_\odot$ and $Z = 0.03$ (representing η Boo). The dashed lines correspond to modes of degree $l = 0$, and the solid lines to $l = 1$. The vertical solid line indicates the location of the model whose frequencies are illustrated in the right panel, the thick solid lines show two of the $l = 1$ π modes and, the thick dashed line shows one of the γ modes (see text). *b)* Échelle diagram using a frequency separation of $\Delta\nu = 40.3$ μHz and a zero-point frequency of $\nu_0 = 735$ μHz. Circles are used for modes with $l = 0$, triangles for $l = 1$, squares for $l = 2$, and diamonds for $l = 3$. The size of the symbols indicates the relative amplitude of the modes, estimated from the mode inertia, with crosses used for modes whose symbols would otherwise be too small. Figures adapted from Christensen-Dalsgaard *et al.* (1995).

(1991); Christensen-Dalsgaard (2004); Miglio *et al.* (2008); Dupret *et al.* (2009), and Aerts *et al.* (2010).

Figure 3.10 shows theoretical oscillations frequencies for a subgiant star whose parameters were chosen to match η Boo. The left panel shows the evolution with time of the model frequencies for modes with $l = 0$ (dashed lines) and $l = 1$ (solid lines). The $l = 0$ frequencies decrease slowly with time as the star expands. At a given moment in time, such as that marked by the vertical line, the radial modes are regularly spaced in frequency, with a large separation of $\Delta\nu \approx 40$ μHz. However, the behavior of the $l = 1$ modes (solid lines) is rather different. They undergo a series of *avoided crossings* (Osaki, 1975; Aizenman *et al.*, 1977), during which an $l = 1$ mode is bumped upwards by the mode below, and in turn it bumps the mode above. The effect is to disturb the regular spacing of the $l = 1$ modes in the vicinity of each avoided crossing. The right panel of Fig. 3.10 shows the frequencies in échelle format for the single model that is marked in the left panel by the vertical line. The bumping from regularity of the $l = 1$ modes (triangles) is obvious and there is one extra $l = 1$ mode at each avoided crossing, where the modes are "squeezed together."

Bumping of $l = 1$ modes has been observed in the subgiant stars η Boo (Kjeldsen *et al.*, 1995, 2003; Carrier *et al.*, 2005), β Hyi (Bedding *et al.*, 2007), the CoRoT target HD 49385 (Deheuvels *et al.*, 2010), and the Kepler target "Gemma" (KIC 11026764 Metcalfe *et al.*, 2010). Many other subgiants stars observed with Kepler also show bumping of $l = 1$ modes, and observations have so far been published for "Tigger" (KIC 11234888) and "Boogie" (KIC 11395018) by Mathur *et al.* (2011), and for "Mulder" (KIC 10273246) and "Scully" (KIC 10920273) by Campante *et al.* (2011). It was also suggested by Bedding *et al.* (2010a) that Procyon has an $l = 1$ mixed mode at low frequency (see Fig. 3.4), based on the narrowness of the peak in the power spectrum. Note that mixed modes are expected to have longer lifetimes (smaller linewidths) than pure p modes because they have larger mode inertias (e.g., Christensen-Dalsgaard, 2004). Mode inertia is a measure of the total interior mass that is affected by the oscillation.

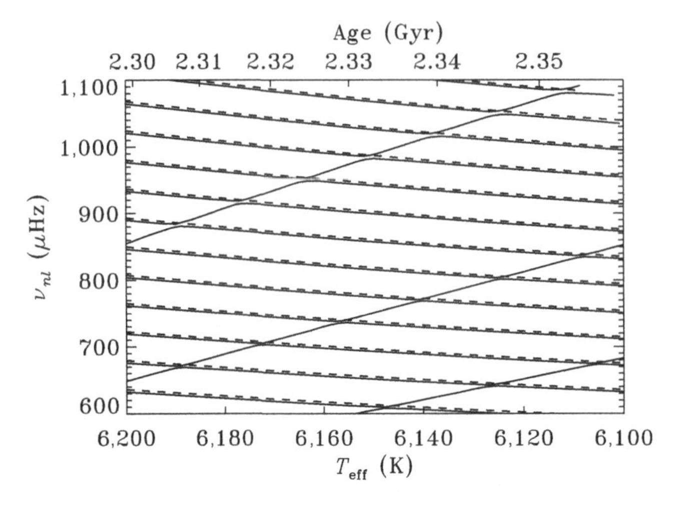

FIG. 3.11. Avoided crossings for $l = 2$ modes in a model of η Boo. This is similar to the left panel of Fig. 3.10, but for a slightly different model and with the solid lines now showing $l = 2$ modes (the dashed lines again show $l = 0$ modes). Adapted from Christensen-Dalsgaard and Houdek (2010).

We now discuss the theoretical aspects of mixed modes in more detail. In particular, what causes the mode bumping and why are $l = 1$ modes the most affected? To answer these questions, we return to the idea of a star with two cavities, as mentioned earlier. We can think of subgiants and red giants as having p modes trapped in the envelope and g modes trapped in the core. Strictly speaking, these are hypothetical modes within each cavity rather than true oscillation modes of the whole star, and so they should probably be called "envelope p modes" and "core g modes". Aizenman *et al.* (1977) referred to these artificially isolated modes as π modes and γ modes, which is a useful abbreviation that we will adopt. They are the modes that would exist if the two cavities were completely decoupled. In reality, there is coupling and this leads to a mixed mode whenever a π mode and a γ mode with the same l are sufficiently close to each other in frequency. This mixed mode has p-mode character in the envelope and g-mode character in the core. Note also that the frequencies of the π modes decrease with time as the star evolves, while those of the γ modes increase with time.

This picture allows us to understand the features seen in Fig. 3.10. For $l = 0$, there are no mixed modes because radial g modes do not exist and so we have pure p modes that are regularly spaced in frequency. For $l = 1$, the γ modes are visible as the three upward-sloping ridges, one of which is marked with a thick dashed line (see also Fig. 1b of Aizenman *et al.*, 1977). The $l = 1$ π modes can also be traced in Fig. 3.10 by connecting the downward-sloping parts of the tracks, as shown for two adjacent π modes by the thick solid lines. Recall that a mixed mode occurs when a π mode and a γ mode are close to each other in frequency, which happens at the intersections. In fact, we see that several π modes are able to couple with each γ mode (see Deheuvels and Michel, 2010b). Indeed, all the $l = 1$ modes are bumped to some extent, with those closest in frequency to each γ mode being bumped the most, as seen in the échelle diagram (right panel of Fig. 3.10).[4]

Figure 3.11 shows the situation for the $l = 2$ modes in a similar set of models. The coupling between the two cavities for $l = 2$ modes is much weaker than for $l = 1$ modes,[5]

[4] "All modes are mixed, but some modes are more mixed than others" (with apologies to George Orwell, 1945).

[5] The coupling strength varies inversely with the width of the evanescent region between the two cavities, and this region turns out to be wider for $l = 2$ modes than for $l = 1$ modes (Christensen-Dalsgaard, 2004).

which means that the π and γ modes must have almost exactly the same frequency in order to couple. In Fig. 3.11 we see that the $l = 2$ modes (solid lines) mostly slope downward (following the π-mode tracks), indicating they are essentially behaving as pure p modes. The upward-sloping segments (which follow the γ-mode tracks) correspond to modes behaving essentially as pure g modes. These g-dominated modes will have very high inertias and very low amplitudes, making them undetectable. As a result, the observable $l = 2$ modes will only be bumped slightly or not at all, as can be seen in the right panel of Fig. 3.10 (squares).

3.8.1 *The p-g diagram*

It has long been recognized that mixed modes have great potential for asteroseismology because their frequencies are very sensitive to stellar interiors and because they change quite rapidly as the star evolves (e.g., Dziembowski and Pamyatnykh, 1991; Christensen-Dalsgaard *et al.*, 1995). One way to exploit this information is by performing a direct comparison of observed frequencies with models. This has been done for η Boo (Christensen-Dalsgaard *et al.*, 1995; Guenther and Demarque, 1996; Di Mauro *et al.*, 2003a; Guenther, 2004; Carrier *et al.*, 2005), β Hyi (Di Mauro *et al.*, 2003b; Fernandes and Monteiro, 2003; Brandão *et al.*, 2011), the CoRoT target HD 49385 (Deheuvels and Michel, 2010b), and the Kepler target "Gemma" (Metcalfe *et al.*, 2010).

However, adjusting the model parameters to fit the observed frequencies can be difficult and time-consuming. We can ask whether the information contained in the mixed modes can be used more elegantly. Based on the preceding discussion of mixed modes and mode bumping, we are led to consider the frequencies of the avoided crossings themselves. For example, the échelle diagram in Fig. 3.10 shows two avoided crossings,[6] at about 730 and 1,050 μHz. We recognize these as the frequencies of the γ modes (recall that these are the pure g modes that would exist in the core cavity if it could be treated in isolation). Much of the diagnostic information contained in the mixed modes can be captured in this way. This is because the overall pattern of the mixed modes is determined by the mode bumping at each avoided crossing, and these patterns are determined by the g modes trapped in the core (the γ modes). For related discussion on this point, see Deheuvels and Michel (2010a).

This discussion suggests a new asteroseismic diagram, inspired by the classical C-D diagram, in which the frequencies of the avoided crossings (the γ modes) are plotted against the large separation of the p modes. This p-g diagram, so named because it plots g-mode frequencies versus p-mode frequencies, could prove to be an instructive way to display results of many stars and to make a first comparison with theoretical models (Campante *et al.*, 2011).

3.9 Solar-like oscillations in red giants

As outlined in Section 3.4, there has recently been tremendous progress in asteroseismology of G- and K-type red giants, with solar-like oscillations detected by CoRoT and Kepler in thousands of stars. There have also been rapid developments in the theory (e.g., Guenther *et al.*, 2000; Dziembowski *et al.*, 2001; Christensen-Dalsgaard, 2004; Dupret *et al.*, 2009; Montalbán *et al.*, 2010; Di Mauro *et al.*, 2011), as discussed in detail by Christensen-Dalsgaard (Chapter 7 of this book).

At first glance, the oscillation spectra of red giants are remarkably solar-like. As first revealed by CoRoT observations (De Ridder *et al.*, 2009), they have radial and non-radial

[6] Strictly speaking, these features in the échelle diagram should not really be called avoided crossings, as that term refers to the close approaches of two modes in the evolution diagram. However, there does not seem to be any harm in using the term.

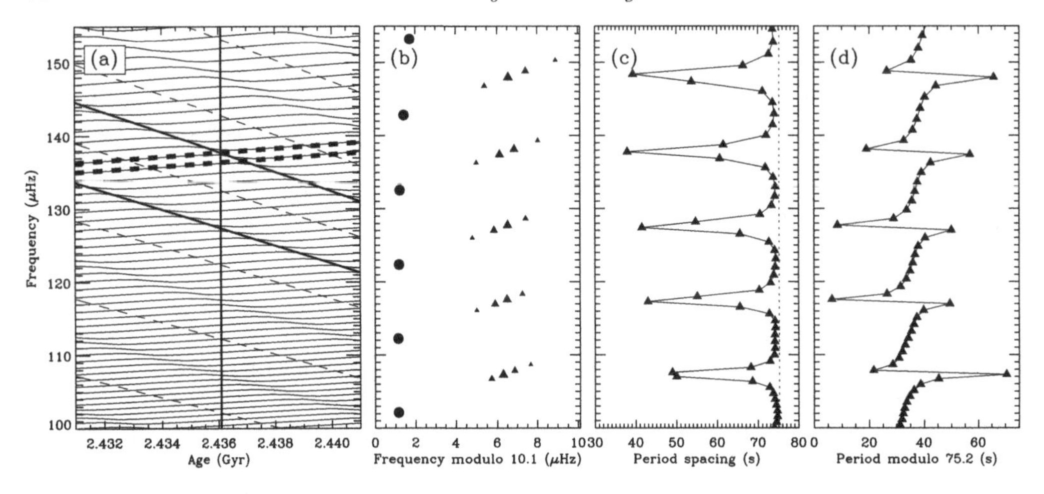

FIG. 3.12 (a) Evolution of oscillation frequencies in a model of a hydrogen-shell-burning red giant with a mass of $1.5M_\odot$ and solar metallicity, calculated using ASTEC and ADIPLS (Christensen-Dalsgaard, 2008b,a). Downward-sloping dashed lines show radial modes ($l = 0$) and solid lines show dipole modes ($l = 1$). The thick downward-sloping solid lines show two of the $l = 1$ π modes and the thick upward-sloping dashed lines show two of the γ modes (see text). The vertical solid line indicates the model whose frequencies are illustrated in the next three panels. (b) Conventional échelle diagram for one model (abscissa shows frequencies modulo the p-mode frequency spacing, $\Delta\nu$). Circles show $l = 0$ modes and triangles show $l = 1$ modes, with symbol size roughly proportional to expected amplitude. (c) Period spacings between adjacent $l = 1$ modes for the same model. The period spacing of the g-dominated modes (ΔP_g) can be seen from the maximum values to be about 75 s. (d) Period échelle diagram of $l = 1$ modes for the same model (the abscissa is the period modulo the g-mode period spacing, ΔP_g).

modes that closely follow the asymptotic relation. In fact, to a large extent their power spectra are simply scaled versions of each other, so that the large and small spacings are in a roughly constant ratio (e.g., Bedding *et al.*, 2010b). This high degree of homology is rather unhelpful for asteroseismology, in the sense that the small separations lack much of the diagnostic potential that makes them so valuable in main-sequence stars. Fortunately, there are other features in the oscillation spectra of red giants that make up for this loss. We now discuss one feature in some detail, namely the presence of mixed modes with g-mode period spacings.

Even after the first month, Kepler data revealed multiple $l = 1$ modes per order that indicated the presence of mixed modes (see Fig. 5 of Bedding *et al.*, 2010b). As with the subgiants discussed in Section 3.8, the large density gradient outside the helium core divides the star into two coupled cavities, with the oscillations behaving like p modes in the envelope and like g modes in the core. The difference with red giants is that the spectrum of g modes in the core (the γ modes) is much denser. It is also important to note that, while p modes are approximately equally spaced in frequency (see Section 3.2.2), an asymptotic analysis shows that g modes are approximately equally spaced in *period* (Tassoul, 1980).

Figure 3.12a shows the time evolution of frequencies of a red giant model for modes with $l = 0$ (dashed lines) and $l = 1$ (solid lines). It is very instructive to compare this with Fig. 3.10. In both cases, we can follow the downward-sloping $l = 1$ π modes (two examples are shown by thick solid lines). They are approximately equally spaced in frequency (with separation $\Delta\nu$) and they run parallel to the $l = 0$ modes (which are pure p modes). In both figures we can also follow the upward-sloping $l = 1$ γ modes (examples shown by thick dashed lines). The difference is that for the subgiant in Fig. 3.10, there are only three γ modes, but for the red giant in Fig. 3.12a there are dozens. As before, we have a mixed mode for every π mode and a mixed mode for every γ mode. Most of these mixed modes are g-dominated (they follow the γ-mode tracks), with very high inertia

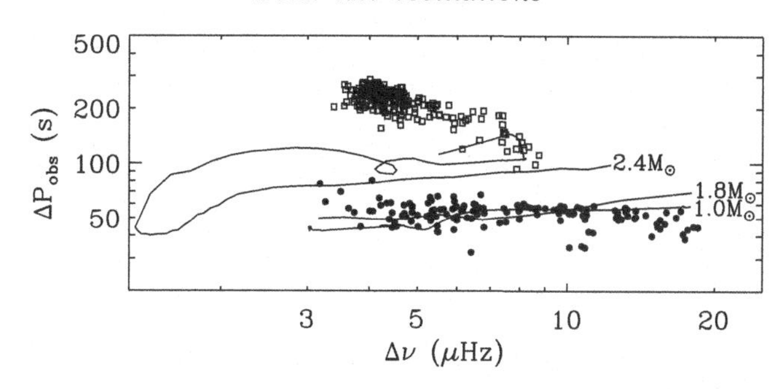

FIG. 3.13. Period spacings of $l = 1$ modes in red giants. The symbols are measurements obtained with Kepler (Bedding *et al.*, 2011). Filled circles indicate hydrogen-shell-burning giants and open squares indicate helium-core-burning stars. The solid lines show model calculations using ASTEC and ADIPLS (Christensen-Dalsgaard, 2008b,a). The period spacings were computed from the models by averaging over the observable modes (see White *et al.*, 2011, for details). The hydrogen-shell-burning giants evolve from right to left as they ascend the red giant branch. The 2.4-$M_\odot$ model continues into the helium-core-burning stage, looping back to the right to join the so-called secondary clump (Girardi, 1999). The lower-mass models stop before the helium-burning stage because they undergo a helium flash that cannot be calculated by ASTEC. Solar metallicity was adopted for all models, which were computed without mass loss.

and very low amplitude. However, some mixed modes fall close enough to a π mode for the coupling to be significant, producing p-dominated mixed modes with sufficient amplitude to be observable. These are shown by the triangles in Fig 3.12b, which is the échelle diagram of the model indicated by the vertical line in Fig. 3.12a. Note that the clusters of observed $l = 1$ modes are centred at the positions expected for pure $l = 1$ p modes, which are about halfway between the $l = 0$ modes.

The pattern of $l = 1$ clusters just described has been observed in red giants using data from Kepler (Beck *et al.*, 2011; Bedding *et al.*, 2011) and CoRoT (Mosser *et al.*, 2011a). The spacings within the $l = 1$ clusters clearly provide valuable information about the gravity modes trapped in the core but to interpret them we must take into account the mode bumping. Each cluster is squeezed together by the presence of one extra mode and so the spacing we measure from the observable modes, which we denote $\Delta P_{\rm obs}$, will be substantially less than the spacing of the core g modes (the γ modes), which we denote $\Delta P_{\rm g}$. This is shown for the model in Fig. 3.12c, which plots the period spacing between consecutive $l = 1$ mixed modes. From the upper envelope we see that $\Delta P_{\rm g} \approx 75\,$s, but this is determined from the g-dominated modes, whereas the observable (p-dominated) modes have period differences that are lower by a factor ranging from 1.2 to 2.

Another way to visualize the frequencies is shown in Fig. 3.12d. This shows the model frequencies for $l = 1$ in échelle format, but setting the horizontal axis to be period modulo the g-mode period spacing (rather than frequency modulo the large frequency separation). Doing this for real stars is difficult because most of the $l = 1$ modes are not observable, so that finding the correct period spacing is equivalent to estimating the exact number of missing $l = 1$ modes in each gap. Bedding *et al.* (2011) were only able to do this unambiguously for those stars with the largest number of detected modes. For the remainder, however, it was still possible to measure the average period spacing of the observed $l = 1$ modes ($\Delta P_{\rm obs}$). This quantity is plotted in Fig. 3.13 against the large frequency separation, and the stars are seen to divide nicely into two groups (see also Mosser *et al.*, 2011a). Theoretical models confirm that these groups correspond to stars still burning hydrogen in a shell (filled circles) and those also burning helium in the core (open squares; for more details, see Bedding *et al.*, 2011; White *et al.*, 2011). These issues are discussed in detail by Christensen-Dalsgaard, in Chapter 7 of this book, who shows that the larger period spacing in He-burning stars is due to the presence of a convective core.

Finally, Simon O'Toole (personal communication) has pointed out that the close agreement between the period spacings found in the He-burning red giants and the hot subdwarf B (sdB) stars from Kepler data (Reed *et al.*, 2011) is nicely consistent with the accepted explanation that sdB stars are the cores of He-burning red giants that have lost their envelopes.

Acknowledgments

Many thanks to Pere Pallé and colleagues for organizing this fantastic Winter School. I am grateful to Dennis Stello, Tim White, and Jørgen Christensen-Dalsgaard for providing figures. I also thank them, as well as Hans Kjeldsen, Daniel Huber, and Othman Benomar, for many helpful discussions.

REFERENCES

Aerts, C., Christensen-Dalsgaard, J., Cunha, M., and Kurtz, D. W. 2008. The current status of asteroseismology. *Sol. Phys.*, **251**, 3.

Aerts, C., Christensen-Dalsgaard, J., and Kurtz, D. W. 2010. *Asteroseismology.* Heidelberg, Germany: Springer.

Aizenman, M., Smeyers, P., and Weigert, A. 1977. Avoided crossing of modes of non-radial stellar oscillations. *A&A*, **58**, 41.

Althaus, L. G., Córsico, A. H., Isern, J., and García-Berro, E. 2010. Evolutionary and pulsational properties of white dwarf stars. *A&A Rev.*, **18**, 471.

Antoci, V., Handler, G., Campante, T. L., *et al.* 2011. Solar-like oscillations as probe of the convective envelope of a delta Scuti star. *Nature.* submitted.

Appourchaux, T., Gough, D. O., Catala, C., *et al.* 1993. Prisma: the first space mission to see inside the stars. T. M. Brown, (ed.), *GONG 1992: Seismic Investigation of the Sun and Stars*, vol. 42., 411, ASP Conf. Ser.

Appourchaux, T., Michel, E., Auvergne, M., *et al.* 2008. CoRoT sounds the stars: p-mode parameters of Sun-like oscillations on HD 49933. *A&A*, **488**, 705.

Arentoft, T., Kjeldsen, H., Bedding, T. R., *et al.* 2008. A multi-site campaign to measure solar-like oscillations in Procyon. I. Observations, data reduction and slow variations. *ApJ*, **687**, 1180.

Baglin, A. 1991. Stellar seismology from space: the EVRIS experiment on board Mars 94. *Sol. Phys.*, **133**, 155.

Baliunas, S. L., Vaughan, A. H., Hartmann, L., Liller, W., and Dupree, A. K. 1981. Short time-scale variability of chromospheric CA II in late-type stars. *ApJ*, **246**, 473.

Ballot, J. 2010. Current methods for analyzing light curves of solar-like stars. *Astron. Nachr.*, **331**, 933.

Ballot, J., Gizon, L., Samadi, R., *et al.* 2011. Accurate p-mode measurements of the G0V metal-rich CoRoT target HD 52265. *A&A*, **530**, A97.

Balona, L. A., Pigulski, A., De Cat, P., *et al.* 2011. Kepler observations of the variability in B-type stars. *MNRAS*, **413**, 2403.

Barban, C., Michel, E., Martić, M., *et al.* 1999. Solar-like oscillations of Procyon A: stellar models and time series simulations versus observations. *A&A*, **350**, 617.

Barban, C., Deheuvels, S., Baudin, F., *et al.* 2009. Solar-like oscillations in HD 181420: data analysis of 156 days of CoRoT data. *A&A*, **506**, 51.

Basu, S., Grundahl, F., Stello, D., *et al.* 2011. Sounding open clusters: asteroseismic constraints from kepler on the properties of ngc 6791 and ngc 6819. *ApJ*, **729**, L10.

Bazot, M., Vauclair, S., Bouchy, F., and Santos, N. C. 2005. Seismic analysis of the planet-hosting star mu Arae. *A&A*, **440**, 615.

Bazot, M., Ireland, M. J., Huber, D., *et al.* 2011. The radius and mass of the close solar twin 18 Scorpii derived from asteroseismology and interferometry. *A&A*, **526**, L4.

Bazot, M., Monteiro, M. J. P. F. G., and Straka, C. W. 2008. Current issues in asteroseismology. L. Gizon, and M. Roth (eds.), *Second HELAS International Conference: Helioseismology, Asteroseismology and MHD Connections*, vol. 118. Journal of Physics Conference Series.

Beck, P. G., Bedding, T. R., Mosser, B., *et al.* 2011. Kepler detected gravity-mode period spacings in a red giant star. *Science*, **332**, 205.

Bedding, T. R. 2003. Solar-like oscillations in semiregular variables. *Ap&SS*, **284**, 61.

Bedding, T. R. and Kjeldsen, H. 2003. Solar-like oscillations. *PASA*, **20**, 203.

Bedding, T. R. and Kjeldsen, H. 2008. Asteroseismology from solar-like oscillations oscillations. Page 21 of: *14th Cambridge Workshop on Cool Stars, Stellar Systems, and the Sun*, vol. 384. ASP Conf. Ser.

Bedding, T. R. and Kjeldsen, H. 2010. Scaled oscillation frequencies and échelle diagrams as a tool for comparative asteroseismology. *Commun. Asteroseismology*, **161**, 3.

Bedding, T. R., Kjeldsen, H., Reetz, J., and Barbuy, B. 1996. Measuring stellar oscillations using equivalent widths of absorption lines. *MNRAS*, **280**, 1155.

Bedding, T. R., Butler, R. P., Kjeldsen, H., *et al.* 2001. Evidence for solar-like oscillations in β Hydri. *ApJ*, **549**, L105.

Bedding, T. R., Kjeldsen, H., Butler, R. P., *et al.* 2004. Oscillation frequencies and mode lifetimes in α Centauri A. *ApJ*, **614**, 380.

Bedding, T. R., Kjeldsen, H., Bouchy, F., *et al.* 2005. The non-detection of oscillations in Procyon by MOST: is it really a surprise? *A&A*, **432**, L43.

Bedding, T. R., Butler, R. P., Carrier, F., *et al.* 2006. Solar-like oscillations in the metal-poor subgiant nu Indi: constraining the mass and age using asteroseismology. *ApJ*, **647**, 558.

Bedding, T. R., Kjeldsen, H., Arentoft, T., *et al.* 2007. Solar-like oscillations in the G2 subgiant β Hydri from dual-site observations. *ApJ*, **663**, 1315.

Bedding, T. R., Kjeldsen, H., Campante, T. L., *et al.* 2010a. A multi-site campaign to measure solar-like oscillations in Procyon. II. Mode frequencies. *ApJ*, **713**, 935.

Bedding, T. R., Huber, D., Stello, D., *et al.* 2010b. Solar-like oscillations in low-luminosity red giants: first results from kepler. *ApJ*, **713**, L176.

Bedding, T. R., Mosser, B., Huber, D., *et al.* 2011. Gravity modes as a way to distinguish between hydrogen- and helium-burning red giant stars. *Nature*, **471**, 608.

Belkacem, K., Samadi, R., Goupil, M.-J., *et al.* 2009. Solar-like oscillations in a massive star. *Science*, **324**, 1540.

Belkacem, K., Dupret, M. A., and Noels, A. 2010. Solar-like oscillations in massive main-sequence stars. I. Asteroseismic signatures of the driving and damping regions. *A&A*, **510**, A6.

Belkacem, K., Goupil, M. J., Dupret, M. A., *et al.* 2011. The underlying physical meaning of the $\nu_{\rm max} - \nu_{\rm c}$ relation. *A&A*, **530**, A142.

Benomar, O., Baudin, F., Campante, T. L., *et al.* 2009. A fresh look at the seismic spectrum of HD49933: analysis of 180 days of CoRoT photometry. *A&A*, **507**, L13.

Bertaux, J.-L., Schmitt, J., Lebrun, J.-C., Bouchy, F., and Guibert, S. 2003. Two-site simultaneous observations of solar-like stellar oscillations in radial velocities with a Fabry-Pérot calibration system. *A&A*, **405**, 367.

Bonanno, A., Benatti, S., Claudi, R., *et al.* 2008. Detection of solar-like oscillations in the G5 subgiant μ Her. *ApJ*, **676**, 1248.

Borucki, W. J., Koch, D., Basri, G., *et al.* 2010. Kepler planet-detection mission: introduction and first results. *Science*, **327**, 977.

Bouchy, F. and Carrier, F. 2001. P-mode observations on α Cen A. *A&A*, **374**, L5.

Bouchy, F. and Carrier, F. 2002. The acoustic spectrum of alpha Cen A. *A&A*, **390**, 205.

Bouchy, F., Bazot, M., Santos, N. C., Vauclair, S., and Sosnowska, D. 2005. Asteroseismology of the planet-hosting star μ Arae. I. The acoustic spectrum. *A&A*, **440**, 609.

Brandão, I. M., Doğan, G., Christensen-Dalsgaard, J., *et al.* 2011. Asteroseismic modeling of the solar-type subgiant star β Hydri. *A&A*, **527**, A37.

Broomhall, A.-M., Chaplin, W. J., Davies, G. R., *et al.* 2009. Definitive Sun-as-a-star p-mode frequencies: 23 years of BiSON observations. *MNRAS*, **396**, L100.

Brown, T. M. and Gilliland, R. L. 1994. Asteroseismology. *ARAandA*, **32**, 37.

Brown, T. M., Gilliland, R. L., Noyes, R. W., and Ramsey, L. W. 1991. Detection of possible *p*-mode oscillations on Procyon. *ApJ*, **368**, 599.

Bruntt, H. 2009. Accurate fundamental parameters of CoRoT asteroseismic targets: the solar-like stars HD 49933, HD 175726, HD 181420, and HD 181906. *A&A*, **506**, 235.

Bruntt, H., Kjeldsen, H., Buzasi, D. L., and Bedding, T. R. 2005. Evidence for granulation and oscillations in Procyon from photometry with the WIRE satellite. *ApJ*, **633**, 440.

Bruntt, H., Bedding, T. R., Quirion, P.-O., *et al.* 2010. Accurate fundamental parameters for 23 bright solar-type stars. *MNRAS*, **405**, 1907.

Buey, J. T., Auvergne, M., Vuillemin, A., and Epstein, G. 1997. Towards Asteroseismology from Space: the EVRIS Experiment. I. The Photometer: Description and calibration. *PASP*, **109**, 140.

Butler, R. P., Bedding, T. R., Kjeldsen, H., *et al.* 2004. Ultra-high-precision velocity measurements of oscillations in α Cen A. *ApJ*, **600**, L75.

Buzasi, D. L. 2002. Asteroseismic results from WIRE. Page 616 of: C. Aerts, T. R. Bedding, and J. Christensen-Dalsgaard (eds.), *IAU Colloqium 185: Radial and Nonradial Pulsations as Probes of Stellar Physics*, vol. 259. ASP Conf. Ser.

Buzasi, D. L., Catanzarite, J., Laher, R., *et al.* 2000. The detection of multimodal oscillations on α Ursae Majoris. *ApJ*, **532**, L133.

Campante, T. L., Handberg, R., Mathur, S., *et al.* 2011. Asteroseismology from *Kepler* ten-month time series: the evolved Sun-like stars KIC 10273246 and KIC 10920273. *A&A*. submitted.

Cantiello, M., Langer, N., Brott, I., *et al.* 2009. Sub-surface convection zones in hot massive stars and their observable consequences. *A&A*, **499**, 279.

Carrier, F., Bouchy, F., Kienzle, F., *et al.* 2001. Solar-like oscillations in β Hydri: Confirmation of a stellar origin for the excess power. *A&A*, **378**, 142.

Carrier, F., Eggenberger, P., and Bouchy, F. 2005. New seismological results on the G0 IV η Bootis. *A&A*, **434**, 1085.

Carrier, F., Kjeldsen, H., Bedding, T. R., *et al.* 2007. Solar-like oscillations in the metal-poor subgiant ν Indi. II. Acoustic spectrum and mode lifetime. *A&A*, **470**, 1059.

Carrier, F., De Ridder, J., Baudin, F., *et al.* 2010. Non-radial oscillations in the red giant HR 7349 measured by CoRoT. *A&A*, **55**, A73.

Chaplin, W. J., Elsworth, Y., Isaak, G. R., *et al.* 1997. Solar p-mode linewidths from recent BiSON helioseismological data. *MNRAS*, **288**, 623.

Chaplin, W. J., Appourchaux, T., Arentoft, T., *et al.* 2008a. AsteroFLAG: First results from hare-and-hounds Exercise #1. *Astron. Nachr.*, **329**, 549.

Chaplin, W. J., Houdek, G., Appourchaux, T., *et al.* 2008b. Challenges for asteroseismic analysis of Sun-like stars. *A&A*, **485**, 813.

Chaplin, W. J., Appourchaux, T., Elsworth, Y., *et al.* 2010. The asteroseismic potential of kepler: first results for solar-type stars. *ApJ*, **713**, L169.

Chaplin, W. J., Kjeldsen, H., Christensen-Dalsgaard, J., *et al.* 2011. Ensemble asteroseismology of solar-type stars with the NASA Kepler mission. *Science*, **332**, 213.

Christensen-Dalsgaard, J. 1984. What will asteroseismology teach us? Page 11 of: A. Mangeney and F. Praderie (eds.), *Workshop on Space Research in Stellar Activity and Variability.* Meudon: Observatoire de Paris.

Christensen-Dalsgaard, J. 2004. Physics of solar-like oscillations. *Sol. Phys.*, **220**, 137.

Christensen-Dalsgaard, J. 2008b. ASTEC: the Aarhus stellar evolution code. *Ap&SS*, **316**, 13.

Christensen-Dalsgaard, J. 2008a. ADIPLS: the Aarhus adiabatic oscillation package. *Ap&SS*, **316**, 113.

Christensen-Dalsgaard, J. and Houdek, G. 2010. Prospects for asteroseismology. *Ap&SS*, **328**, 51.

Christensen-Dalsgaard, J., Däppen, W., and Lebreton, Y. 1988. Solar oscillation frequencies and the equation of state. *Nat*, **336**, 634.

Christensen-Dalsgaard, J., Bedding, T. R., and Kjeldsen, H. 1995. Modeling solar-like oscillations in η Bootis. *ApJ*, **443**, L29.

Christensen-Dalsgaard, J., Kjeldsen, H., and Mattei, J. A. 2001. Solar-like oscillations of semiregular variables. *ApJ*, **562**, L141.

Cunha, M. S., Aerts, C., Christensen-Dalsgaard, J., *et al.* 2007. Asteroseismology and interferometry. *AandAR*, **14**, 217.

De Ridder, J., Barban, C., Baudin, F., *et al.* 2009. Non-radial oscillation modes with long lifetimes in giant stars. *Nature*, **459**, 398.

Degroote, P., Briquet, M., Catala, C., *et al.* 2009. Evidence for nonlinear resonant mode coupling in the β Cephei star HD 180642 (V1449 Aquilae) from CoRoT photometry. *A&A*, **506**, 111.

Degroote, P., Briquet, M., Auvergne, M., *et al.* 2010. Detection of frequency spacings in the young O-type binary HD 46149 from CoRoT photometry. *A&A*, **519**, A38.

Deheuvels, S. and Michel, E. 2010a. Constraints on the core μ-gradient of the solar-like star HD 49385 via low-degree mixed modes. *Astron. Nachr.*, **331**, 929.

Deheuvels, S. and Michel, E. 2010b. New insights on the interior of solar-like pulsators thanks to CoRoT: the case of HD 49385. *Ap&SS*, **328**, 259.

Deheuvels, S., Bruntt, H., Michel, E., *et al.* 2010. Seismic and spectroscopic characterization of the solar-like pulsating CoRoT target HD 49385. *A&A*. submitted.

Di Mauro, M. P., Christensen-Dalsgaard, J., Kjeldsen, H., Bedding, T. R., and Paternò, L. 2003a. Convective overshooting in the evolution and seismology of η Bootis. *A&A*, **404**, 341.

Di Mauro, M. P., Christensen-Dalsgaard, J., and Paternò, L. 2003b. A study of the solar-like properties of β Hydri. *Ap&SS*, **284**, 229.

Di Mauro, M. P., Cardini, D., Catanzaro, G., *et al.* 2011. Solar-like oscillations from the depths of the red-giant star KIC 4351319 observed with *Kepler*. *MNRAS*. submitted.

Dupret, M., Belkacem, K., Samadi, R., *et al.* 2009. Theoretical amplitudes and lifetimes of non-radial solar-like oscillations in red giants. *A&A*, **506**, 57.

Dziembowski, W. A. and Pamyatnykh, A. A. 1991. A potential asteroseismological test for convective overshooting theories. *A&A*, **248**, L11.

Dziembowski, W. A., Paternó, L., and Ventura, R. 1988. How comparison between observed and calculated p-mode eigenfrequencies can give information on the internal structure of the sun. *AandA*, **200**, 213.

Dziembowski, W. A., Gough, D. O., Houdek, G., and Sienkiewicz, R. 2001. Oscillations of α UMa and other red giants. *MNRAS*, **328**, 601.

Edmonds, P. D. and Gilliland, R. L. 1996. K giants in 47 Tucanae: detection of a new class of variable stars. *ApJ*, **464**, L157.

Elsworth, Y., Howe, R., Isaak, G. R., McLeod, C. P., and New, R. 1990. Evidence from solar seismology against non-standard solar-core models. *Nature*, **347**, 536.

Fernandes, J. and Monteiro, M. J. P. F. G. 2003. HR diagram and asteroseismic analysis of models for beta Hydri. *A&A*, **399**, 243.

Fletcher, S. T., Chaplin, W. J., Elsworth, Y., Schou, J., and Buzasi, D. 2006. Frequency, splitting, linewidth and amplitude estimates of low-l p modes of α Cen A: analysis of wide-field infrared explorer photometry. *MNRAS*, **371**, 935.

Frandsen, S., Jones, A., Kjeldsen, H., *et al.* 1995. CCD photometry of the δ-Scuti star κ^2 Bootis. *A&A*, **301**, 123.

Frandsen, S., Carrier, F., Aerts, C., *et al.* 2002. Detection of solar-like oscillations in the G7 giant star xi Hya. *A&A*, **394**, L5.

Fridlund, M., Gough, D. O., Jones, A., *et al.* 1994. STARS: a proposal for a dedicated space mission to study stellar structure and evolution. Page 416 of: R. K. Ulrich, Jr. E. J. Rhodes, and W. Däppen (eds.), *GONG '94: Helio- and Astero-seismology from Earth and Space*. ASP Conf. Ser.

García, R. A., Jiménez, A., Mathur, S., *et al.* 2008. Update on g-mode research. *Astronomische Nachrichten*, **329**, 476.

García, R. A., Regulo, C., Samadi, R., *et al.* 2009. Solar-like oscillations with low amplitude in the CoRoT target HD 181906. *A&A*, **506**, 41.

Gastine, T. and Dintrans, B. 2011. Convective quenching of stellar pulsations. *A&A*, **528**, A6.

Gilliland, R. L. 2008. Photometric oscillations of low-luminosity red giant stars. *AJ*, **136**, 566.

Gilliland, R. L., Brown, T. M., Christensen-Dalsgaard, J., *et al.* 2010b. Kepler asteroseismology program: introduction and first results. *PASP*, **122**, 131–43.

Gilliland, R. L., Brown, T. M., Kjeldsen, H., *et al.* 1993. A search for solar-like oscillations in the stars of M67 with CCD ensemble photometry on a network of 4 m telescopes. *AJ*, **106**, 2441.

Gilliland, R. L., Jenkins, J. M., Borucki, W. J., *et al.* 2010a. Initial Characteristics of Kepler Short Cadence Data. *ApJ*, **713**, L160.

Gilliland, R. L., McCullough, P. R., Nelan, E. P., *et al.* 2011. Asteroseismology of the Transiting Exoplanet Host HD 17156 with Hubble Space Telescope Fine Guidance Sensor. *ApJ*, **726**, 2.

Girardi, L. 1999. A secondary clump of red giant stars: why and where. *MNRAS*, **308**, 818–32.

Gough, D. O. 1986. EBK quantization of stellar waves. Page 117 of: Y. Osaki (ed.), *Hydrodynamic and Magnetodynamic Problems in the Sun and Stars*. Tokyo, Japan: University of Tokyo Press.

Gough, D. O. 2001. The birth of asteroseismology. *Science*, **291**, 2325.

Gough, D. O. 2003. On the principal asteroseismic diagnostic signatures. *Ap&SS*, **284**, 165.

Grec, G., Fossat, E., and Pomerantz, M. A. 1983. Full-disk observations of solar oscillations from the geographic South Pole: latest results. *Sol. Phys.*, **82**, 55.

Grigahcène, A., Antoci, V., Balona, L., *et al.* 2010. Hybrid γ Doradus-δ Scuti pulsators: new insights into the physics of the oscillations from Kepler observations. *ApJ*, **713**, L192.

Grundahl, F., Kjeldsen, H., Frandsen, S., *et al.* 2006. SONG: Stellar Oscillations Network Group: a global network of small telescopes for asteroseismology and planet searches. *Mem. Soc. Astron. Ital.*, **77**, 458.

Guenther, D. B. 2004. Quantitative analysis of the oscillation spectrum of η bootis. *ApJ*, **612**, 454.

Guenther, D. B. and Demarque, P. 1996. Seismology of η Bootis. *ApJ*, **456**, 798.

Guenther, D. B., Demarque, P., Buzasi, D., *et al.* 2000. Evolutionary model and oscillation frequencies for Ursae Majoris: a comparison with observations. *ApJ*, **530**, L45.

Heasley, J. N., Janes, K., Labonte, B., *et al.* 1996. The prospects for asteroseismology from ground-based sites. *PASP*, **108**, 385.

Hekker, S. 2010. Observations and interpretation of solar-like oscillations in red-giant stars. *Astron. Nachr.*, **331**, 1004.

Hekker, S., Kallinger, T., Baudin, F., *et al.* 2009. Characteristics of solar-like oscillations in red giants observed in the CoRoT exoplanet field. *A&A*, **506**, 465.

Hekker, S., Barban, C., Baudin, F., *et al.* 2010. Oscillation mode lifetimes of red giants observed during the initial and first anticentre long run of CoRoT. *A&A*, **520**, A60.

Hekker, S., Basu, S., Stello, D., *et al.* 2011a. Asteroseismic inferences on red giants in open clusters NGC 6791, NGC 6819, and NGC 6811 using Kepler. *A&A*, **530**, A100.

Hekker, S., Gilliland, R. L., Elsworth, Y., *et al.* 2011b. Characterization of red giant stars in the public Kepler data. *MNRAS*, **414**, 2594-2601.

Hekker, S., Elsworth, Y., De Ridder, J., *et al.* 2011c. Solar-like oscillations in red giants observed with Kepler: comparison of global oscillation parameters from different methods. *A&A*, **525**, A131.

Huber, D., Bedding, T. R., Stello, D., *et al.* 2010. Asteroseismology of red giants from the first four months of Kepler data: global oscillation parameters for 800 stars. *ApJ*, **723**, 1607.

Huber, D., Bedding, T. R., Stello, D., *et al.* 2011b. Testing scaling relations for solar-like oscillations from the main-sequence to red giants using Kepler data. *ApJ*. submitted.

Huber, D., Bedding, T. R., Arentoft, T., *et al.* 2011a. Solar-like oscillations and activity in Procyon: a comparison of the 2007 MOST and ground-based radial velocity campaigns. *ApJ*, **731**, 94.

Jenkins, J. M., Caldwell, D. A., Chandrasekaran, H., *et al.* 2010. Initial characteristics of Kepler long cadence data for detecting transiting planets. *ApJ*, **713**, L120.

Kallinger, T. and Matthews, J. M. 2010. Evidence for granulation in early A-type stars. *ApJ*, **711**, L35.

Kallinger, T., Zwintz, K., Pamyatnykh, A. A., Guenther, D. B., and Weiss, W. W. 2005. Pulsation of the K 2.5 giant star GSC 09137-03505? *A&A*, **433**, 267.

Kallinger, T., Mosser, B., Hekker, S., *et al.* 2010a. Asteroseismology of red giants from the first four months of Kepler data: fundamental stellar parameters. *A&A*, **522**, A1.

Kallinger, T., Weiss, W. W., Barban, C., *et al.* 2010b. Oscillating red giants in the CoRoT exo-field: asteroseismic radius and mass determination. *A&A*, **509**, A77.

Kambe, E., Ando, H., Sato, B., *et al.* 2008. Development of iodine cells for the Subaru HDS and the Okayama HIDES. III. An improvement on radial velocity measurement technique. *PASJ*, **60**, 45.

Kiss, L. L., Szabo, G. M., and Bedding, T. R. 2006. Variability in red supergiant stars: pulsations, long secondary periods and convection noise. *MNRAS*, **372**, 1721.

Kjeldsen, H. and Bedding, T. R. 1995. Amplitudes of stellar oscillations: the implications for asteroseismology. *A&A*, **293**, 87.

Kjeldsen, H., Bedding, T. R., Viskum, M., and Frandsen, S. 1995. Solar-like oscillations in η Boo. *AJ*, **109**, 1313.

Kjeldsen, H., Bedding, T. R., Frandsen, S., and Dall, T. H. 1999. A search for solar-like oscillations and granulation in α Cen A. *MNRAS*, **303**, 579.

Kjeldsen, H., Bedding, T. R., and Christensen-Dalsgaard, J. 2000. MONS: measuring oscillations in nearby stars. Page 73 of: L. Szabados and D. Kurtz (eds.), *IAU Colloqium 176: The Impact of Large-Scale Surveys on Pulsating Star Research*, vol. 203. ASP Conf. Ser.

Kjeldsen, H., Bedding, T. R., Baldry, I. K., *et al.* 2003. Confirmation of Solar-Like Oscillations in η Bootis. *AJ*, **126**, 1483.

Kjeldsen, H., Bedding, T. R., Butler, R. P., *et al.* 2005. Solar-like oscillations in α Centauri B. *ApJ*, **635**, 1281.

Kjeldsen, H., Bedding, T. R., and Christensen-Dalsgaard, J. 2008. Correcting stellar oscillation frequencies for near-surface effects. *ApJ*, **683**, L175.

Li, L. H., Robinson, F. J., Demarque, P., Sofia, S., and Guenther, D. B. 2002. Inclusion of turbulence in solar modeling. *ApJ*, **567**, 1192.

Martić, M., Schmitt, J., Lebrun, J.-C., *et al.* 1999. Evidence for global pressure oscillations on Procyon. *A&A*, **351**, 993.

Martić, M., Lebrun, J.-C., Appourchaux, T., and Korzennik, S. G. 2004. p-mode frequencies in solar-like stars. I. Procyon A. *A&A*, **418**, 295.

Mathur, S., García, R. A., Catala, C., *et al.* 2010. The solar-like CoRoT target HD 170987: spectroscopic and seismic observations. *A&A*, **518**, A53.

Mathur, S., Handberg, R., Campante, T. L., *et al.* 2011. Solar-like oscillations in KIC 11395018 and KIC 11234888 from 8 months of Kepler data. *ApJ*, **733**, 95.

Matthews, J. M., Kuschnig, R., Guenther, D. B., *et al.* 2004. No stellar p-mode oscillations in space-based photometry of Procyon. *Nat*, **430**, 51. (Erratum: **430**, 921.)

Metcalfe, T. S., Monteiro, M. J. P. F. G., Thompson, M. J., *et al.* 2010. A precise asteroseismic age and radius for the evolved sun-like star kic 11026764. *ApJ*, **723**, 1583.

Michel, E., Baglin, A., Auvergne, M., *et al.* 2008. CoRoT measures solar-like oscillations and granulation in stars hotter than the Sun. *Science*, **322**, 558.

Miglio, A., Montalbán, J., Eggenberger, P., and Noels, A. 2008. Gravity modes and mixed modes as probes of stellar cores in main-sequence stars: From solar-like to β Cep stars. *Astronomische Nachrichten*, **329**, 529–34.

Miglio, A., Montalbán, J., Baudin, F., *et al.* 2009. Probing populations of red giants in the galactic disk with CoRoT. *A&A*, **503**, L21.

Montalbán, J., Miglio, A., Noels, A., Scuflaire, R., and Ventura, P. 2010. Seismic diagnostics of red giants: first comparison with stellar models. *ApJ*, **721**, L182.

Mosser, B., Maillard, J. P., Mékarnia, D., and Gay, J. 1998. New limit on the p-mode oscillations of Procyon obtained by Fourier transform seismometry. *A&A*, **340**, 457.

Mosser, B., Michel, E., Appourchaux, T., *et al.* 2009. The CoRoT target HD 175726: an active star with weak solar-like oscillations. *A&A*, **506**, 33.

Mosser, B., Belkacem, K., Goupil, M., *et al.* 2010. Red giant seismic properties analyzed with CoRoT. *A&A*, **517**, A22.

Mosser, B., Barban, C., Montalban, J., *et al.* 2011a. Mixed modes in red-giant stars observed with CoRoT. *A&A*, **525**, L9.

Mosser, B., Belkacem, K., Goupil, M. J., *et al.* 2011b. The universal red-giant oscillation pattern. An automated determination with CoRoT data. *A&A*, **525**, L9.

Orwell, G. 1945. *Animal Farm*. London, United Kingdom: Secker and Warburg.

Osaki, J. 1975. Nonradial oscillations of a 10 solar mass star in the main-sequence stage. *PASJ*, **27**, 237–58.

Pérez Hernández, F. and Christensen-Dalsgaard, J. 1998. The phase function for stellar acoustic oscillations. IV. Solar-like stars. *MNRAS*, **295**, 344.

Pijpers, F. P. 2006. *Methods in helio- and asteroseismology*. Imperial College Press.

Reed, M. D., Baran, A., Quint, A. C., *et al.* 2011. First Kepler results on compact pulsators. VIII. Mode identifications via period spacings in g-mode pulsating Subdwarf B stars. *MNRAS*, **414**, 2885–92.

Retter, A., Bedding, T. R., Buzasi, D., Kjeldsen, H., and Kiss, L. L. 2003. Arcturus. *ApJ*, **591**, L151.

Rosenthal, C. S., Christensen-Dalsgaard, J., Nordlund, Å., Stein, R. F., and Trampedach, R. 1999. Convective contributions to the frequencies of solar oscillations. *AandA*, **351**, 689.

Roxburgh, I. W. 2010. Asteroseismology of solar and stellar models. *Ap&SS*, **328**, 3.

Roxburgh, I. and Favata, F. 2003. The Eddington Mission. *Ap&SS*, **284**, 17.

Samadi, R., Goupil, M.-J., and Houdek, G. 2002. Solar-like oscillations in δ Scuti stars. *A&A*, **395**, 563.

Schou, J. and Buzasi, D. L. 2001. Observations of p-modes in α Cen. Page 391 of: *Helio- and Asteroseismology at the Dawn of the Millenium, Proc. SOHO 10/GONG 2000 Workshop.* ESA SP-464.

Scuflaire, R. 1974. The non radial oscillations of condensed polytropes. *A&A*, **36**, 107.

Smith, M. A. 1982. Precise radial velocities. I. A preliminary search for oscillations in Arcturus. *ApJ*, **253**, 727.

Stello, D. and Gilliland, R. L. 2009. Solar-like oscillations in a metal-poor globular cluster with the hubble space telescope. *ApJ*, **700**, 949.

Stello, D., Bruntt, H., Preston, H., and Buzasi, D. 2008. Oscillating K giants with the WIRE satellite: determination of their asteroseismic masses. *ApJ*, **674**, L53.

Stello, D., Chaplin, W. J., Basu, S., Elsworth, Y., and Bedding, T. R. 2009. The relation between $\Delta\nu$ and $\nu_{\rm max}$ for solar-like oscillations. *MNRAS*, **400**, L80.

Stello, D., Basu, S., Bruntt, H., *et al.* 2010. Detection of solar-like oscillations from Kepler photometry of the open cluster NGC 6819. *ApJ*, **713**, L182.

Tabur, V., Bedding, T. R., Kiss, L. L., *et al.* 2010. Period-luminosity relations of pulsating M giants in the solar neighbourhood and the Magellanic Clouds. *MNRAS*, **409**, 777.

Tarrant, N. J., Chaplin, W. J., Elsworth, Y., Spreckley, S. A., and Stevens, I. R. 2007. Asteroseismology of red giants: photometric observations of Arcturus by SMEI. *MNRAS*, **382**, L48.

Tarrant, N. J., Chaplin, W. J., Elsworth, Y., Spreckley, S. A., and Stevens, I. R. 2008. Oscillations in β Ursae Minoris: observations with SMEI. *A&A*, **483**, L43.

Tassoul, M. 1980. Asymptotic approximations for stellar nonradial pulsations. *ApJS*, **43**, 469.

Traub, W. A., Mariska, J. T., and Carleton, N. P. 1978. A search for stellar oscillations. *ApJ*, **223**, 583.

Vauclair, S., Laymand, M., Bouchy, F., *et al.* 2008. The exoplanet-host star ι Horologii: an evaporated member of the primordial Hyades cluster. *A&A*, **482**, L5.

Verner, G. A., Elsworth, Y., Chaplin, W. J., *et al.* 2011. Global asteroseismic properties of solar-like oscillations observed by Kepler: a comparison of complementary analysis methods. *MNRAS*, **415**, 3539–51.

Walker, G., Matthews, J., Kuschnig, R., *et al.* 2003. The MOST asteroseismology mission: ultraprecise photometry from space. *PASP*, **115**, 1023.

White, T. R., Bedding, T. R., Stello, D., *et al.* 2011. Calculating asteroseismic diagrams for solar-like oscillations. *ApJ*. submitted.

Zwintz, K., Kuschnig, R., Weiss, W. W., Gray, R. O., and Jenkner, H. 1999. Hubble deep field guide star photometry. *A&A*, **343**, 899.

4. Studying stars through frequency inversions

SARBANI BASU

4.1 Introduction

Helioseismology, the study of the Sun using solar oscillations, has provided us with the means to probe the solar interior. Since the discovery of the oscillations in 1962 (Leighton *et al.*, 1962) and their interpretation as global oscillation modes by Ulrich (1970) and Leibacher and Stein (1971), helioseismology has been used extensively to study the interior of the Sun, mainly through inversions of solar frequencies. With space missions such as CoRoT (Baglin *et al.*, 2006) and *Kepler* (Borucki *et al.*, 2010) now observing oscillations of other stars, inversions of stellar frequencies may soon be feasible. There are two ways by which we could use seismic data to make inferences about the stars. The first way involves trying to find models whose frequencies match the observed frequencies, usually referred to as "forward modeling." This is essentially what is done in most fields of astronomy. The end result of the process is a model that is the best match to the observations. The second way is to invert the data. Inversions use the data directly to make inferences about the star. In the case of inversions, we can make a distinction between the structure of the star and the structure of the best-fit model. These days inversions are used to study the solar interior, while forward modeling is used to study other stars.

It is not possible to do an inverse analysis unless we can do the forward analysis. Thus, to invert frequencies to determine stellar structure, we need to know how stellar frequencies depend on stellar properties. The stellar oscillations we will be concerned with have very small amplitudes and the periods are much smaller than heat-transport time scales. Such oscillations can be described using the theory of linear, adiabatic, oscillations. The oscillations are usually labeled by three "quantum" numbers – the radial order n, the degree ℓ, and the azimuthal order m. The radial order n can have any integer value and is the number of nodes in the radial direction. Positive values of n are used to denote acoustic modes, that is, the so-called p modes (p for pressure, since the dominant restoring force for these modes is provided by the pressure gradient). Negative values of n are used to denote modes for which buoyancy provides the main restoring force. These are usually referred to as g modes (g for gravity). Modes with $n = 0$ are the so-called fundamental or f modes. The f modes are essentially surface gravity modes whose frequencies are largely independent of the stratification of the stellar interior. As a result, f mode frequencies have not usually been used to determine the structure of the Sun. However, these have been used to draw inferences about the solar radius (e.g., Schou *et al.*, 1997; Antia *et al.*, 1998; Lefebvre *et al.*, 2007). The degree ℓ denotes the number of nodal planes that intersect the surface of a star and m is the number of nodes along the equator.

For a spherically symmetric star, all modes with the same degree ℓ and order n have the same frequency. Asphericities such as rotation lift this degeneracy and "split" the frequencies, causing them to become m-dependent. Very often, the frequency $\nu_{n\ell m}$ of a mode is written in term of "splitting coefficients" as

$$\frac{\omega_{n\ell m}}{2\pi} = \nu_{n\ell m} = \nu_{n\ell} + \sum_{j=1}^{j_{\max}} a_j(n\ell)\mathcal{P}_j^{n\ell}(m), \tag{4.1}$$

where a_j are the splitting or "a" coefficients and $\mathcal{P}$ are polynomials related to Clebsch-Gordon coefficients. It can be shown that for slow rotation the central frequency $\nu_{n\ell}$

depends only on structure, the odd-order a coefficients depend on rotation, and the even-order a coefficients depend on structural asphericities, magnetic fields, and second-order effects of rotation. In this chapter, we shall only be dealing with results obtained with the central frequencies, $\nu_{n\ell}$.

The rest of the chapter is organized as follows: We derive the equations that relate stellar frequencies to stellar structure in Section 4.2 and put the equation in a form that can be inverted in Section 4.3. The two popular techniques used to invert solar frequencies are described in Section 4.4. We discuss some inversion results and how we may interpret them in Section 4.5. In Section 4.6 we try to determine what we need to invert frequencies of other stars. Unless explicitly mentioned, results shown in the figures were obtained specifically for this chapter.

4.2 The oscillation equation

The derivation of the equations of stellar oscillations have been discussed extensively in literature; see, for example, Cox (1980), Unno *et al.* (1989), Christensen-Dalsgaard, and Berthomieu (1991), Gough (1993), Christensen-Dalsgaard (2002, 2003). We give a brief overview of the steps that allow us to determine how frequencies of stellar oscillation are related to stellar structure.

Derivation of the equations of linear adiabatic stellar oscillations starts with the basic equations of fluid dynamics. These are the continuity equation and the momentum equation, together with the Poisson's equation to describe the gravitational field, that is,

$$\frac{\partial \rho}{\partial t} + \nabla \cdot (\rho \mathbf{v}) = 0, \tag{4.2}$$

$$\rho \left(\frac{\partial \mathbf{v}}{\partial t} + \mathbf{v} \cdot \nabla \mathbf{v} \right) = -\nabla P - \rho \nabla \Phi, \tag{4.3}$$

$$\nabla^2 \Phi = 4\pi G \rho, \tag{4.4}$$

where $\mathbf{v}$ is the velocity of the fluid element, Φ is the gravitational potential, and G the gravitational constant. Also needed is the heat equation, which can be written in the form

$$\frac{dq}{dt} = \frac{1}{\rho(\Gamma_3 - 1)} \left(\frac{dp}{dt} - \frac{\Gamma_1 P}{\rho} \frac{d\rho}{dt} \right), \tag{4.5}$$

where

$$\Gamma_1 = \left(\frac{\partial \ln p}{\partial \ln \rho} \right)_{\rm ad}, \quad \text{and,} \quad \Gamma_3 - 1 = \left(\frac{\partial \ln T}{\partial \ln \rho} \right)_{\rm ad}. \tag{4.6}$$

We assume that the equilibrium structure is static and the oscillations are small perturbations to the equilibrium structure. Thus, all time derivatives of equilibrium quantities vanish.

The equations describing solar oscillations are obtained by a linear perturbation analysis of Equations 4.2–4.5. Since we are assuming adiabaticity, $dq/dt = 0$, and thus Equation 4.5 can be used to determine how changes in pressure are related to changes in density.

We can write the perturbation to pressure as:

$$P(r, \theta, \phi, t) = P_0(r) + P'(r, \theta, \phi, t). \tag{4.7}$$

The subscript "0" denotes the equilibrium quantity which, by definition, does not depend on time. The superscript "′" denotes the perturbation. We have assumed that the equilibrium structure is spherically symmetric in addition to being static and hence P_0 is only a function of r. Perturbations to other quantities, such as density, can be expressed in the same form. These are the Eulerian perturbations, which are evaluated at a specified point described by coordinates (r, θ, ϕ). The Lagrangian perturbation, on the other hand, is given by

$$\delta P(r,t) = P(\vec{r} + \vec{\xi}(\vec{r},t)) - P(\vec{r}) = P'(\vec{r},t) + \vec{\xi}(\vec{r},t) \cdot \nabla P, \tag{4.8}$$

where $\vec{\xi}$ is the displacement from the equilibrium position. The perturbations to the other quantities can be written in exactly the same way.

By definition, the equilibrium state has no velocities; the velocity $\mathbf{v}$ in the context of stellar oscillations is simply the time derivative of the displacement. Since we are looking for oscillatory solutions, we search for solutions with a time dependence of the form $\exp(-i\omega t)$, ω being the frequency of the oscillation. Such a choice is possible because the equations do not explicitly depend on t.

Substituting the perturbed quantities (Eulerian) in the basic equations (4.2–4.4), and keeping only linear terms in the perturbations, we get:

$$\rho' + \nabla \cdot (\rho_0 \vec{\xi}) = 0, \tag{4.9}$$

$$-\omega^2 \rho \vec{\xi} = -\nabla P' - \rho_0 \nabla \Phi' - \rho' \nabla \Phi_0, \tag{4.10}$$

$$\nabla^2 \Phi' = 4\pi G \rho'. \tag{4.11}$$

It can be shown (see Christensen-Dalsgaard, 2003) that if we assume a spherically symmetric equilibrium state, we can write the angular dependence of the perturbations in the form of spherical harmonics, $Y_\ell^m(\theta, \phi)$, where ℓ and m are the degree and the azimuthal order that we described earlier.

It is easier to consider the Lagrangian perturbation for the heat equation since under the assumptions that have been made, the time derivative of the various quantities is simply the time derivative of the Lagrangian perturbation of those quantities. Thus, from Equation 4.5 we get

$$\frac{\partial \delta q}{\partial t} = \frac{1}{\rho_0(\Gamma_{3.0} - 1)} \left(\frac{\partial \delta P}{\partial t} - \frac{\Gamma_{1,0} P_0}{\rho_0} \frac{\partial \delta \rho}{\partial t} \right). \tag{4.12}$$

In the adiabatic limit, where energy loss is negligible, $\partial \delta q / \partial t = 0$ and hence,

$$P' + \vec{\xi} \cdot \nabla P = \frac{\Gamma_{1,0} P_0}{\rho_0} (\rho' + \vec{\xi} \cdot \nabla \rho). \tag{4.13}$$

The term $\Gamma_{1,0} P_0 / \rho_0$ in Equation 4.13 is nothing but the squared, unperturbed sound-speed c_0^2. Eliminating P' and ρ', and expressing the the gravitational potential as an integral, we can combine Equations 4.9 through 4.11 using Equation 4.13 to get one equation:

$$-\omega^2 \rho \vec{\xi} = \nabla(c^2 \rho \nabla \cdot \vec{\xi} + \nabla P \cdot \vec{\xi}) - \vec{g} \nabla \cdot (\rho \vec{\xi}) - G \rho \nabla \left(\int_V \frac{\nabla \cdot (\rho \vec{\xi})\, dV}{|\vec{r} - \vec{r'}|} \right). \tag{4.14}$$

This equation describes how the mode frequencies ω depend on the structure of the star. In Equation 4.14 we have dropped the subscript 0 from the equilibrium quantities for convenience, because the perturbations, ρ', P', and so on, do not occur. This is the starting point of most inversions.

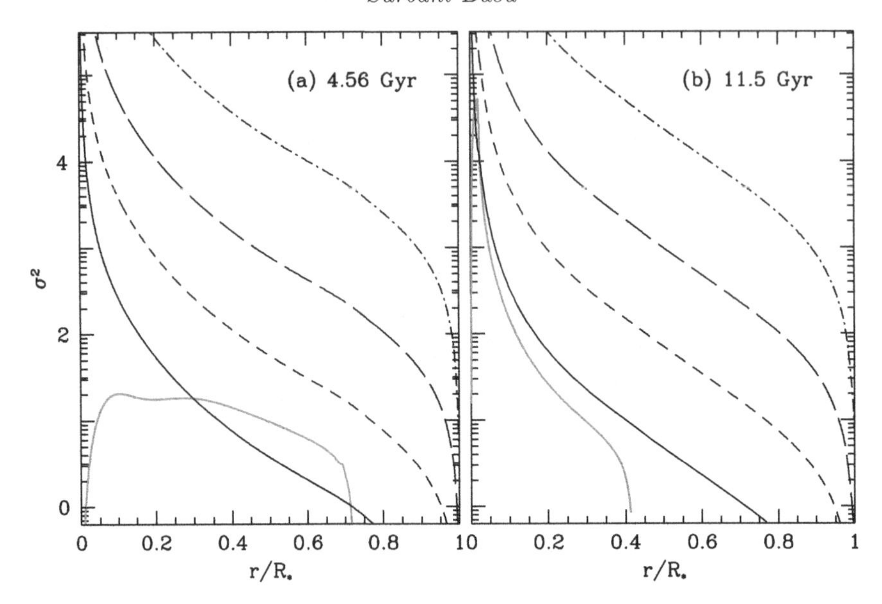

FIG. 4.1. The propagation diagram for two 1 $M_\odot$ stars: *a)* one on the main sequence and *b)* one evolved. The ordinate is the squared dimensionless frequency $\sigma^2 = (R^2/GM)\omega^2$, R being total radius, M the total mass, and G the gravitational constant. The gray line shows the squared Brunt-Väisälä frequency. The black lines show the squared Lamb frequency for $\ell = 1$ (solid), $\ell = 5$ (small dashes), $\ell = 20$ (large dashes), and $\ell = 100$ (dot dashed line). Note that while the dimensionless Lamb frequency does not change much with age, the Brunt-Väisälä changes drastically. Models were constructed using the YREC code (Demarque *et al.*, 2008).

While it is not obvious from Equation 4.14, different oscillation modes penetrate to different depths in a star and hence carry information from different parts of a star. This property allows us to invert frequencies and build up a picture of the star. A somewhat different analysis of Equations 4.9 through 4.11 (see Unno *et al.*, 1989; Christensen-Dalsgaard, 2003) shows that there are two fundamental frequencies that determine the behavior of a mode. The first is the buoyancy frequency, known as the *Brunt-Väisälä* frequency, defined as

$$N^2 = g\left(\frac{1}{\Gamma_1 P}\frac{dP}{dr} - \frac{1}{\rho}\frac{d\rho}{dr}\right). \tag{4.15}$$

This is the frequency with which a small element of fluid will oscillate when it is disturbed from its equilibrium position. The fluid is unstable to convection when $N^2 < 0$. The other relevant frequency is the so-called *Lamb frequency*, which depends on the degree ℓ of the modes and is defined as

$$S_\ell^2 = \frac{\ell(\ell+1)c^2}{r^2}. \tag{4.16}$$

Figure 4.1 shows the Brunt-Väisälä frequency for a 1 $M_\odot$ stellar model at two states of evolution. Also shown is the Lamb frequency for a few values of ℓ. Diagrams like Fig. 4.1 are often known as propagation diagrams since they help us determine the region of a star where different types of modes propagate. Modes with frequency ω for which $\omega^2 > N^2$ and $\omega^2 > S_\ell^2$ are the p modes. These have highest amplitudes at the surface. Modes for which $\omega^2 < N^2$ and $\omega^2 < S_\ell^2$ are the g modes or gravity modes and these have the highest amplitudes toward the core. In evolved stars, the Brunt-Väisälä frequency can become larger than the Lamb frequency in certain regions (see Fig. 4.1b) and in these cases, a mode may be a "mixed" mode, that is, it can have g mode-like characteristics near the core and p mode-like characteristics near the surface. The lower turning point of a p mode, that is, the radius to which can it penetrate, can be determined using the Lamb frequency. The lower turning point, r_t, is the point where $\omega^2 = S_\ell^2$. As can be seen from Fig. 4.1, low-ℓ p modes penetrate deeper than high-ℓ p modes of the same frequency, while high-frequency p modes penetrate deeper than low-frequency p modes of the same degree.

4.3 The inversion equations

Although Equation 4.14 is the starting point of inversions, it is clear that we cannot use the equation as it stands. The problem with using it is as follows: we can determine the oscillation frequencies ω from observation and we want to determine c^2 and ρ (pressure P is related to ρ through the equation of hydrostatic equilibrium; similarly, $\vec{g}$ can be derived from ρ). However, we do not have any means of determining the displacement vector $\vec{\xi}$. We may be able to determine $\vec{\xi}$ at the surface, but we cannot find $\vec{\xi}$ as a function of radius. The way out of this impasse is to recognize that Equation 4.14 defines an eigenvalue problem of the form

$$\mathcal{L}(\vec{\xi}_{n,\ell}) = -\omega_{n,\ell}^2 \vec{\xi}_{n,\ell}\,, \tag{4.17}$$

$\mathcal{L}$ being a differential operator in Equation 4.14, and under specific boundary conditions, namely $\rho = P = 0$ at the outer boundary, the eigenvalue problem defined by Equation 4.14 is Hermitian (Chandrasekhar, 1964).

If $\mathcal{O}$ be a differential Hermitian operator, and x and y be two eigenfunctions of the operator, then

$$\int x^* \mathcal{O}(y) dV = \int y \mathcal{O}(x^*) dV, \tag{4.18}$$

where * indicates a complex conjugate. This is often written as

$$< x, \mathcal{O}(y) > = < \mathcal{O}(x), y >, \tag{4.19}$$

where the operation $<>$ is usually called an inner product. The inner product needs to be defined appropriately for each operator. Hermitian operators have the properties of their eigenvalues being real and the variational principle applying. This means that if a is the eigenvalue corresponding to the eigenfunction x, then

$$a = \frac{< x, \mathcal{O}(x) >}{< x, x >}. \tag{4.20}$$

Chandrasekhar (1964) showed that the oscillation equation (Equation 4.14, 4.17) constitutes a Hermitian, eigenvalue problem under the boundary condition $\rho = P = 0$ at $r = R$ if the inner product is defined as:

$$< \vec{\xi}, \vec{\eta} > = \int_V \rho \vec{\xi}^* \cdot \vec{\eta} d^3\vec{r} = 4\pi \int_0^R [\xi_r^*(r)\eta_r(r) + L^2 \xi_t^*(r)\eta_t(r)] r^2 \rho dr, \tag{4.21}$$

where the subscripts r and t, respectively, denote the radial and tangential components of the displacement eigenfunctions $\vec{\xi}$ and $\vec{\eta}$. Thus, if the frequency $\omega_{n,\ell}$ is the eigenvalue corresponding to the displacement eigenfunction $\vec{\xi}_{n,\ell}$, then

$$-\omega_{n,\ell}^2 \vec{\xi}_{n,\ell} = \frac{\int_V \rho \vec{\xi}_{n,\ell}^* \cdot \mathcal{L}(\vec{\xi}_{n,\ell}) d^3\vec{r}}{\int_V \rho \vec{\xi}_{n,\ell}^* \cdot \vec{\xi}_{n,\ell} d^3\vec{r}}. \tag{4.22}$$

Note that the denominator is nothing but the mode inertia (Christensen-Dalsgaard 2003):

$$E_{n,\ell} = \int_V \rho \vec{\xi}_{n,\ell} \cdot \vec{\xi}_{n,\ell}\, d^3\vec{r} = \int_0^R \rho[\xi^2_{r,n,\ell} + \ell(\ell+1)\xi^2_{t,n,\ell}]r^2 dr. \tag{4.23}$$

Modes that penetrate deeper into the star (low-degree modes) have higher mode inertia than those that do not penetrate as deep (higher-degree modes). Additionally, for modes of a given degree, higher frequency modes have larger inertia than lower frequency modes.

To obtain the equation that we will be able to invert, we linearize Equation 4.17 around a known model, usually called the *reference* model. Thus, if $\mathcal{L}$ is the operator for a model, we assume that the operator for the Sun or other star can be expressed as $\mathcal{L} + \delta\mathcal{L}$; and if $\vec{\xi}$ is the displacement eigenfunction for the model, then the displacement eigenfunction for the Sun is $\vec{\xi} + \delta\vec{\xi}$; similarly, if ω is the frequency of a given model of the model, then that for the Sun is $\omega + \delta\omega$. Thus, we can write:

$$(\mathcal{L} + \delta\mathcal{L})(\vec{\xi} + \delta\vec{\xi}) = -(\omega + \delta\omega)(\vec{\xi} + \delta\vec{\xi}), \tag{4.24}$$

which on expansion gives

$$\begin{aligned}\int \vec{\xi}^* \mathcal{L}\vec{\xi} dV + \int \vec{\xi}^* \delta\mathcal{L}\vec{\xi} dV + \int \vec{\xi}^* \mathcal{L}\delta\vec{\xi} dV + \int \vec{\xi}^* \delta\mathcal{L}\delta\vec{\xi} dV = -\omega^2 \int \vec{\xi}^* \vec{\xi} dV \\ - \omega^2 \int \vec{\xi}^* \delta\vec{\xi} dV - (\delta\omega^2) \int \vec{\xi}^* \vec{\xi} dV - 2\omega\delta\omega \int \vec{\xi}^* \vec{\xi} dV\end{aligned} \tag{4.25}$$

if we only keep terms linear in the perturbations. Because $\mathcal{L}$ is a Hermitian operator, all but two terms in Equation 4.25 cancel out to give

$$\int \vec{\xi}^* \delta\mathcal{L}\vec{\xi} dV = -2\omega\delta\omega \int \vec{\xi}^* \vec{\xi} dV, \tag{4.26}$$

or

$$\frac{\delta\omega}{\omega} = -\frac{\int_V \rho\vec{\xi} \cdot \delta\mathcal{L}\vec{\xi}\, d^3\vec{r}}{2\omega^2 \int_V \rho\vec{\xi} \cdot \vec{\xi}\, d^3\vec{r}} = -\frac{\int_V \rho\vec{\xi} \cdot \delta\mathcal{L}\vec{\xi}\, d^3\vec{r}}{2\omega^2 E}. \tag{4.27}$$

One such equation can be written for each mode, and these are the equations that we invert. Note that Equation 4.27 shows that for a given perturbation, modes with lower inertia will be changed more than modes with higher mode inertia.

From Equation 4.14 we can see that

$$\begin{aligned}\delta\mathcal{L}\vec{\xi} = \nabla(\delta c^2 \nabla \cdot \vec{\xi} + \delta\vec{g} \cdot \vec{\xi}) + \nabla\left(\tfrac{\delta\rho}{\rho}\right) c^2 \nabla \cdot \vec{\xi} + \tfrac{1}{\rho}\nabla\rho\delta c^2 \nabla \cdot \vec{\xi} \\ + \delta\vec{g}\nabla \cdot \vec{\xi} - G\nabla \int_V \tfrac{\nabla\cdot(\delta\rho\vec{\xi})}{|\vec{r}-\vec{r}'|} d^3\vec{r}'\end{aligned} \tag{4.28}$$

where δc^2, $\delta\vec{g}$, and $\delta\rho$ are the differences in sound speed and acceleration due to gravity and density between the model and the Sun or star. The quantity $\delta\vec{g}$ can be expressed in terms of $\delta\rho$.

Substituting Equation 4.28 in Equation 4.27 and rearranging the terms, we can write

$$\frac{\delta\omega}{\omega} = -\frac{1}{2\omega^2 E_k}(I_1 + I_2 + I_3 + I_4), \tag{4.29}$$

where

$$I_1 = -\int_0^R \rho(\nabla\cdot\vec{\xi})^2\delta c^2 r^2\,dr \tag{4.30}$$

$$I_2 = \int_0^R \xi_r[\rho\nabla\cdot\vec{\xi} + \nabla\cdot(\rho\vec{\xi})]\delta g r^2\,dr \tag{4.31}$$

$$\delta g(r) = \frac{4\pi G}{r^2}\int_0^r \delta\rho(s)s^2\,ds \tag{4.32}$$

$$I_3 = \int_0^R \rho c^2\xi_r\nabla\cdot\vec{\xi}\frac{d}{dr}\left(\frac{\delta\rho}{\rho}\right)r^2\,dr \tag{4.33}$$

and

$$I_4 = \frac{4\pi G}{2\ell+1}\int_0^R \nabla\cdot(\rho\vec{\xi})r^2\,dr\left[\frac{1}{r^{\ell+1}}\int_0^r s^{\ell+2}\left(\rho\nabla\cdot\vec{\xi} - \rho\frac{d\xi_r}{ds} - \frac{\ell+2}{s}\rho\xi_r\right)\frac{\delta\rho}{\rho}\,ds\right] \tag{4.34}$$

(see Antia and Basu, 1994).

Equation 4.30 can be rewritten in terms of $\delta c^2/c^2$, the relative difference between the squared sound speed of the Sun/star and the reference model. Similarly, Equations 4.31–4.34 can be rewritten (in some cases after changing the order of the integrals, see, e.g., Gough, 1991; Gough and Thompson, 1991) in terms of the relative density difference $\delta\rho/\rho$. Thus, Equation 4.29 for each mode i can be rewritten as:

$$\frac{\delta\omega_i}{\omega_i} = \int K^i_{c^2,\rho}(r)\frac{\delta c^2}{c^2}(r)\,dr + \int K^i_{\rho,c^2}(r)\frac{\delta\rho}{\rho}(r)\,dr \tag{4.35}$$

The terms $K^i_{c^2,\rho}(r)$ and $K^i_{\rho,c^2}(r)$ are known functions of the reference model and represent the change in frequency in response to changes in sound speed and density, respectively. These two functions are known as the "kernels" of the inversion.

4.3.1 *The inversion kernels*

We show sound-speed and density kernels for a few modes in Fig. 4.2. Note that sound-speed kernels are positive at all radii, while density kernels oscillate between positive and negative values. This oscillation reduces the contribution of the second term of the RHS of Equation 4.35 and explains why density inversions are usually much more difficult that sound-speed inversions. We can also see that modes of higher ℓ are restricted closer to the surface, while modes of lower ℓ penetrate deeper. This is exactly what we expect from our earlier discussion about propagation of these modes. We also see that higher frequency modes of a given ℓ penetrate deeper than lower frequency modes of the same ℓ. However, notice that the amplitude of the sound-speed kernels of very low frequency, low-ℓ, modes is actually larger closer to the core than that of modes of higher frequency. This is the reason behind the attempts to detect lower and lower frequency solar p modes. In the absence of definitive g mode frequencies, low-frequency, low-degree p modes can increase our knowledge of the Sun (see, e.g., Basu *et al.*, 2009).

We are not constrained to inverting for the sound-speed and density profiles only. It is not very difficult to convert one set of kernels to another. Gough (1991) discusses how this may be done. It is relatively easy to change the kernels pair for (c^2, ρ) into that for (Γ_1, ρ). Note that $c^2 = \Gamma_1 P/\rho$, and hence

$$\frac{\delta c^2}{c^2} = \frac{\delta\Gamma_1}{\Gamma_1} + \frac{\delta P}{P} - \frac{\delta\rho}{\rho}. \tag{4.36}$$

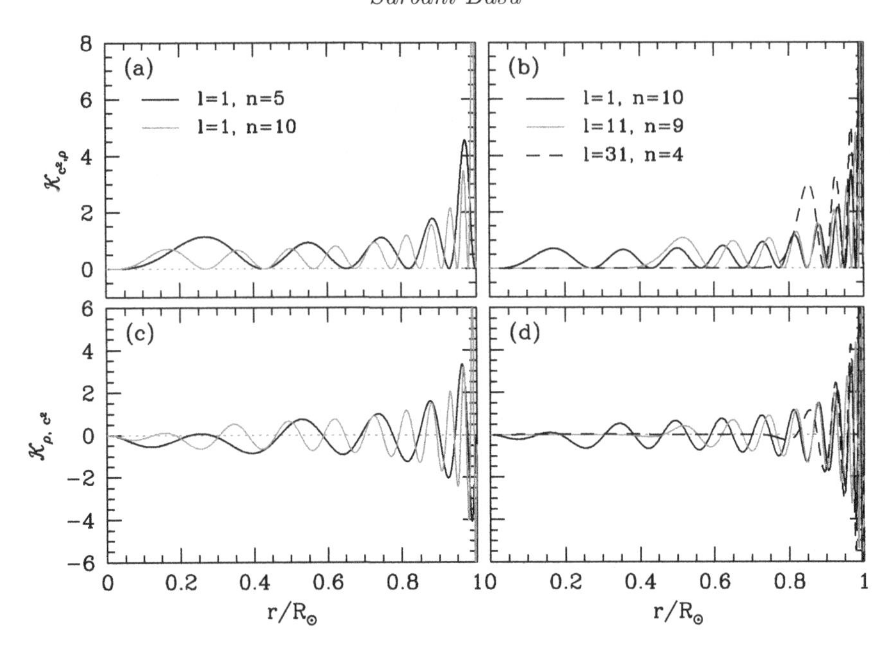

FIG. 4.2. The upper panels show the sound-speed kernel $K_{c^2,\rho}$ and the lower panels show the density kernels K_{ρ,c^2} for a solar model. *(a)* and *(c)* show kernels for modes with the same degree but different radial order (hence, different frequency); *(b)* and *(d)* show kernels for modes with different degrees but similar frequencies.

Substituting Equation 4.36 in Equation 4.35, we get

$$\frac{\delta\omega_i}{\omega_i} = \int K^i_{c^2,\rho}(r)\frac{\delta\Gamma_1}{\Gamma_1}\, dr + \int K^i_{c^2,\rho}\frac{\delta P}{P}dr + \int \left(K^i_{\rho,c^2} - K^i_{c^2,\rho}\right)\frac{\delta\rho}{\rho}dr. \tag{4.37}$$

Thus, we see that $K_{\Gamma_1,\rho} = K_{c^2,\rho}$. To obtain the K_{ρ,Γ_1}, we have to use the equation of hydrostatic equilibrium, $dP/dr = -g\rho$, to write $\int K_{c^2,\rho}(\delta P/P)dr$ in terms of $\delta\rho/\rho$ though an integral. Doing that, and changing the order of integrations, we can write $\int K_{c^2,\rho}(\delta P/P)dr - \int K_{c^2,\rho}(r)(\delta\rho/\rho)dr + \int K_{\rho,c^2}(\delta\rho/\rho)dr$ as $\int B(r)\delta\rho/\rho$, where $B(r)$ is the kernel K_{ρ,Γ_1}.

It is usually not possible to get a closed form for kernels other than (c^2,ρ) and (Γ_1,ρ). For instance, going from (c^2,ρ) to (u,Γ_1) $(u \equiv P/\rho)$, we find that $K_{\Gamma_1,u} \equiv K_{c^2,\rho}$ (vide Eq 4.36). But the second kernel of the pair, K_{u,Γ_1}, does not have a closed form solution. It can be written as

$$K_{u,\Gamma_1} = K_{c^2,\rho} - P\frac{d}{dr}\left(\frac{\psi}{P}\right), \tag{4.38}$$

where, ψ is a solution of the equation

$$\frac{d}{dr}\left(\frac{P}{r^2\rho}\frac{d\psi}{dr}\right) - \frac{d}{dr}\left(\frac{Gm}{r^2}\psi\right) + \frac{4\pi G\rho}{r^2}\psi = -\frac{d}{dr}\left(\frac{P}{r^2}F\right) \tag{4.39}$$

for $F = K_{\rho,c^2}$.

It has also been common to use the knowledge of the equation of state of stellar matter to inject some more information into the system and convert (Γ_1,ρ) kernels to kernels of (u,Y), Y being the helium abundance. Kernels for Y are nonzero only in the ionization zone. This, as we shall show later, can be advantageous for inversions. The adiabatic index Γ_1 can be expressed as

$$\frac{\delta\Gamma_1}{\Gamma_1} = \left(\frac{\partial \ln \Gamma_1}{\partial \ln P}\right)_{\rho,Y} \frac{\delta P}{P} + \left(\frac{\partial \ln \Gamma_1}{\partial \ln \rho}\right)_{P,Y} \frac{\delta\rho}{\rho} + \left(\frac{\partial \ln \Gamma_1}{\partial Y}\right)_{\rho,P} \delta Y$$
$$\equiv \Gamma_{1,P}\frac{\delta P}{P} + \Gamma_{1,\rho}\frac{\delta\rho}{\rho} + \Gamma_{1,Y}\delta Y \qquad (4.40)$$

where the partial derivatives can be calculated using the equation of state. Once we know this, it is easy to see that

$$K_{Y,u} \equiv \Gamma_{1,Y} K_{\Gamma_1,\rho}, \quad \text{and} \quad K_{u,Y} \equiv \Gamma_{1,P} K_{\Gamma_1,\rho} - P\frac{d}{dr}\left(\frac{\psi}{P}\right), \qquad (4.41)$$

with ψ being the solution of Equation 4.39 with $F \equiv (\Gamma_{1,P} + \Gamma_{1,\rho})K_{\Gamma_1,\rho} + K_{\rho,\Gamma_1}$.

Some kernel calculations may be even more involved. Elliott (1996) discusses how one may calculate kernels for (A^*, Γ_1), where A^* is the convective stability parameter

$$A^* = \frac{1}{\Gamma_1}\frac{d\ln P}{d\ln r} - \frac{d\ln\rho}{d\ln r} \qquad (4.42)$$

starting from kernels of (Γ_1, ρ). Because the stability parameter is very small in the convection zone but rises sharply at the convective zone (CZ) boundary, inversions for A^* have been done to determine the solar CZ boundary. This kernel pair can also be useful to study Γ_1 differences in the ionization zones of low mass stars.

4.3.2 *The surface term*

Unfortunately, Equation 4.35 is not enough to represent all the differences between the Sun and the models. There is an additional term that arises due to uncertainties in modeling layers just below the solar surface, and deviations from adiabaticity.

Equation 4.27 implies that we can invert the frequencies provided we know how to model the Sun properly, and provided that the frequencies can be described by the equations for adiabatic oscillations. Neither of these two assumptions is completely correct. For instance, our treatment of convection in stars is known to be approximate. Simulations of stellar convection indicate that there are significant departures between the temperature gradients obtained from the mixing length approximation and those from a full treatment of convection. The deviations mainly occur close to the solar surface (see, e.g., Robinson *et al.*, 2003, and references therein). This results in differences in the density and pressure profiles. Not using a full treatment of convection also means that several physical effects, such as turbulent pressure and turbulent kinetic energy, are missing from the models. The adiabatic approximation that is used to calculate the frequencies of the models also breaks down near the surface where the thermal time scale is comparable to the period of oscillations. All such factors imply that the RHS of Equation 4.35 does not fully account for the frequency difference $\delta\nu/\nu$ between the Sun and the model.

Fortunately, the major uncertainties are located in a thin layer near the surface. For modes that are not of very high degree ($\ell \simeq 200$ and lower), the structure of the wavefront near the surface is almost independent of the degree of the modes – the wave vector of these modes is almost completely radial in the near-surface layers. This implies that any additional difference in frequency due to errors in the surface structure has to be a function of frequency alone once differences in mode inertia has been taken into account. It can also be shown (e.g., Gough, 1990) that surface perturbations cause the difference in frequency to be a slowly varying function of frequency that can be modeled as a sum of low degree polynomials. This effect is illustrated in Fig. 4.3, where we show the frequency differences between the Sun and two solar models that differ only near the surface because of differences in the convection formalism used to construct the models.

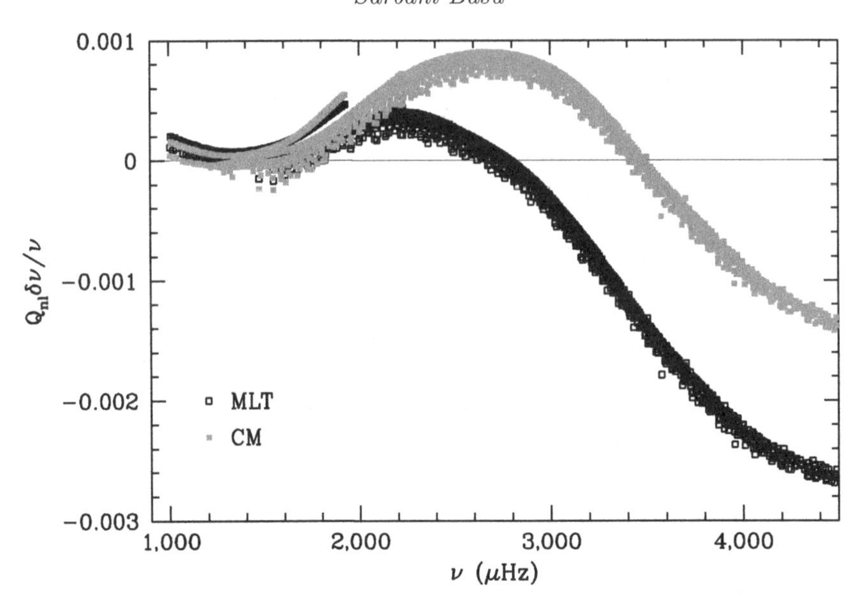

FIG. 4.3. The scaled frequency differences between the Sun and two standard models, one constructed with the mixing length approximation (black points) and the other with the Canuto and Mazzitelli (1991) formulation (gray points). The difference between the two models lie in the near-surface regions. (Data from Basu *et al.*, 1999.)

The frequencies have been plotted after being scaled by the quantity

$$Q_{nl} = E_{n\ell}(\omega_{n\ell})/\bar{E}_0(\omega_{n\ell}). \tag{4.43}$$

Here, $E_{n\ell}(\omega_{n\ell})$ is the mode inertia of a model of degree ℓ, order n, and frequency $\omega_{n\ell}$. The quantity $\bar{E}_0(\omega_{n\ell})$ is the mode inertia of a radial mode (i.e., $\ell = 0$) interpolated to the frequency $\omega_{n\ell}$. This scaling corrects for the fact that for a given perturbation, low-inertia modes undergo a larger frequency shift than high-inertia modes. Note from the figure that for each model, the scaled frequency differences for modes of all degrees tend to fall on a single curve, and that the curve is a slowly varying function of frequency. Thus, we can modify Equation 4.35 to represent the difference between the Sun and the reference model as

$$\frac{\delta\omega_i}{\omega_i} = \int \mathcal{K}^i_{c^2,\rho}(r)\frac{\delta c^2}{c^2}(r)\, dr + \int \mathcal{K}^i_{\rho,c^2}(r)\frac{\delta\rho}{\rho}(r)\, dr + \frac{F(\omega_i)_{\rm surf}}{E_i} \tag{4.44}$$

where $F(\omega_{n\ell})$ is a slowly varying function of frequency that arises due to the errors in modeling the near-surface regions (see also Dziembowski *et al.*, 1990; Antia and Basu, 1994).

When we invert oscillation data to determine the density profile, we need to ensure that mass is conserved, that is,

$$\int \frac{\delta\rho}{\rho}\rho(r)r^2 dr = 0 \tag{4.45}$$

This is usually done by defining an additional mode with:

$$\omega = 0, \quad K_{c^2,\rho} = 0, \quad \text{and} \quad K_{\rho,c^2} = \rho(r)r^2 \tag{4.46}$$

4.4 Techniques for inverting oscillation frequencies

For N observed modes, Equation 4.44 represents N equations that need to be "inverted." The data $\delta\omega_i/\omega_i$ or equivalently $\delta\nu_i/\nu_i$ are known. The kernels $K^i_{c^2,\rho}$ and

K^i_{ρ,c^2} are known functions of the reference model. The task of the inversion process is to estimate $\delta c^2/c^2$ and $\delta\rho/\rho$ after somehow accounting for the surface term $F_{\rm surf}$. There is one inherent problem: no matter how many modes we observe, we can only have a finite amount of data and hence a finite amount of information. The two unknowns, $\delta c^2/c^2$ and $\delta\rho/\rho$, being functions, have an infinite amount of information. Thus, there is really no way that we can recover the two functions from the data. The best we may hope for is to find localized averages of the two functions at a finite number of points. How localized the averages are depends on the data we have. During the inversion process, we often have to make additional assumptions, such as the sound speed and density of a star being positive definite functions, in order to get physically valid solutions.

Gough (1985) gave a prescription for the numerical inversion of solar oscillation frequencies using Equation 4.44. Since then, many groups have inverted available solar frequencies to build up a detailed picture of the Sun. A short description of the type of inversion methods discussed here can be found in Gough and Thompson (1991) and Christensen-Dalsgaard *et al.* (2003). There are two popular ways of inverting Equation 4.44: (1) the regularised least squares (RLS) method, and (2) the method of optimally localized averages (OLA). The two inversion techniques have two different aims. In the case of RLS, the aim is to find the $\delta c^2/c^2$ and $\delta\rho/\rho$ profiles that give the best fit to the data (i.e., give the smallest residuals) while keeping the errors small. The aim of OLA, on the other hand, is not to fit the data at all, but to find linear combinations of the frequency differences in such a way that the corresponding combination of kernels provides a localized average of the unknown function, again while keeping the errors small.

4.4.1 *The regularized least squares technique*

RLS inversions start with expressing the three unknown functions in Equation 4.44 in terms of well-defined basis functions. Thus,

$$
\begin{aligned}
F(\omega) &= \sum_{i=1}^{m} a_i \psi_i(\omega), \\
\frac{\delta c^2}{c^2} &= \sum_{i=1}^{n} b_i \phi_i(r), \\
\frac{\delta\rho}{\rho} &= \sum_{i=1}^{n} c_i \phi_i(r)
\end{aligned}
\tag{4.47}
$$

where $\psi_i(\omega)$ are suitable basis functions in frequency ω, and $\phi_i(r)$ are suitable basis functions in radius r. Thus, for N observed modes, Equation 4.44 represents N equations of the form:

$$
\begin{aligned}
& b_1 \int \phi_1(r) K^i_{c^2} dr + b_2 \int \phi_2(r) K^i_{c^2} dr + \cdots + b_n \int \phi_n(r) K^i_{c^2} dr \\
& + c_1 \int \phi_1(r) K^i_{\rho} dr + c_2 \int \phi_2(r) K^i_{\rho} dr + \cdots + c_n \int \phi_n(r) K^i_{\rho} dr \\
& + a_1 \psi_1(\omega_i)/E_i + a_2 \psi_2(\omega_i)/E_i + \cdots + a_m \psi_m(\omega_i)/E_i = \delta\omega_i/\omega_i,
\end{aligned}
\tag{4.48}
$$

where, for ease of writing, we have denoted $K^i_{c^2,\rho}$ as $K^i_{c^2}$ and K^i_{ρ,c^2} as K^i_{ρ}. What we need to do is find coefficients a_i, b_i, and c_i such that we get the best fit to the data ω_i/ω_i. This is done by minimizing

$$
\chi^2 = \sum_{i=1}^{N} \left(\frac{\frac{\delta\omega_i}{\omega_i} - \Delta\omega_i}{\sigma_i} \right)^2
\tag{4.49}
$$

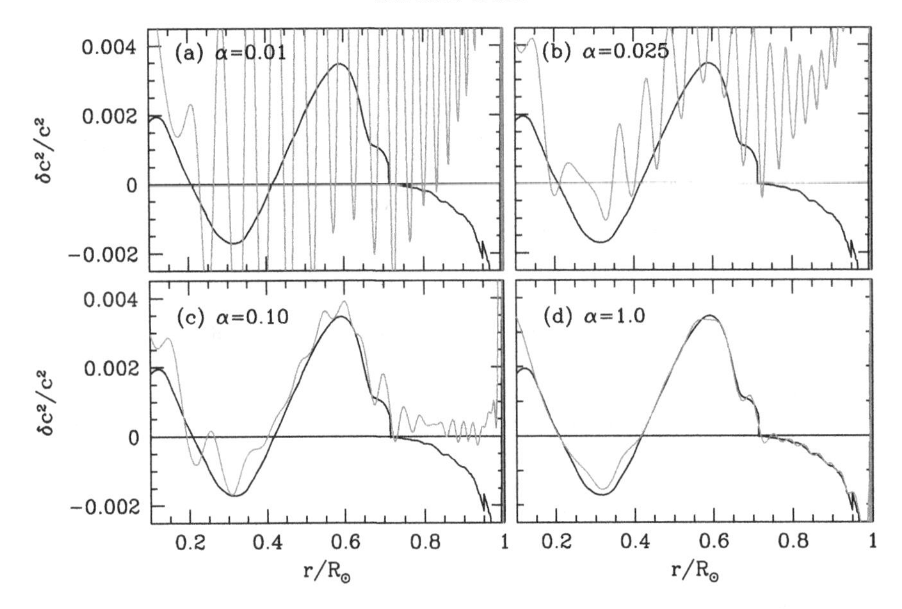

FIG. 4.4. Sound speed differences between two solar models obtained by inverting their frequency differences using the RLS technique. In each panel the black line is the true difference between the models and the gray line is the inversion results. The different panels show the result using different values of the smoothing parameter α. The reference model used is model BP04 of Bahcall *et al.* (2005), and the test model is model BSB(GS98) of Bahcall *et al.* (2006). We only used those modes that have been observed in the Sun; specifically, we used the mode-set BiSON-13 of Basu *et al.* (2009). Random errors, corresponding to the uncertainties in the observed mode-set, were added to the data of the test model.

where $\Delta\omega_i$ represents the LHS of Equation 4.48. While such a minimization does give us a solution for the three unknown functions, the solutions often are extremely oscillatory in nature (see Fig. 4.4). The main reason for this is that the data have errors and the oscillatory solution is a result of error propagation into the unknown function. The phenomenon is also related to the so-called Gibbs' Phenomenon in mathematics and is a result of the fact that we only have a finite amount of data but are trying to recover infinite information in the form of three functions. To ensure a more physical profile, "regularization" or "smoothing" techniques are applied (Gough, 1984; Craig and Brown, 1986). This is done by demanding that while trying to find the least squares solution, we also try to minimize the second derivative of the unknown functions. Thus, instead of minimizing Equation 4.49, we minimize

$$\begin{aligned}\chi^2_{\rm reg} &= \chi^2 + ||L||^2 \\ &= \sum_{i=1}^{N} \left(\frac{\frac{\delta\omega_i}{\omega_i} - \Delta\omega_i}{\sigma_i}\right)^2 + \alpha^2 \int_0^R \left[\left(\frac{d^2}{dr^2}\frac{\delta\rho}{\rho}\right)^2 + \left(\frac{d^2}{dr^2}\frac{\delta c^2}{c^2}\right)^2\right]\end{aligned} \tag{4.50}$$

where α is the regularization or smoothing parameter. We could, if we wanted to, have different smoothing parameters for the the two functions. The influence of the regularization parameter on the solution can be seen in Fig. 4.4. Smoothing is not applied to the surface term $F(\omega)$; instead, the number and type of basis functions selected are such that it is a slowly varying function of frequency. Often, it is desirable to smooth $\delta c^2/c^2$ and $\delta\rho/\rho$ preferentially in one part of the star. To do this, a function of r is included in the smoothing term so that smoothing is higher wherever the function is larger. Thus:

$$\chi^2_{\rm reg} = \sum_{i=1}^{N} \left(\frac{\frac{\delta\omega_i}{\omega_i} - \Delta\omega_i}{\sigma_i}\right)^2 + \alpha^2 \int_0^R q(r)^2 \left[\left(\frac{d^2}{dr^2}\frac{\delta\rho}{\rho}\right)^2 + \left(\frac{d^2}{dr^2}\frac{\delta c^2}{c^2}\right)^2\right] \tag{4.51}$$

where $q(r)$ is the smoothing profile.

4.4.1.1 Implementing RLS

While many different methods can be used to determine the coefficients a_i, b_i, and c_i, we shall concentrate here on the implementation used by Antia and Basu (1994) and Basu and Thompson (1996).

Equation 4.48 clearly gives us N equations in $k = 2n + m$ unknowns. These equations can this be represented as:

$$A\mathbf{x} = \mathbf{d} \tag{4.52}$$

where A is a $N \times (2n + m)$ matrix; $\mathbf{x}$ is a vector of length $(2n + m)$ consisting of the unknown coefficients a_i, b_i, and c_i; and $\mathbf{d}$ is the vector with the data, that is, $\delta\omega/\omega$. To take data errors into account, each row i of each matrix is divided by σ_i, where σ_i is the error for the ith data point. We have to determine the elements of vector $\mathbf{x}$ and that can be done with the help of a matrix operation called singular value decomposition (SVD).

An SVD of matrix A results in the decomposition of A into three matrices, $A = U\Sigma V^T$. U and V have the property that $U^TU = I$, $V^TV = VV^T = I$, and $\Sigma = \mathrm{diag}(s_1, s_2, \ldots, s_{2n+m})$, $s_1 \geq s_2 \geq \cdots s_{2n+m}$ being the singular values of A. Thus, the given set of equation can be reduced to

$$U\Sigma V^T\mathbf{x} = \mathbf{d}\,, \tag{4.53}$$

and hence,

$$\mathbf{x} = V\Sigma^{-1}U^T\mathbf{d}\,. \tag{4.54}$$

Since the equations have already been normalized by the errors, the standard error on any component x_k of vector $\mathbf{x}$ is given by

$$e^2 = \sum_{i=1}^{2n+m} \frac{v_{ki}^2}{s_i^2}\,. \tag{4.55}$$

There is still the issue of implementing the smoothing condition. We can do this by replacing the integrals in Equation 4.51 by a sum over M uniformly spaced points and adding the corresponding equations to the system of equations to be solved. Thus, we add the following terms:

$$\frac{\alpha}{\sqrt{M}} q(r_j) \left(\frac{d^2}{dr^2} \frac{\delta c^2}{c^2} \right)_{r=r_j} = \frac{\alpha}{\sqrt{M}} q(r_j) \sum_i^n b_i \frac{d^2}{dr^2} \phi_i(r_j) = 0 \tag{4.56}$$

for $i = 1 \ldots n$ and $j = 1 \ldots M$; and

$$\frac{\alpha}{\sqrt{M}} q(r_j) \left(\frac{d^2}{dr^2} \frac{\delta \rho}{\rho} \right)_{r=r_j} = \frac{\alpha}{\sqrt{M}} q(r_j) \sum_i^n c_i \frac{d^2}{dr^2} \phi_i(r_j) = 0 \tag{4.57}$$

for $i = n + 1 \ldots 2n$ and $j = M + 1 \ldots 2M$. We now have $N' = N + 2M$ equations in $k = 2n + m$ unknowns and we can again use the SVD decomposition technique to solve the equations.

4.4.1.2 Choosing inversion parameters

The discussion to follow assumes that we are doing a sound-speed inversion with density as the second parameter. It easily generalizes to all other combinations of functions. There are at least three parameters that need to be specified to do an RLS inversion: (a) the number of basis functions in frequency used to define the surface term, (b) the number of basis functions in radius to describe the two functions, and (c) the regularization or smoothing parameter α. We have the option of increasing the number of parameters by having different numbers of basis functions for c^2 and ρ and/or different values of α for the two functions. We do not discuss those cases here. Basu and Thompson (1996) have

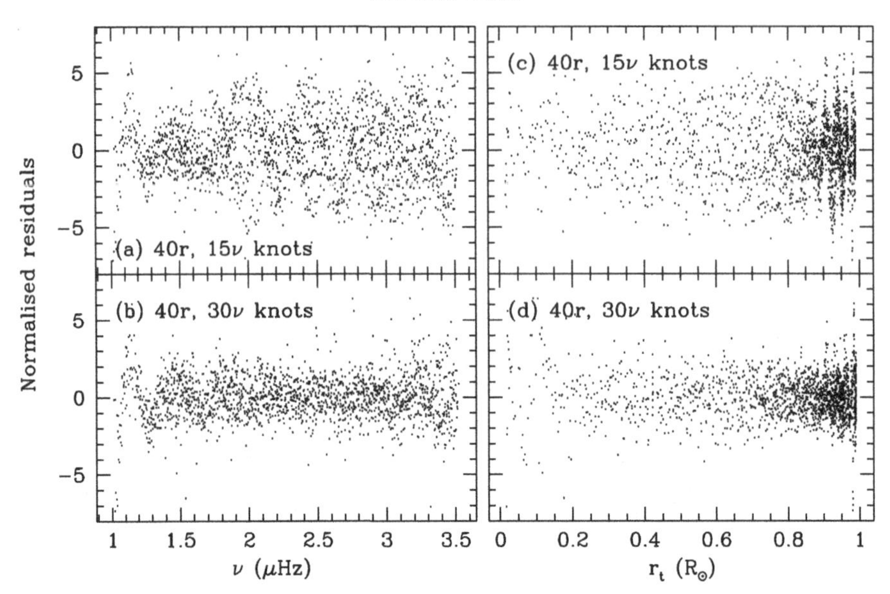

FIG. 4.5. *(a), (b)* Residuals plotted as a function of mode frequency and *(c), (d)* the lower turning point of the modes for RLS inversions with different numbers of ν knots. Note that there is considerable structure in the residuals for the case with 15 knots in frequency. The structure is reduced when 30 frequency knots are used. The remaining structure can be eliminated by increasing the number of r knots. The models and mode set used are the same as those in Fig. 4.4.

described the influence of the parameters on the inversion results and how those can be determined. We take the same approach here, although the results shown were obtained specifically for this chapter.

The number of basis functions depends on the type of basis functions used. It is usually advantageous to have well-localized basis functions so that the solution at one radius is not correlated much with the solution at other radii. The basis functions that I have used extensively are the so-called cubic B-spline (see Antia and Basu, 1994; Basu and Thompson, 1996). These functions have the advantage that they are localized (unlike normal cubic splines). The other advantage is that it is easy to apply smoothing to them because they have continuous first and second derivatives. The positions where B-splines are defined are known as knots and each spline is defined over five knots. For m basis functions, $m + 2$ knots need to be defined. A description of these splines can be found in de Boor (2001). For inversions of solar frequencies, knots in frequency are usually kept equidistant in frequency. Knots in r are defined so that they are equidistant in the acoustic depth τ. This takes into account the fact that modes spend more time in regions of lower sound speed, making those regions easier to resolve.

It is usual to determine the number of frequency knots first. This is done by fixing the number of r knots to a reasonable value and then examining the residuals left as the number of frequency knots is changed. The aim is to eliminate structure in the residuals when they are plotted against mode frequency. Figure 4.5 demonstrates this. Tests such as the "run test" (Miller and Freund, 1965) can be used to test for randomness.

In order to determine the number of r knots, one uses a combination of the reduced χ^2 obtained as a function of the number of r knots for reasonable smoothing and of the "condition" number of the matrix A (Equation 4.52) as a function of the number of r knots. The condition number is the ratio of the largest singular value to the smallest singular value. An example of this analysis is shown in Fig. 4.6. Note that there is a sharp decrease in χ^2_ν as the number of r knots is increased. The number of knots selected should lie after the decrease, on the flatter part of the curve. The behavior of

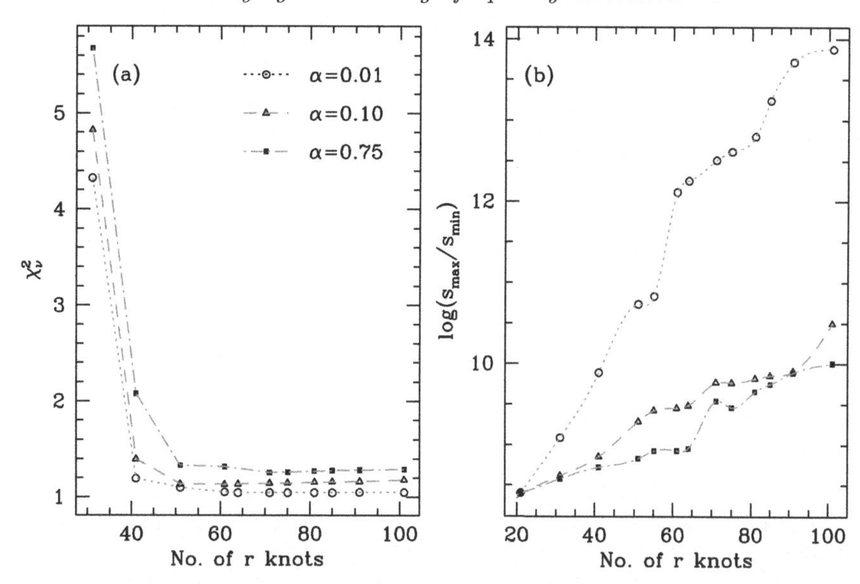

FIG. 4.6. *(a)* The χ^2 per degree of freedom and *(b)* the condition number of matrix A plotted as a function of r knots for three values of smoothing. The number of frequency knots was held fixed at 30. The models and mode set used are the same as those shown in Fig. 4.4.

the condition number is the opposite, and there is a steep rise beyond a certain number of knots. Experience shows that it is best to select the number of knots from the part of the curve just after the steep jump in the condition number. The figure shows that the position of the jump depends on the smoothing. Thus, the selection of knots and smoothing have to be done in a somewhat iterative manner.

The smoothing parameter α is usually determined by inspecting the so-called L-curve (Hansen, 1992) that shows the balance between how smooth the solution is and how large the residuals are by plotting the smoothing constraint $||L||^2$ (Equation 4.50) as a function of the χ^2 per degree of freedom. Since the smoothing parameter increases the mismatch between the data and the solution, one immediately infers that the lowest possible χ^2_ν will be obtained for $\alpha = 0$. However, in this case we will also get the highest possible $||L||^2$ because the oscillatory nature of the solution is not constrained at all. A large value of α, on the other hand, would imply a linear solution (as we are minimizing the second derivative) and hence low $||L||^2$, but the fit to the data would be bad and therefore the χ^2_ν obtained would be large. The optimum value of smoothing should lie somewhere in between. An example of an L-curve is shown in Fig. 4.7. Note that there is a definite elbow in this curve. The value of smoothing is usually chosen to be somewhere in the elbow.

The smoothing parameter not only influences how smooth the solution is, but also determines the extent of uncertainties in the solution caused by data errors. The higher the smoothing, the lower the uncertainties and *vice versa.* A plot of the uncertainties in sound-speed inversion results as a function of radius for a few values of the smoothing parameter is shown in Fig. 4.8.

It should be noted that the number of knots and the smoothing parameter selected depend on the type of basis functions used. They also depend on the number of modes available for inversion and the errors in the data. Hence, the parameters obtained for one set of modes cannot automatically be used for another set. It should also be noted that the selection of parameters is not as well defined as the description may lead one to believe. Determining the smoothing is the most tricky – very often the L-curve does not have a well defined elbow. Determining the number of knots is usually easier, however, there can be cases where the fall in χ^2_ν is gentle, as is the rise in the condition number,

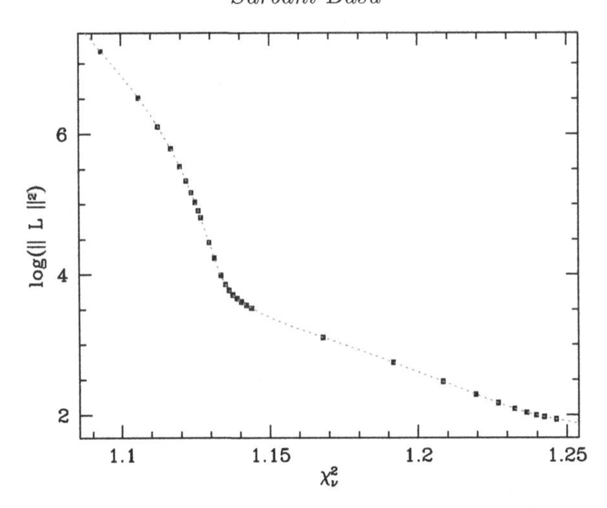

FIG. 4.7. Plot of the smoothing constraint $||L||^2$ against χ^2_ν when doing a sound-speed inversion with 70 r knots and 30 frequency knots. The models and mode set used are the same as those shown in Fig. 4.4.

making the selection difficult. In all cases, it is good practice to try and determine the parameters for a few pairs of models using only the observed modes and errors before inverting observations.

4.4.2 *Optimally localized averages*

The method of optimally localized averages (OLA) was developed by Backus and Gilbert (1968, 1970). Gough (1985) described its use for solar data. Our discussion of OLA inversions of Equation 4.44 assumes that we want to invert for sound-speed differences. For an OLA inversion, we need to determine coefficients c_i such that the sum

$$\mathcal{K}(r_0, r) = \sum_i c_i(r_0) K^i_{c^2,\rho}(r) \tag{4.58}$$

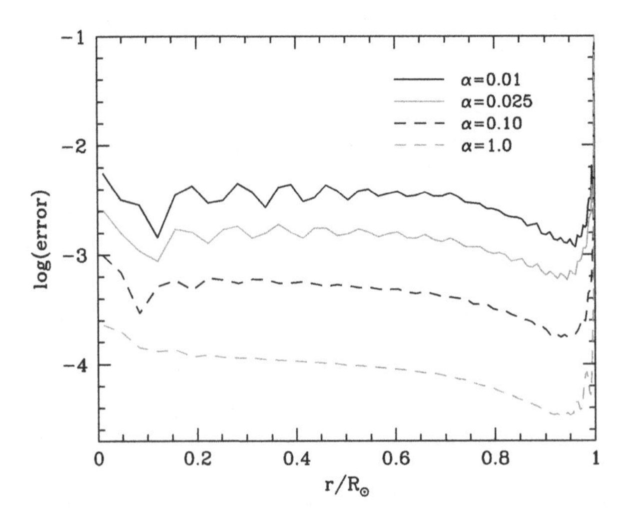

FIG. 4.8. The propagated errors in sound-speed inversion results plotted as a function of radius for a few values of the smoothing parameter α. The models and mode set used are the same as those shown in Fig. 4.4.

is well localized around $r = r_0$. If $\int \mathcal{K}(r_0, r)dr = 1$, then

$$\left\langle \frac{\delta c^2}{c^2} \right\rangle (r_0) = \sum_i c_i(r_0) \frac{\delta \omega_i}{\omega_i} \tag{4.59}$$

is the inversion result at radius r_0 as long as

$$\mathcal{C}(r_0, r) = \sum_i c_i(r_0) K^i_{\rho,c^2} \tag{4.60}$$

is small, and the surface term contribution

$$\mathcal{F} = \sum_i c_i(r_0) \frac{F_{\rm surf}(\omega_i)}{E_i} \tag{4.61}$$

is small too. As is clear, the error in the solution will be $e^2(r_0) = \sum_i c_i^2(r_0)\sigma_i^2$, where σ_i is the error associated with the frequency of mode i.

The function $\mathcal{K}(r_0, r)$ is usually referred to as the *averaging kernel* or the *resolution kernel* at $r = r_0$. The term $\mathcal{C}(r_0, r)$ is the *cross-term kernel* and measures the contribution of the second variable (in this case density) on the inversion results. The coefficients c_i are known as the *inversion coefficients.* The resolution kernel derives its name from the fact that the width of the resolution kernel is a measure of the resolution of the inversions because the kernels represent the region over which the underlying function is averaged. Thus the aim of OLA is to obtain the narrowest possible resolution kernels that the data allow, without increasing the errors in the solution to an unacceptable level.

There are two variants of OLA that are widely used these days. The first is the original one, which, for reasons that will be clear later, is known as *multiplicative* optimally localized averages, or MOLA. This implementation is fairly computation-intensive because it requires a large matrix to be set up and inverted at every radius at which we need a solution (i.e., every r_0). Pijpers and Thompson (1992) developed an alternative implementation, commonly referred to as *subtractive* optimally localized averages, or SOLA. SOLA requires just one inversion. The idea behind this innovation is that one can decide what the averaging kernels should look like (the *target* kernel) and then determine the inversion coefficients such that the sum in Equation 4.58 results in averaging kernels that are similar to the target.

4.4.2.1 Implementing OLA

The inversion coefficients for MOLA are determined by minimizing

$$\int \left(\sum_i c_i K^i_{c^2,\rho} \right)^2 J(r_0, r)dx + \beta \int \left(\sum_i K^i_{\rho,c^2} \right)^2 dr + \mu \sum_{i,j} c_i c_j E_{ij} \tag{4.62}$$

where $J = (r - r_0)^2$ (often, J is defined as $12(r - r_0)^2$) and E_{ij} are elements of the error-covariance matrix. Usually errors in mode frequencies are assumed to be uncorrelated, and hence $E_{ij} = \sigma_i \delta_{ij}$. This implementation is called the multiplicative OLA because the function J multiplies the averaging kernel in the first term of Equation 4.62. The parameter β ensures that the cross-term kernel is small, and μ ensures that the error in the solution is small. The constraint of unimodularity of the averaging kernel $\mathcal{K}$ is applied through a Lagrange multiplier. We also need to apply the constraint that the surface term be small. If the surface term is expressed as $F_{\rm surf}(\omega) = \sum_{j=1}^{\Lambda} a_j \Psi_j(\omega)$, $\Psi_j(\omega)$ being suitable basis functions, we get surface-term constraints of the form

$$\sum_i c_i \frac{\Psi_j(\omega_i)}{E_i} = 0, \; j = 1, \ldots, \Lambda \tag{4.63}$$

SOLA proceeds through minimizing

$$\int \left(\sum_i c_i K^i_{c^2,\rho} - \mathcal{T} \right)^2 dr + \beta \int \left(\sum_i K^i_{\rho,c^2} \right)^2 dr + \mu \sum_{i,j} c_i c_j E_{ij} \tag{4.64}$$

where $\mathcal{T}$ is the target averaging kernel. The difference between the averaging and target kernels in the first term of Equation 4.64 is the reason why this implementation is called subtractive OLA. Again, we have to apply the condition of unimodularity for the averaging kernels and the surface constraints. The choice of the target kernel is often dictated by what the purpose of the inversion is. It is usual to use a Gaussian, as was done by Pijpers and Thompson (1992) in their original paper. They used

$$\mathcal{T}(r_0, r) = \frac{1}{f\Delta} \exp\left(- \left[\frac{r - r_0}{\Delta(r_0)} \right]^2 \right) \tag{4.65}$$

where f is a normalization factor that ensures that $\int \mathcal{T} dr = 1$. However, this form has the disadvantage that for small r_0, $\mathcal{T}$ may not be equal to 0 at $r = 0$ where the kernels $K_{c^2,\rho}$ and K_{ρ,c^2} are zero. This leads to a forced mismatch between the target and the resultant averaging kernel. One way out is to force the target kernels to go to zero at $r = 0$ using a modified Gaussian such as

$$\mathcal{T}(r_0, r) = Ar \exp\left(- \left[\frac{r - r_0}{\Delta(r_0)} + \frac{\Delta(r_0)}{2r_0} \right]^2 \right) \tag{4.66}$$

where A is the normalization factor that ensures that the target is unimodular.

The width of the averaging kernel at a given target radius r_0 depends on the amount of information present for that radius; one usually finds that the widths are smaller toward the surface than toward the core. This is a reflection of the fact that (for p modes at least) the amplitudes of the kernels are larger toward the surface. Thus, the target kernels need to be defined such that they have a variable width. The usual practice is to define the width Δ_f at a fiducial target radius $r_0 = r_f$ and specify the widths at other locations as $\Delta(r_0) = \Delta_f c(r_0)/c(r_f)$, where c is the speed of sound. This inverse variation of the width with sound speed reflects the ability of models to resolve stellar structure (see Thompson, 1993).

The equations that result from the minimization of Equations 4.62 and 4.64 and applying the constraints can be written as a set of linear equations of the form

$$A\mathbf{c} = \mathbf{v}\,. \tag{4.67}$$

For M observed modes the matrix elements A_{ij} for SOLA inversions are given by

$$A_{ij} = \begin{cases} \int K^i_{c^2} K^j_{c^2} dx + \beta \int K^i_\rho K^j_\rho dx + \mu E_{ij} & (i, j \le M) \\ \int K^i_{c^2} dx & (i \le M, j = M+1) \\ \int K^j_{c^2} dx & (j \le M, i = M+1) \\ 0 & (i = j = M+1) \\ \psi_j(\omega_i)/E_i & (i \le M, M+1 < j \le M+1+\Lambda) \\ \psi_i(\omega_j)/E_j & (M+1 < i \le M+1+\Lambda, j \le M) \\ 0 & (\text{otherwise}) \end{cases} \tag{4.68}$$

The vectors $\mathbf{c}$ and $\mathbf{v}$ have the form

$$\mathbf{c} = \begin{pmatrix} c_1 \\ c_2 \\ \cdot \\ \cdot \\ \cdot \\ c_M \\ \lambda \\ 0 \\ \cdot \\ \cdot \\ 0 \end{pmatrix}, \quad \text{and} \quad \mathbf{v} = \begin{pmatrix} \int K_{c^2}^1 \mathcal{T} dx \\ \int K_{c^2}^2 \mathcal{T} dx \\ \cdot \\ \cdot \\ \cdot \\ \int K_{c^2}^M \mathcal{T} dx \\ 1 \\ 0 \\ \cdot \\ \cdot \\ 0 \end{pmatrix}. \tag{4.69}$$

For MOLA, the elements of A for $i, j \leq M$ change, and are given by

$$A_{ij} = (1 - x_0)^2 \int K_{c^2}^i K_{c^2}^j dx + \beta \int K_\rho^i K_\rho^j dx + \mu E_{ij}. \tag{4.70}$$

And as far as the vector $\mathbf{v}$ is concerned, the $(M+1)$th element for MOLA is the same as that for SOLA, but all other elements are equal to 0. The column vector $\mathbf{c}$ is identical in both cases.

The main difference between MOLA and SOLA is that in MOLA the matrix A depends on the target radius r_0 and hence it needs to be set up and inverted at each r_0. Matrix A for SOLA, though, is independent of r_0 and hence has to be set up and inverted just once. For SOLA, it is the vector $\mathbf{v}$ that contains the r_0 dependence. This makes SOLA less intensive computationally. The matrix A, for both MOLA and SOLA, is a symmetric matrix and standard specialized routines can be used to invert it.

4.4.2.2 Choosing inversion parameters

Choosing inversion parameters for OLA is a bit more difficult than that for RLS. For MOLA inversions, we need to determine at least three parameters – the number of surface terms Λ, the error suppression parameter μ, and the cross-term suppression parameter β. A fourth parameter, the width of the target averaging kernels $\Delta(r_f)$ at a fiducial radius of r_f, is needed for SOLA. There is an extensive discussion of SOLA and MOLA parameter selection in Rabello-Soares *et al.* (1999) and to some extent we follow that approach here, although the results shown were obtained specifically for this article. Pijpers and Thompson (1994) have a discussion of parameter selection for rotation inversions using SOLA. However, rotation inversions do not need a surface term or a cross-term suppression parameter and, hence, the problem is easier.

As in the case of RLS inversions, the surface term is the first parameter to be determined. This is done by fitting different numbers of basis functions Ψ to the frequency differences that are to be inverted, but scaled with Q_{nl}, and increasing the number of terms till there is no large-scale structure left in the residuals when plotted as a function of frequency. An example of this is shown in Fig. 4.9.

In the case of SOLA, the width of the target averaging kernel is such that the mismatch between the obtained and target averaging kernel is minimized. The mismatch can be defined as:

$$\chi(r_0) = \int \left[\mathcal{K}(r_0, r) - \mathcal{T}(r_0, r)\right]^2 dr. \tag{4.71}$$

The target kernels as defined by either Equation 4.65 or Equation 4.66 are all positive. However, if the selected target width is too small, the resultant averaging kernels have negative side-bands (Gibbs' phenomenon again). Figure 4.10 shows how the obtained averaging kernel can change with the target width. At small Δ_f one can see negative

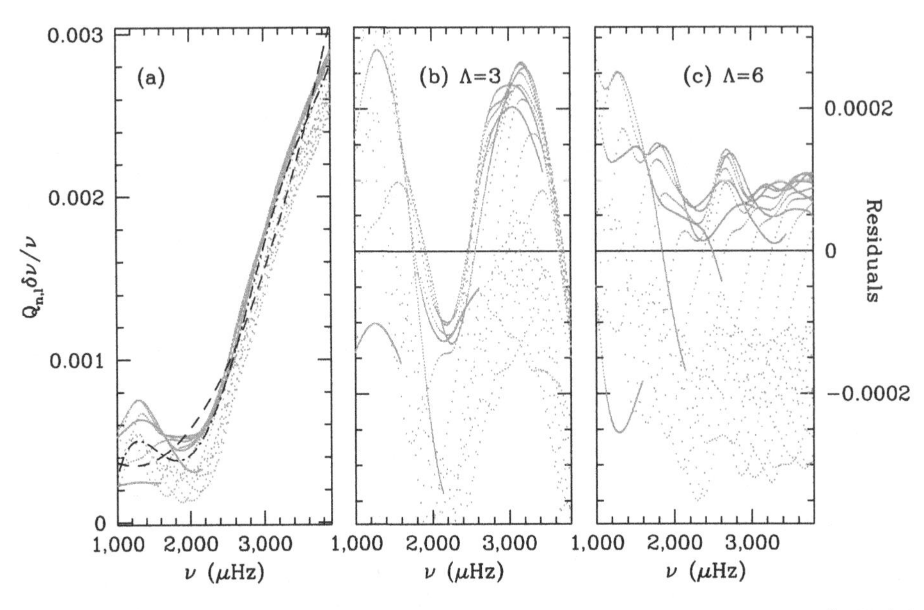

FIG. 4.9. *(a)* The scaled frequency differences between the models BP04 and BSB(GS96) fitted with the first three Legendre polynomials (dashed line) and the first six Legendre polynomials (dot-dashed lines). The residuals of the fits are shown in panels *(b)* and *(c)*. The residuals have not been normalized with the errors. Note the large-scale structure present in the residuals when only three polynomials are used.

side-bands for the low widths, which increase the mismatch $\chi(r_0)$. As will be seen next, narrow averaging kernels also increase the error in the solution for a given error suppression parameter μ. One generally chooses the smallest width that avoids negative side-bands. Choosing a larger width than is necessary would mean an unnecessary degradation of the resolution of the inversions.

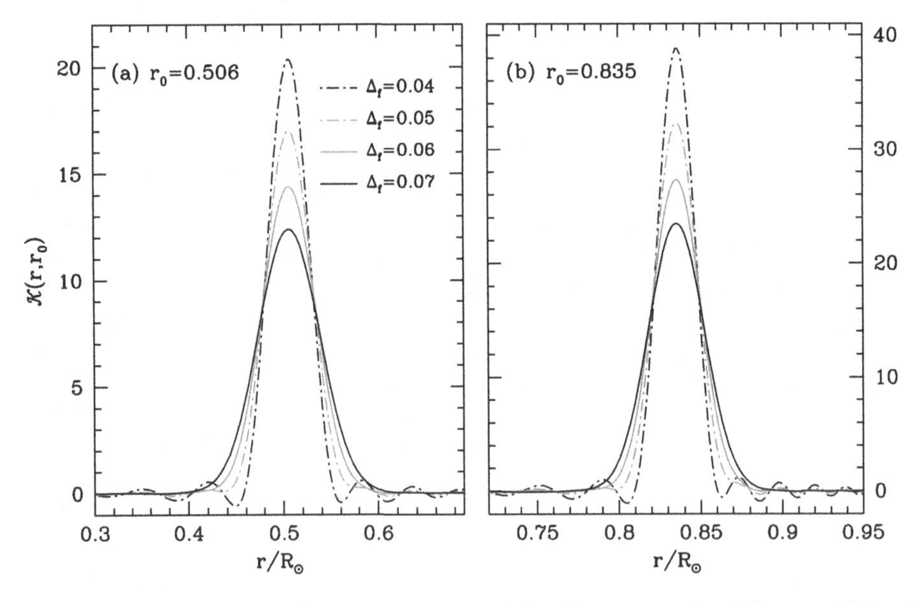

FIG. 4.10. The averaging kernels obtained for sound-speed inversions for different widths of the target averaging kernels at two target radii. The widths are defined at the fiducial radius of 0.2 R$_\odot$. Values of μ and β have been kept fixed. The models and mode-set used are the same as those shown in Fig. 4.4.

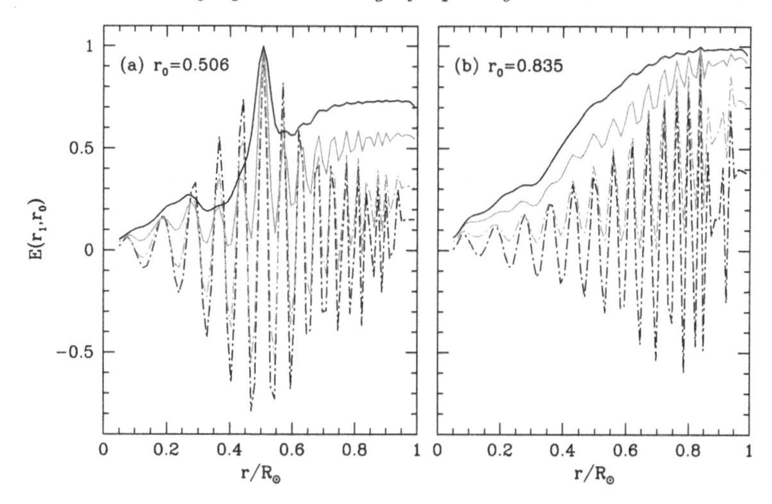

FIG. 4.11. The error correlation between the solution at two target radii and the solutions at other target radii for a few values of the target width. The values of μ and β have been kept fixed. The line types are the same as in Fig. 4.10. The models and mode set used are the same as those shown in Fig. 4.4.

In the case of MOLA, the error parameter μ and the cross-term parameter β determine the averaging kernel. While no mismatch can be defined there, one can try to minimize the negative sidebands by ensuring that the quantity

$$\chi'(r_0) = \int_0^{r_A} \mathcal{K}^2(r_0, r)dr + \int_{r_B}^1 \mathcal{K}^2(r_0, r)dr \tag{4.72}$$

is small. Here, r_A and r_B are defined in such a way that the averaging kernel $\mathcal{K}$ has its maximum at $(r_A + r_B)/2$ and its full width at half maximum is $(r_B - r_A)/2$.

In practice, small negative side lobes can be an advantage. Because the solutions at all radii are obtained from the same set of data, the error in the inversion result at one radius is correlated with that at another. The error correlation between solutions at radii r_1 and r_2 are given by

$$E(r_1, r_2) = \frac{\sum c_i(r_1)c_i(r_2)\sigma_i^2}{[\sum c_i^2(r_1)\sigma_i^2]^{1/2}[\sum c_i^2(r_2)\sigma_i^2]^{1/2}} \tag{4.73}$$

The error correlation has values between ± 1. A value of $+1$ implies complete correlation; a value of -1 implies complete anticorrelation. Correlated errors can introduce features into the solution on the scale of the order of the correlation function width (Howe and Thompson, 1996). Although wide averaging kernels reduce $\chi(r_0)$ and $\chi'(r_0)$ and also reduce uncertainties in the results, they can increase error correlations, as can be seen in Fig. 4.11.

It should be noted that it is almost impossible to reduce error correlations for density inversions. The conservation of mass condition forces the solution in one part of the star to be sensitive to the solution at other parts of the star. There is an anticorrelation of errors between the core (where density is the highest) and the outer layers (where density is the lowest), and the cross-over occurs around the radius at which $r^3\rho$ has the largest value. Figure 4.12 shows the correlation function for density inversions at a few target radii.

For a given error-suppression parameter β, changing the width of the averaging kernel also changes the cross-term kernels, and hence the contribution of the second function on the inversion results of the first function. The contribution of the cross-term kernels

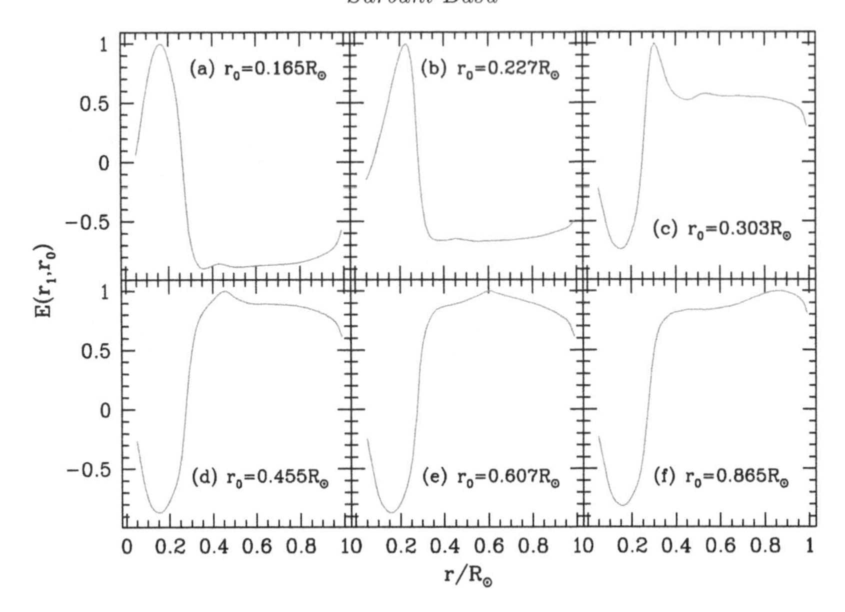

FIG. 4.12. The error correlation function for density inversions at a few target radii. Note the difference in shape of these functions compared with those shown in Fig. 4.11. The models and mode set used are the same as those shown in Fig. 4.4.

can be gauged using the quantity

$$C(r_0) = \sqrt{\int \mathcal{C}^2(r_0, r)dr}. \tag{4.74}$$

We need to aim for small $C(r_0)$ in order to get a good inversion. Figure 4.13 shows how the error in the solution, the mismatch of the averaging kernels, and the measure of the cross term change when the width of the target kernel is changed. As can be seen

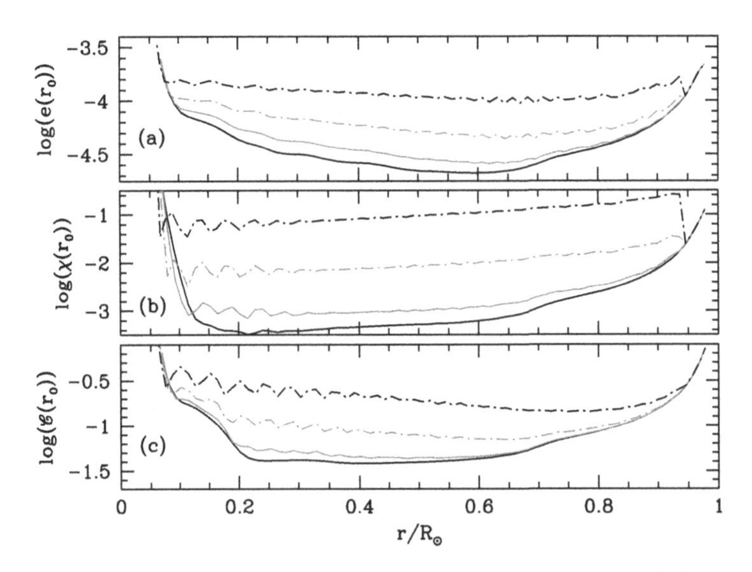

FIG. 4.13. The error in the solution, the mismatch between target and obtained averaging kernels, and the influence of the cross-term kernels plotted as a function of target radius for a few values of the target width. The values of μ and β have been kept fixed. The line types are the same as in Fig. 4.10. The models and mode set used are the same as those shown in Fig. 4.4.

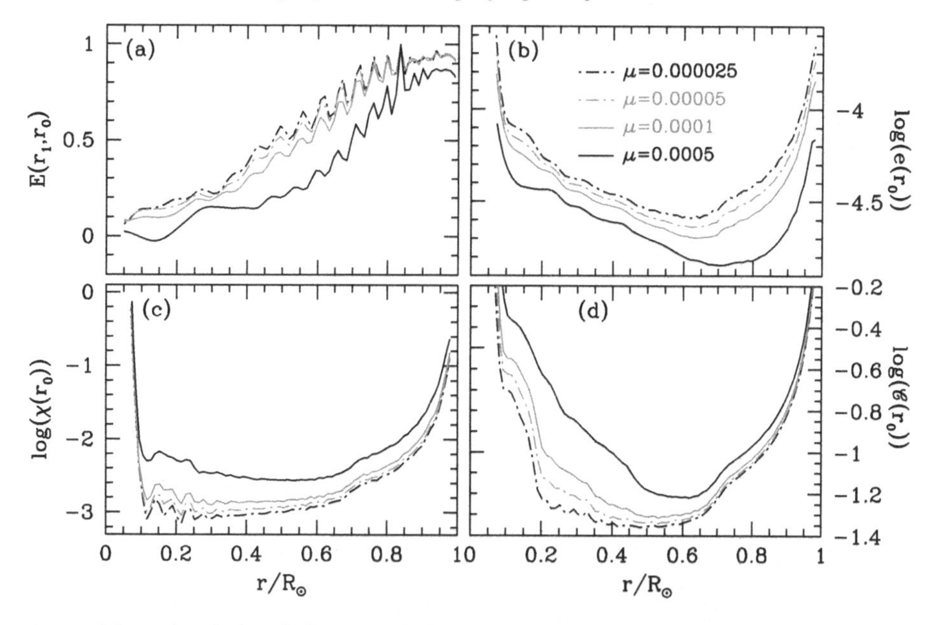

FIG. 4.14. $E(r_1, r_0)$, $e(r_0)$, $\chi(r_0)$, and $C(r_0)$ as a function of radius plotted for a few values of the error-suppression parameter μ. $E(r_1, r_0)$ is only shown for $r_0 = 0.835$. The models and mode set used are the same as those shown in Fig. 4.4.

from this figure and Fig. 4.11, the requirement that the error correlation be small acts against the other requirements. Thus, the choice of Δ needs to be a balance of all these quantities.

There are two other parameters that need to be determined: the error-suppression parameter μ and the cross term suppression parameter β. Increasing μ decreases the error correlation $E(r_1, r_0)$ and the uncertainty in the solution $e(r_0)$, but increases $\chi(r_0)$ the mismatch between the target kernels and the averaging kernel. The influence of the cross-term as measured by $C(r_0)$ also increases. This can be seen in Fig. 4.14. Increasing β increases $E(r_1, r_0)$, $e(r_0)$, and $\chi(r_0)$, but decreases $C(r_0)$ (see Fig. 4.15). MOLA inversions show the same behavior of $E(r_1, r_0)$, $e(r_0)$, and $C(r_0)$ as μ and β are changed. Although $\chi(r_0)$ is undefined for MOLA, we can use $\chi'(r_0)$ instead and this quantity has the same variation with μ and β as $\chi(r_0)$ does for SOLA inversions.

Determining SOLA and MOLA inversion parameters is thus more complicated than determining RLS parameters. Given that the cross-term parameter reduces systematic errors while the error-suppression parameter reduces random errors, it is usual to try and reduce $C(r_0)$ even if that results in a somewhat larger uncertainty in the solution. As with RLS, it is best if one looks at the behaviour of the solution obtained for models first.

4.4.3 *Some miscellaneous points*

It is customary to talk of RLS inversions in terms of residuals and OLA in terms of averaging kernels. However, it is possible to calculate averaging kernels for RLS inversions too. The first step is to determine the inversion coefficients explicitly. Using the SVD decomposition of matrix A of Equation 4.52, the inversion coefficients are defined as elements of the matrix C, where

$$C = V\Sigma^{-1}U^T. \tag{4.75}$$

These coefficients can then be used to determine the averaging kernels. RLS averaging kernels usually have large side-lobes and structure far away from the target radius.

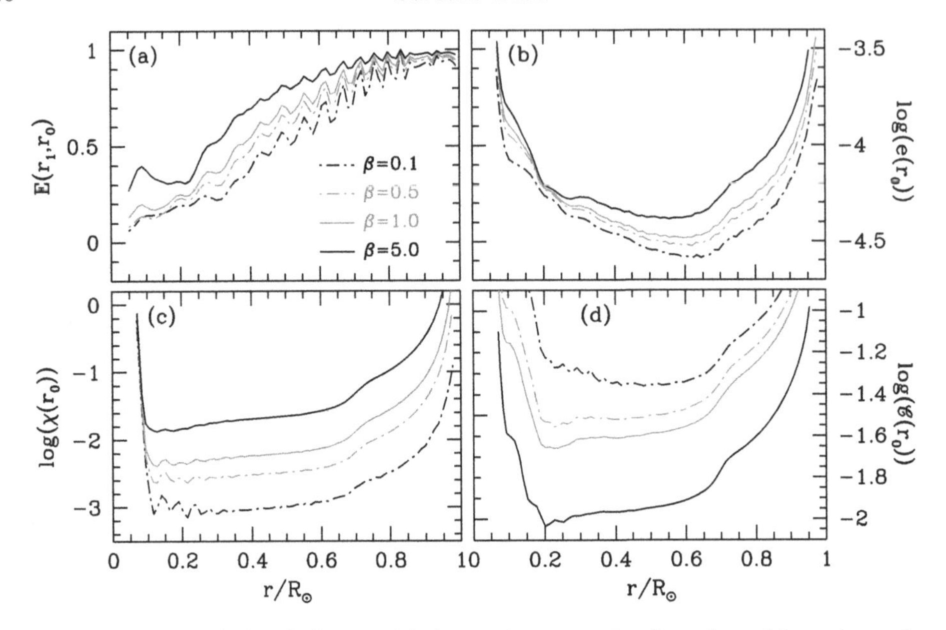

FIG. 4.15. $E(r_1, r_0)$, $e(r_0)$, $\chi(r_0)$, and $C(r_0)$ as a function of radius plotted for a few values of the cross-term suppression parameter β. $E(r_1, r_0)$ is only shown for $r_0 = 0.835$. The models and mode set used are the same as those shown in Fig. 4.4.

Because inversion results do depend on the parameters chosen, the most reliable results are obtained by doing both RLS and OLA inversions and comparing the two results. Agreement between the two results generally implies that the solution is trustworthy, as RLS and OLA are complementary techniques (see, Sekii, 1997 for a discussion). An agreement between the results of the two techniques would be an indication that the inversion result is reasonably correct. OLA inversions are notoriously prone to systematic errors caused by outliers, while RLS results are relatively insensitive. Unless both types of inversions are done, one often cannot detect the error due to bad data points. An example of this is shown in Fig. 4.16. The figure shows the inverted density difference between the Sun and model BP04 of Bahcall *et al.* (2005). The mode set used is a combination of modes of $\ell = 0$, 1, 2, and 3 with frequencies less than 1800 μHz from Bertello *et al.* (2000) obtained by the Global Oscillations at Low Frequencies (GOLF) instrument on board the Solar and Heliospheric Observatory (SOHO) supplemented with modes obtained with Michelson Doppler Imager (MDI) data during the first year of its operation (Schou, 1998). This is the GOLF1low set of Basu *et al.* (2009). RLS inversions show extremely large residuals for the $\ell = 0, n = 3$ and $\ell = 0, n = 5$ modes. As can be seen from the figure, almost identical RLS inversion results are obtained whether or not these two modes are used. SOLA inversion results, on the other hand, change drastically, especially at very low radii. This discrepancy would not have been noticed if both RLS and SOLA inversions had not been done.

As discussed earlier, structure inversions rely on a linearization of the oscillations around a reference model. This often leads to questions about the influence of the reference model on helioseismic inferences. Basu *et al.* (2000) investigated the uncertainties caused by the reference model in determining the sound-speed, density, and Γ_1 profiles of the Sun. They found that the effect is negligible for modern, standard solar models.

4.5 Interpreting inversion results

The first inversions of solar oscillation frequencies were done to determine the structure of the Sun. These days, inversions of solar frequencies are used to study the physics of

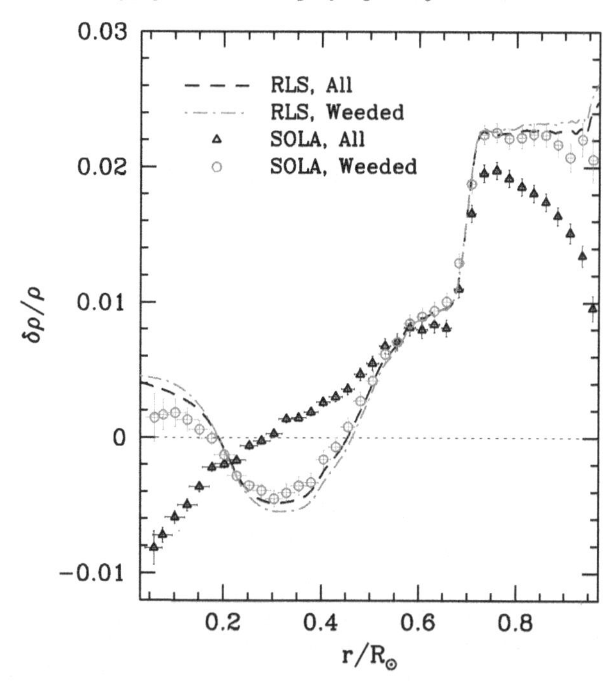

FIG. 4.16. Relative density differences between the Sun and model BP04 obtained using both RLS and SOLA techniques. The results marked "All" were obtained using the mode set GOLF1low while the results marked "Weeded" where obtained when the $\ell = 0, n = 3$, and $\ell = 3, n = 5$ modes were removed from the GOLF1low set. Note that RLS results remain the same, but SOLA results change drastically. (Data for this figure were obtained from Basu *et al.*, 2009.)

the solar interior. By comparing models constructed with different inputs, one can make inferences about the physical processes inside the Sun. A case in point is that of the diffusion and gravitational settling of helium and heavy elements. Before 1993, diffusion and settling were not routinely included in solar and stellar models, as the processes were believed to be too slow to make a difference. However, Christensen-Dalsgaard *et al.* (1993) showed that models with diffusion agree well with the Sun, while models without diffusion do not. This can be seen from Fig. 4.17. The main reason for the disagreement between the no-diffusion model and the Sun is that it has a shallower convection zone than the Sun. Detailed helioseismic studies have shown that the base of the solar convection zone is at 0.713 ± 0.001 $R_\odot$ (Christensen-Dalsgaard *et al.*, 1991; Basu and Antia, 1997). For models without diffusion, the convection zone base is at around 0.726 $R_\odot$. Models without diffusion also have extremely high convection-zone helium abundance. Helioseismic estimates of the solar convection-zone helium abundance are 0.248 ± 0.003 (Basu, 1998); in contrast, the no-diffusion model shown has a convection-zone helium abundance of 0.265 compared with 0.245 for the model with diffusion.

It is clear from Fig. 4.17 that in most of the interior the sound speed of the model with diffusion is within about 0.1% of that in the Sun. The most striking discrepancy is the finger-like structure at the base of the convection zone. The strong peak in the relative sound-speed difference has been identified to be the result of the sharp gradient in the helium and heavy-element abundances just below the base of the convection zone in solar models (Christensen-Dalsgaard *et al.*, 1996; Basu and Antia, 1997). This discrepancy can be alleviated if some mixing is included in the region just below the convection-zone base.

Sound-speed and density differences between the Sun and models have also been used to shed light on the recent controversy about the heavy metal abundance of the Sun. Grevesse and Sauval (1998) had found $Z/X = 0.023$ for the Sun. This value was challenged by Asplund *et al.* (2005), who claim that Z/X for the Sun is low, and fine

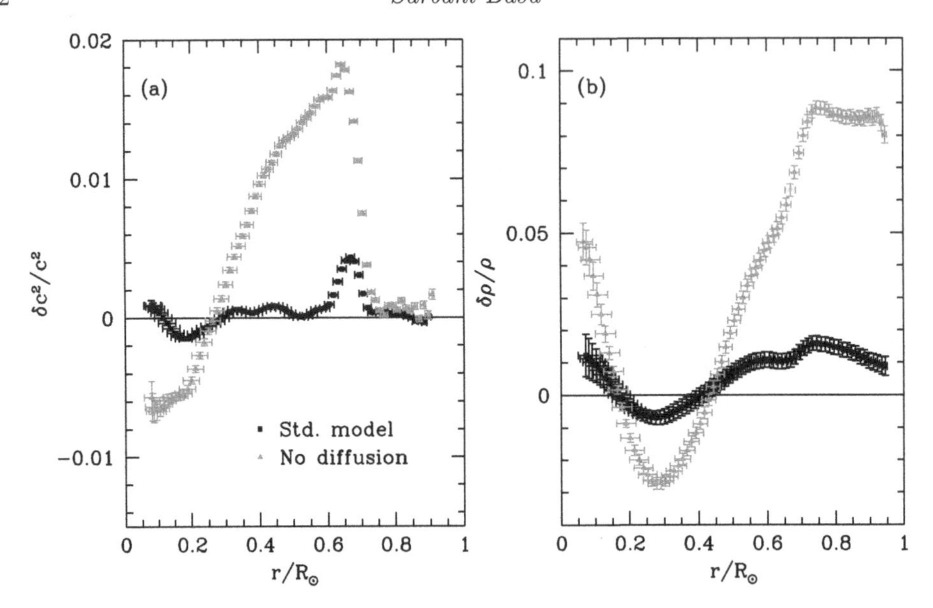

FIG. 4.17. Relative sound speed and density differences between two standard solar models and the Sun. The model marked "Std." is model STD of Basu *et al.* (2000), while the model marked "No diffusion" is model NODIF of the same paper. The former includes diffusion and gravitational settling of helium and heavy elements, the latter does not. The horizontal error bars are a measure of the resolution of the inversion. (Data for this figure were obtained from Basu *et al.*, 2000.)

$Z/X = 0.0165$. However, as can be seen in Fig. 4.18, the model with the Grevesse and Sauval (1998) abundances (marked GS98) agrees better with the Sun than the model with the Asplund *et al.* (2005) abundances (marked AGS05). The large difference between the low-abundance model and the Sun arises because of its shallow convection zone (its convection-zone base is at 0.728 $R_\odot$). There is also a discrepancy in the convection-zone helium abundance, this model, unlike the no-diffusion model in Fig. 4.17, has a lower convection-zone helium abundance than the Sun, with $Y_{CZ} = 0.229$. Thus, sound-speed and density differences are not always enough to determine all the ways in which a model differs from the Sun; we need other information, such as the position of the base of the convection zone and the helium abundance, too. More about the abundance controversy can be read in Basu and Antia (1998) and Antia and Basu (2011).

Inversions to determine Γ_1 differences between the Sun and solar models constructed with different equations of state have been used to test equations of state used in models. The Γ_1 differences between the Sun and models is expected to be zero in the deep interior where conditions are such that gas is completely ionized. Elliott and Kosovichev (1998) determined the Γ_1 difference between the Sun and two models, one constructed with the so-called MHD (Mihalas, Hummer, and Däppen) equation of state (Däppen *et al.*, 1987, 1988; Hummer and Mihalas, 1988; Mihalas *et al.*, 1988) and the other with the OPAL equation of state (Rogers *et al.*, 1996). Much to their surprise, they found significant differences in Γ_1 between the Sun and the two models in the core. The models had identical Γ_1 profiles in the core. This discrepancy was traced back to a deficiency in the two equations of state at high-temperature and high-density regimes. The cause of the discrepancy was identified as the use of the nonrelativistic approximation, instead of the relativistic Fermi-Dirac integrals, to describe partially degenerate electrons. The deficiency has since been corrected in both the OPAL equation of state (Rogers and Nayfonov 2002) and the MHD equation of state (Gong *et al.*, 2001).

The inversion results are, however, only as correct as the kernels used for the inversion. In the early days of helioseismology, it was customary to determine density-difference

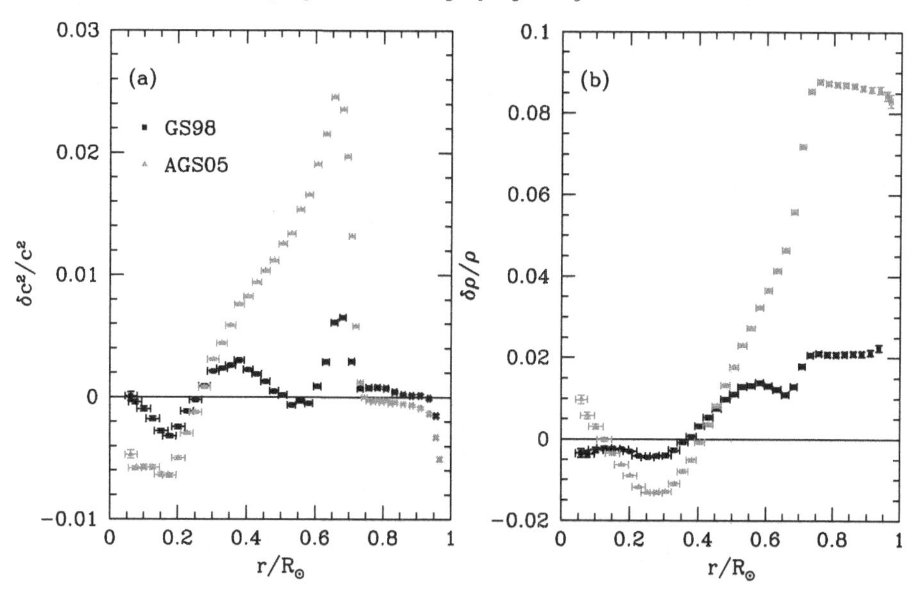

FIG. 4.18. Relative sound speed and density differences between the Sun and two models, one constructed with the Grevesse and Sauval (1998) abundances and the other with the Asplund *et al.* (2005) abundances. The models and data used in this figure are from Serenelli *et al.* (2009).

profiles between the Sun and models using kernels of (ρ, Y) that were derived from kernels of (ρ, Γ_1) using the expansion given in Equation 4.40. This expansion assumes that the equation of state of the Sun is the same as the equation of state of the model. Basu and Christensen-Dalsgaard (1997) realized that this could give rise to systematic errors in the results that are much larger than the uncertainty in the results caused by data errors. The systematic errors become more significant as the uncertainties in the data become smaller. An example is shown in Fig. 4.19, where we show the density differences between the Sun and model S of Christensen-Dalsgaard *et al.* (1996) using both (ρ, Γ_1) and (ρ, Y) kernels. Note that the results are completely different in the core; one result implies that the solar core is denser than that of model S (ρ, Γ_1), the other implies that model S has a denser core than the Sun.

(ρ, Y) and (u, Y) kernels had been used extensively for inversions because Y kernels are nonzero only in the helium ionization zone, making it is very easy to suppress the cross term while keeping the error low. The price paid is the systematic error caused by differences between the equation of state of solar material and that used to construct solar models. Basu and Christensen-Dalsgaard (1997) found that it is possible to use (ρ, Y) and (u, Y) kernels without introducing errors by adding another term in Equation 4.40 that accounts for the intrinsic differences between the equations of state. Thus,

$$\begin{aligned}\frac{\delta\Gamma_1}{\Gamma_1} &= \left(\frac{\delta\Gamma_1}{\Gamma_1}\right)_{\rm int} + \left(\frac{\partial\ln\Gamma_1}{\partial\ln P}\right)_{\rho,Y}\frac{\delta P}{P} + \left(\frac{\partial\ln\Gamma_1}{\partial\ln\rho}\right)_{P,Y}\frac{\delta\rho}{\rho} + \left(\frac{\partial\ln\Gamma_1}{\partial Y}\right)_{\rho,P}\delta Y \\ &= \left(\frac{\delta\Gamma_1}{\Gamma_1}\right)_{\rm int} + \Gamma_{1,P}\frac{\delta P}{P} + \Gamma_{1,\rho}\frac{\delta\rho}{\rho} + \Gamma_{1,Y}\delta Y, \end{aligned} \tag{4.76}$$

where $(\delta\Gamma_1/\Gamma_1)_{\rm int}$ is the intrinsic difference between the equations of state. If the influence of this term is minimized in the inversions, the systematic error due to uncertainties in the equation of state can be made negligibly small. However, the propagated error in the solution becomes as large as the errors obtained when (ρ, Γ_1) or (u, Γ_1) kernels are used, eliminating any advantage that using Y kernels may have given. As we shall see

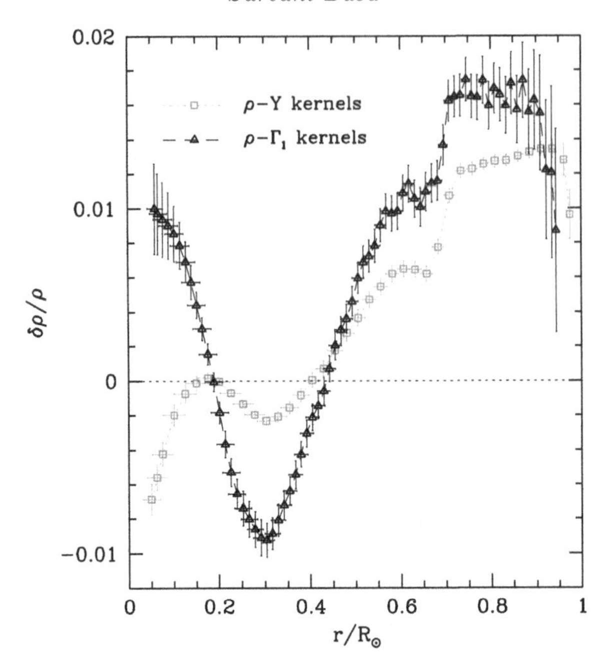

FIG. 4.19. Relative density differences between the Sun and model S obtained with two different sets of kernels. (Data for this figure were obtained from Basu *et al.*, 2009.)

in Section 4.6, using Y as the second variable may still be the best way to do stellar inversions because the data errors involved are large.

The ability to filter out the intrinsic Γ_1 differences gives us the means to examine how deficient the equations of state used to construct solar models really are. To this end, Basu and Christensen-Dalsgaard (1997) determined the intrinsic Γ_1 differences between the Sun and four models – one constructed with the OPAL equation of state, one with the MHD equation of state, one with the simple equation of state proposed by Eggleton *et al.* (1973; henceforth, EFF), and the fourth the EFF equation of state with Coulomb corrections (CEFF; Christensen-Dalsgaard and Däppen, 1992). Updated inversion results with these models are shown in Fig. 4.20. Note that the EFF equation of state is the most deficient, while are others are quite similar. There are, of course, differences in the details.

4.6 Inverting stellar oscillation frequencies

Unlike inversions of solar oscillation frequencies, inversion of stellar frequencies is still in the realm of theoretical studies. Inversions will be difficult since we can only get data for low-degree modes. Other possible problems include the relatively large observational errors, problems with mode identification, and a poor knowledge of the mass and radius of the star. All these factors make stellar inversions difficult. Since mode identification is essential for inversions, in the subsequent discussion we shall assume that the modes can be identified correctly.

There have been several attempts at inverting artificial sets of low-degree modes. Gough and Kosovichev (1993) inverted a set of 64 modes of $\ell = 0$–2. They were attempting to invert for the u profile of a 1.1 $M_\odot$ star with $M/R^3 = 0.468$ with a solar model ($M/R^3 = 1$) as the reference model. They assumed frequency errors of 0.1 μHz. Their results were quite encouraging. Henceforth, we shall refer to this set as the GK set and shall later examine its influence on inversion results. Roxburgh *et al.* (1998) inverted a set of 120 modes of degrees 0–3 (which we shall call the IWR *et al.* set). The error on

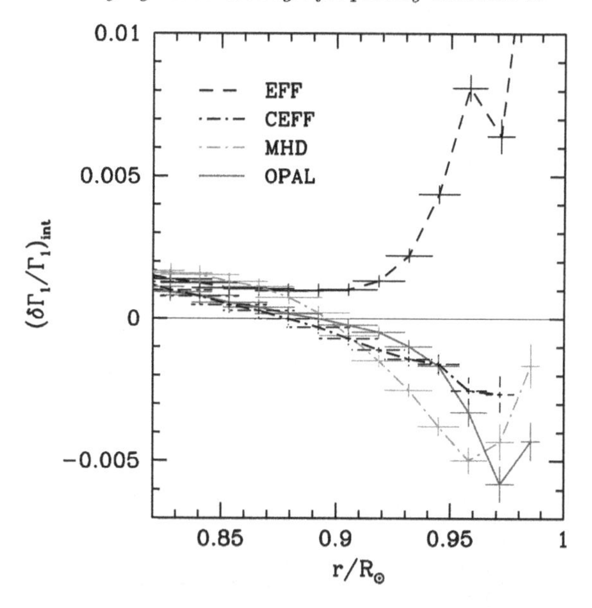

FIG. 4.20. Intrinsic Γ_1 differences between the Sun and solar models constructed with different equations of state. (Data for this figure were obtained from Basu and Antia, 2008.)

each mode frequency was assumed to be 0.3 μ Hz. Their aim was to find the u profile of a 0.8 $M_\odot$ star using a reference model of the same mass but a different age (and hence a different radius and M/R^3). The u profile was then used to determine the density profile of the star. The results were encouraging, but there was some disagreement that can be traced to mismatches in mass and radius, as shall be discussed later. Basu *et al.* (2001) tried to invert for the sound-seed profile of a star using two mode sets, Set 1, which had 47 modes for degrees $\ell = 0$–2, and Set 2, with 65 modes with degrees $\ell = 0$–3. They did not succeed in inverting for c^2 but Basu *et al.* (2002) did succeed in getting a reasonable result for u. The four mode-sets used in the studies are plotted in Fig. 4.21. The IWR *et al.* set is probably too optimistic. For one, current asteroseismology missions (CoRoT and *Kepler*) will not be able to provide $\ell = 3$ frequencies; for another, the n range covered by each degree is too large – we do not get such a coverage even for the Sun! Set 1 is probably a bit too pessimistic. The GK set seems to be similar to what CoRoT and *Kepler*

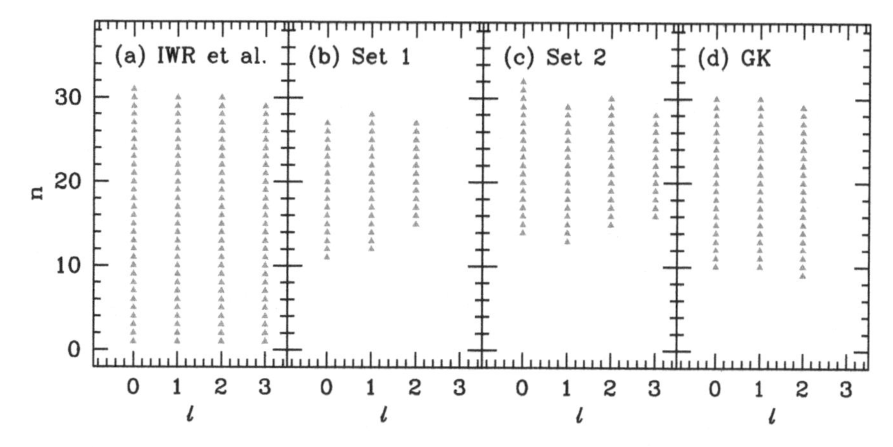

FIG. 4.21. *(a)* The low-degree modes used by Roxburgh *et al.* (1998), *(b) and (c)* Basu *et al.* (2001, 2002) and *(d)* Gough and Kosovichev (1993). We show the position of the modes on the ℓ-n plane.

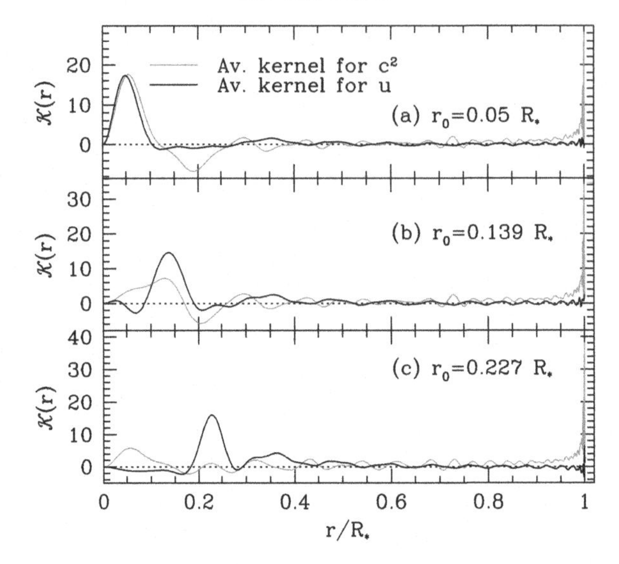

FIG. 4.22. The averaging kernels for c^2 and u inversions obtained by Basu *et al.* (2002) with Set 2. Note that the u averaging kernels are reasonably well localized, but c^2 kernels are not.

are giving. Obtaining the Set 2 mode set will require a Doppler measurement network. Although somewhat unrealistic, these sets allow us to study the effect of the mode set. In particular, the IWR *et al.* set will tell us what is the best that we can do. Basu (2003) reported results from tests of what happens when different mode-sets are used; just we follow the same approach below. It should be noted that the works, mentioned, as well as the results that will be discussed, use the linearized inversion techniques described earlier in this article. There have been proposals for other techniques, and in particular Roxburgh and Vorontsov (2003) have proposed a nonlinear technique for inversions for stellar structure.

All previous researchers have succeeded in inverting for u with Y as the second variable, but not for c^2 with ρ as the second variable. It does not seem possible to localize the (c^2, ρ) averaging kernels, as can be seen in Fig. 4.22. This has to do with the fact that ρ kernels are large, particularly at the surface. In the absence of high-degree modes, it is difficult to suppress that term while simultaneously localizing the averaging kernel. In contrast, Y kernels are nonzero only in the ionization zones and hence it is easy to suppress that term even with a small value of the cross-term suppression parameter β. This, of course, poses a dilemma; the (u, Y) kernel pair is calculated assuming that we know the equation of state (see Section 4.3.1), yet, in Section 4.5 we showed that such an assumption can lead to systematic errors in the result. In the case of stellar inversions we might have to tolerate the uncertainty. And in any case, the random errors will be larger than those in the solar data while the mode set will be much smaller, and hence, the systematic error may be less important.

Figure 4.23 shows the u inversion results between two solar models using the mode-sets shown in Fig. 4.21. The reference model was model S. The test star was model MIX of Basu *et al.* (2000). This is a nonstandard solar model with an artificially mixed core. Mode frequencies were assumed to have an error of 0.3 μHz. The IWR *et al.* set, of course, gives the best results and shows that that if enough modes are available, we can invert the frequency differences successfully. Set 1 gives the worst results. Note that the two mode sets that have $\ell = 3$ modes give much better results than the ones without, particularly at smaller r. This may seem surprising, as $\ell = 3$ modes do not penetrate to those depths. However, the presence of these modes helps in localizing the averaging kernels, as they add information about the outer layers. This is information that needs

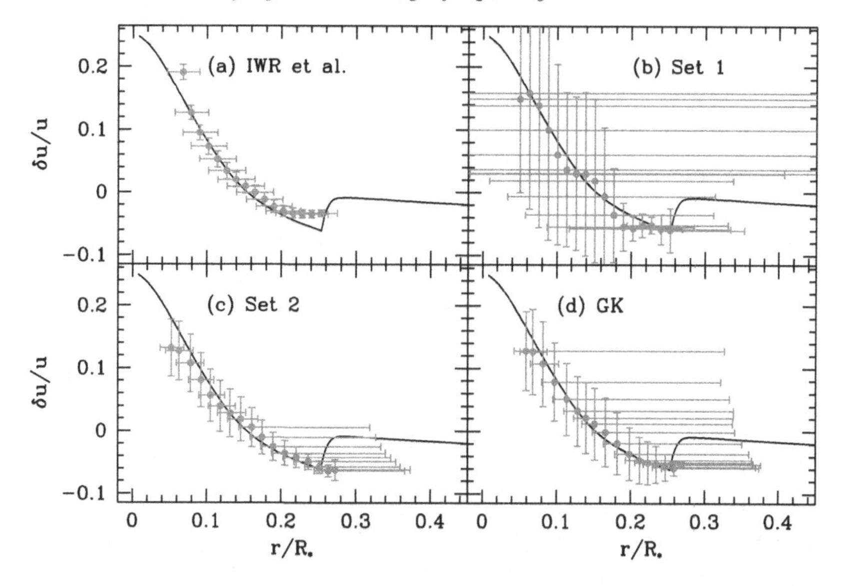

FIG. 4.23. The relative u difference between two solar models obtained by inverting the mode-sets shown in Fig. 4.21. The black line indicates the true difference between the modes, and the points are the inversion results. (Data from Basu, 2003.)

to be removed from the lower-degree kernels in order to filter out information about the core.

We also find that although the GK set gives reasonable results, the resolution of the inversions is poor (as can be seen by the large horizontal error bars). The inversions can be improved considerably if the errors in the frequencies are reduced. This can be seen in Fig. 4.24, where decreasing errors in mode frequency from 0.3 μHz to 0.1 μHz improves the resolution considerably. Because frequency errors in long time-series *Kepler* data are expected to be of this order or even smaller, we have some hopes for doing actual inversions. Note that Set 2 results also improve substantially.

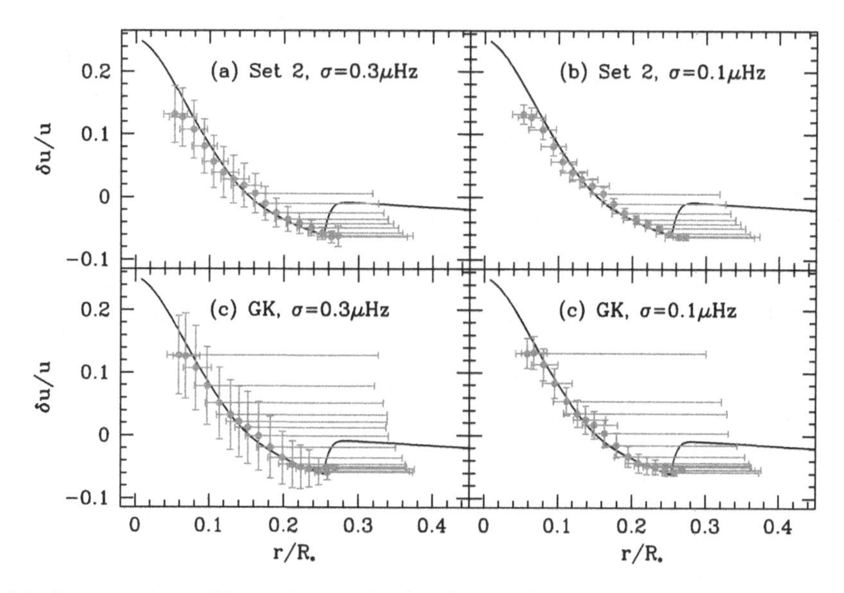

FIG. 4.24. A comparison of inversion results for the same models as in Fig. 4.23 obtained with the Set 2 and GK mode-sets for different frequency errors. (Data from Basu, 2003.)

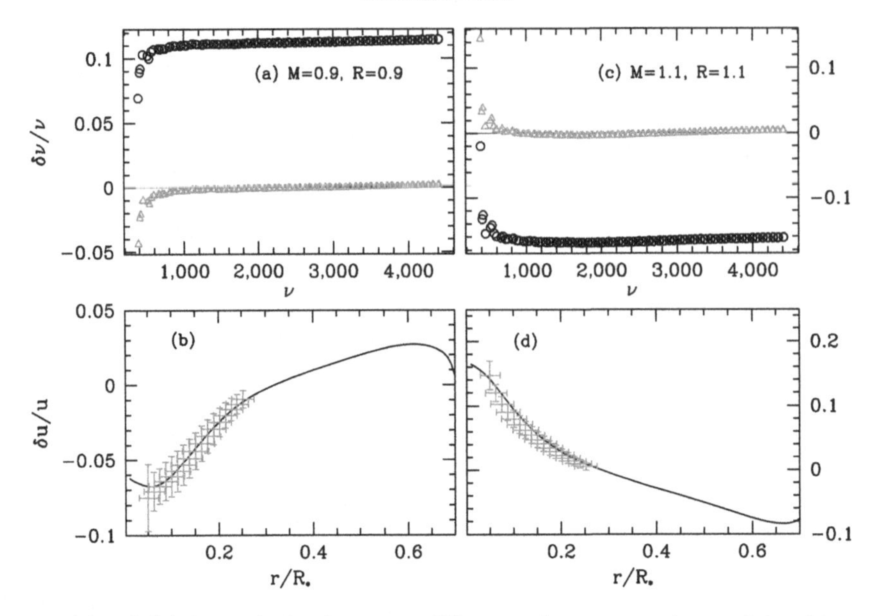

FIG. 4.25. *(a)* and *(c)* show relative frequency differences between a solar model and two models with different mass and radii. The black points denote the actual differences, while the gray points in the panels are the residuals when the offset between the zero difference line and the black points is removed. *(b)* shows the inversion results obtained using the gray points in *(a)*. *(d)* Shows the inversion results for the gray points in *(c)*. The black line denotes the true difference between the models. (Data from Basu, 2003.)

These examples were obtained using solar models and hence the mass and radius of both the reference model and the test model were identical. We cannot expect to know the mass and radius of other stars as well as we know them for the Sun. It is, therefore, quite likely that our reference models will not have the correct mass and radius. Figure 4.25 shows the frequency differences and inversion results between a solar model and two other models, those of a 0.9 $M_\odot$, 0.9 $R_\odot$ star and of a 1.1 $M_\odot$, 1.1 $R_\odot$ star. In both cases, the relative frequency difference shows an almost constant offset. This is merely a manifestation of the fact that the test and reference models have different large frequency separations and that the frequencies scale differently. Recall that large frequency separations scale as $\sqrt{(}M/R^3)$. This constant offset can be fitted and removed, and the residuals then used in the inversion process. The inversion results are surprisingly good. The IWR *et al.* set has been used in these inversions to minimize uncertainties because of other factors.

However, the good results in Fig. 4.25 are an exception rather than the rule. We show in Fig. 4.26 results where the inverted u differences do not agree with the true u differences. There appears to be a constant offset between the actual and inverted differences. The reason for the difference is actually not a profound one: inversion kernels are usually calculated using the dimensionless form of variables, and the inverted result obtained in the process is the relative difference between the dimensionless form of the variables. The squared sound speed c^2 and the squared isothermal sound speed u can be made dimensionless by multiplying the dimensional quantity with R/GM. No offset is seen in the results shown in Fig. 4.25 because all models involved have the same R/M. In contrast, the test model shown in Fig. 4.26 has $R/M = 1.05$, whereas the reference model has $R/M = 1$. If we compare the inversion results with the differences between the dimensionless u (Fig. 4.26c), we see that the results agree. Thus, we can conclude that unless we have independent measures of the mass and radius of a star, inversions will only give us localized information about the dimensionless form of the variables. If the

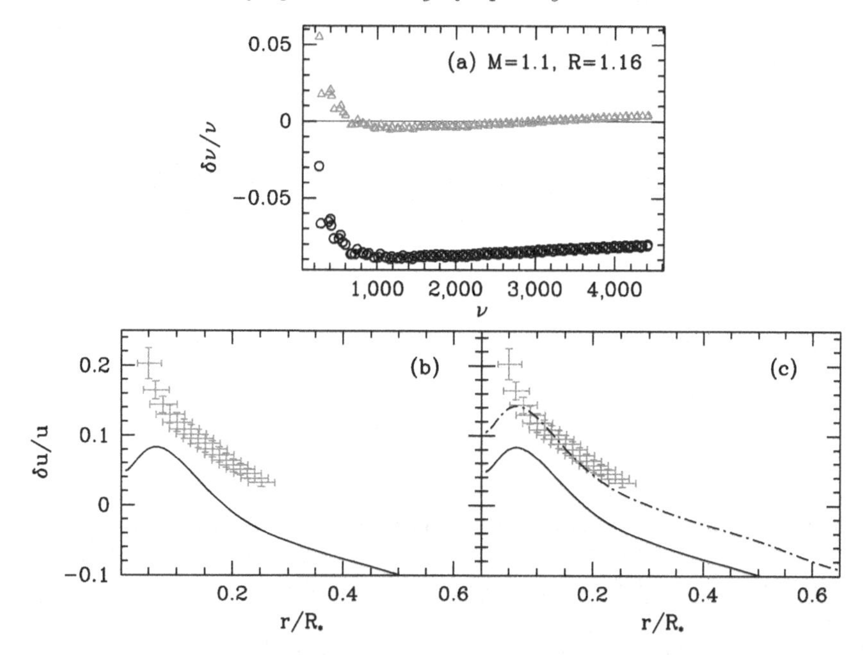

FIG. 4.26. *(a)* The relative frequency differences between a solar model and a 1.1 $M_\odot$, 1.16 $R_\odot$ model. The black points are the actual difference, the gray points are obtained by removing the offset of the black points from the zero difference line. *(b)* The inversion results (gray points) and the true differences (black line). *(c)* The inversion results (gray points) and the true differences (black line) and the differences when the dimensionless u profiles are used to calculate the difference (dot-dashed line). (Data from Basu, 2003.)

inversion results are used to derive other quantities, the differences will propagate. This is most likely to be the cause of the discrepancy between the real and inverted results of Roxburgh *et al.* (1998). Given that we are unlikely to know the exact mass and radius of observed stars, interpreting the results will pose challenges.

In conclusion, inverting stellar frequencies will prove to be much more challenging than inverting solar frequencies. The problems are not merely because of the small mode-set that will be available. The real problem in interpreting the result lies in the uncertainty in the mass and radius of the star. In the solar case, even a difference of 0.03% in the radius of the reference model causes a statistically significant shift in the results (Basu, 1998). Of course, the uncertainties in solar oscillation frequencies are small and the mode sets are large compared with what we can expect from other stars. Nevertheless, these systematic uncertainties need to be kept in mind when inverting stellar data.

Acknowledgments

The author thanks the organizers of the XXII Canary Island Winter School for their hospitality and travel support. Some of the results described in this work was supported by the U.S. National Science Foundation grants ATM-0206130 and ATM-0348837.

REFERENCES

Antia, H. M. and Basu, S. 1994. Nonasymptotic helioseismic inversion for solar structure. *A&AS*, **107**(Nov.), 421–44.

Antia, H. M. and Basu, S. 2011. Are recent solar heavy element abundances consistent with helioseismology? *Journal of Physics Conference Series*, **271**(Jan.), 012034.

Antia, H. M., Basu, S., and Chitre, S. M. 1998. Solar internal rotation rate and the latitudinal variation of the tachocline. *MNRAS*, **298**(Aug.), 543–56.

Asplund, M., Grevesse, N., and Sauval, A. J. 2005. The solar chemical composition. *Cosmic Abundances as Records of Stellar Evolution and Nucleosynthesis*, **336**(Sep.), 25.

Backus, G. E. and Gilbert, J. F. 1968. The resolving power of gross earth data. *Geophysical Journal*, **16**, 169–205.

Backus, G. E. and Gilbert, J. F. 1970. Uniqueness in the inversion of inaccurate gross Earth data. *Royal Society of London Philosophical Transactions Series A*, **266**(Mar.), 123–92.

Baglin, A., Auvergne, M., Barge, P., Deleuil, M., Catala, C., Michel, E., Weiss, W., and the COROT team 2006. Scientific objectives for a Minisat: CoRoT. *ESA Special Publication*, **1306**(Nov.), 33.

Bahcall, J. N., Basu, S., Pinsonneault, M., and Serenelli, A. M. 2005. Helioseismological implications of recent solar abundance determinations. *ApJ*, **618**(Jan.), 1049–56.

Bahcall, J. N., Serenelli, A. M., and Basu, S. 2006. 10,000 standard solar models: a monte carlo simulation. *ApJS*, **165**(Jul.), 400–31.

Basu, S. 1998. Effects of errors in the solar radius on helioseismic inferences. *MNRAS*, **298**(Aug.), 719–28.

Basu, S. 2003. Stellar Inversions. *Ap&SS*, **284**, 153–64.

Basu, S. and Antia, H. M. 1997. Seismic measurement of the depth of the solar convection zone. *MNRAS*, **287**(May.), 189–98.

Basu, S. and Antia, H. M. 2008. Helioseismology and solar abundances. *Phys. Rep.*, **457**(Mar.), 217–83.

Basu, S. and Christensen-Dalsgaard, J. 1997. Equation of state and helioseismic inversions. *A&A*, **322**(Jun.), L5–L8.

Basu, S. and Thompson, M. J. 1996. On constructing seismic models of the Sun. *A&A*, **305**(Jan.), 631.

Basu, S., Däppen, W., and Nayfonov, A. 1999. Helioseismic analysis of the hydrogen partition function in the solar interior. *ApJ*, **518**(Jun.), 985–93.

Basu, S., Pinsonneault, M. H., and Bahcall, J. N. 2000. How much do helioseismological inferences depend on the assumed reference model? *ApJ*, **529**(Feb.), 1084–1100.

Basu, S., Christensen-Dalsgaard, J., Monteiro, M. J. P. F. G., and Thompson, M. J. 2001. Seismology of solar-type stars. *SOHO 10/GONG 2000 Workshop: Helio- and Asteroseismology at the Dawn of the Millennium*, **464**(Jan.), 407–10.

Basu, S., Christensen-Dalsgaard, J., and Thompson, M. J. 2002. SOLA inversions for the core structure of solar-type stars. *Stellar Structure and Habitable Planet Finding*, **485**(Jan.), 249–52.

Basu, S., Chaplin, W. J., Elsworth, Y., New, R., and Serenelli, A. M. 2009. Fresh Insights on the Structure of the Solar Core. *ApJ*, **699**(Jul.), 1403–17.

Bertello, L., Varadi, F., Ulrich, R. K., Henney, C. J., Kosovichev, A. G., García, R. A., and Turck-Chièze, S. 2000. Identification of solar acoustic modes of low angular degree and low radial order. *ApJ*, **537**(Jul.), L143–L146.

Borucki, W. J., and 70 colleagues. 2010. Kepler planet-detection mission: introduction and first results. *Science*, **327**(Feb.), 977.

Canuto, V. M. and Mazzitelli, I. 1991. Stellar turbulent convection: a new model and applications. *ApJ*, **370**(Mar.), 295–311.

Chandrasekhar, S. 1964. A general variational principle governing the radial and the non-radial oscillations of gaseous masses. *ApJ*, **139**(Feb.), 664.

Christensen-Dalsgaard, J. 2002. Helioseismology. *Reviews of Modern Physics*, **74**(Nov.), 1073–1129.

Christensen-Dalsgaard, J. 2003. Lecture notes on stellar oscillations. *http://users-phys.au.dk/jcd/oscilnotes/*.

Christensen-Dalsgaard, J. and Berthomieu, G. 1991. Theory of solar oscillations. *Solar interior and atmosphere (A92-36201 14-92)*. Tucson, AZ, University of Arizona Press. Research supported by SNFO and CNRS. 401–78.

Christensen-Dalsgaard, J. and Daeppen, W. 1992. Solar oscillations and the equation of state. *A&A Rev.*, **4**, 267–361.

Christensen-Dalsgaard, J., Gough, D. O., and Thompson, M. J. 1991. The depth of the solar convection zone. *ApJ*, **378**(Sep.), 413–37.

Christensen-Dalsgaard, J., Proffitt, C. R., and Thompson, M. J. 1993. Effects of diffusion on solar models and their oscillation frequencies. *ApJ*, **403**(Feb.), L75–8.

Christensen-Dalsgaard, J., and 32 colleagues 1996. The current state of solar modeling. *Science*, **272**(May), 1286–92.

Cox, J. P. 1980. Theory of stellar pulsation. *Research supported by the National Science Foundation.* Princeton, NJ. Princeton University Press.

Craig, I. J. D. and Brown, J. C. 1986. Inverse problems in astronomy: A guide to inversion strategies for remotely sensed data. *Research supported by SERC.* Bristol, England and Boston, MA. Adam Hilger, Ltd.

Daeppen, W., Anderson, L., and Mihalas, D. 1987. Statistical mechanics of partially ionized stellar plasma: the Planck-Larkin partition function, polarization shifts, and simulations of optical spectra. *ApJ*, **319**(Aug.), 195–206.

Daeppen, W., Mihalas, D., Hummer, D. G., and Mihalas, B. W. 1988. The equation of state for stellar envelopes. III – Thermodynamic quantities. *ApJ*, **332**(Sep.), 261–70.

de Boor, C. 2001. *A practical guide to Splines.* New York: Springer Verlag.

Demarque, P., Guenther, D. B., Li, L. H., Mazumdar, A., and Straka, C. W. 2008. YREC: the Yale rotating stellar evolution code. Non-rotating version, seismology applications. *Ap&SS*, **316**(Aug.), 31–41.

Dziembowski, W. A., Pamyatnykh, A. A., and Sienkiewicz, R. 1990. Solar model from helioseismology and the neutrino flux problem. *MNRAS*, **244**(Jun.), 542–50.

Eggleton, P. P., Faulkner, J., and Flannery, B. P. 1973. An approximate equation of state for stellar material. *A&A*, **23**(Mar.), 325.

Elliott, J. R. 1996. Equation of state in the solar convection zone and the implications of helioseismology. *MNRAS*, **280**(May), 1244–56.

Elliott, J. R. and Kosovichev, A. G. 1998. The adiabatic exponent in the solar core. *ApJ*, **500**(Jun.), L199.

Gong, Z., Däppen, W., and Zejda, L. 2001. MHD equation of state with relativistic electrons. *ApJ*, **546**(Jan.), 1178–82.

Gough, D. O. 1984. On the rotation of the sun. *Royal Society of London Philosophical Transactions Series A*, **313**(Nov.), 27–38.

Gough, D. O. 1985. Inverting helioseismic data. *Sol. Phys.*, **100**(Oct.), 65–99.

Gough, D. O. 1990. Comments on helioseismic inference. *Progress of Seismology of the Sun and Stars*, **367**, 283.

Gough, D. O. 1993. Linear adiabatic stellar pulsation. *Astrophysical Fluid Dynamics – Les Houches 1987*, 399–560.

Gough, D. O. and Kosovichev, A. G. 1993. Initial asteroseismic inversions. *IAU Colloq. 137: Inside the Stars*, **40**(Jan.), 541.

Gough, D. O. and Thompson, M. J. 1991. The inversion problem. *Solar interior and atmosphere (A92-36201 14-92).* Tucson, AZ. University of Arizona Press, 519–61.

Grevesse, N. and Sauval, A. J. 1998. Standard Solar Composition. *Space Sci. Rev.*, **85**(May), 161–74.

Hansen, P. C. 1992. Numerical tools for analysis and solution of Fredholm integral equations of the first kind. *Inverse Problems*, **8**(Dec.), 849–72.

Howe, R. and Thompson, M. J. 1996. On the use of the error correlation function in helioseismic inversions. *MNRAS*, **281**(Aug.), 1385.

Hummer, D. G. and Mihalas, D. 1988. The equation of state for stellar envelopes. I. An occupation probability formalism for the truncation of internal partition functions. *ApJ*, **331**(Aug.), 794–814.

Lefebvre, S., Kosovichev, A. G., and Rozelot, J. P. 2007. Helioseismic test of nonhomologous solar radius changes with the 11 year activity cycle. *ApJ*, **658**(Apr.), L135–L138.

Leighton, R. B., Noyes, R. W., and Simon, G. W. 1962. Velocity fields in the solar atmosphere. I. Preliminary report. *ApJ*, **135**(Mar.), 474.

Leibacher, J. W. and Stein, R. F. 1971. A new description of the solar five-minute oscillation. *Astrophys. Lett.*, **7**, 191–2.

Mihalas, D., Dappen, W., and Hummer, D. G. 1988. The equation of state for stellar envelopes. II. Algorithm and selected results. *ApJ*, **331**(Aug.), 815–25.

Miller, I., Freund, J. E. 1965. *Probability and statistics for engineers.* Prentice-Hall Mathematics Series, Englewoods Cliffs.

Pijpers, F. P. and Thompson, M. J. 1992. Faster formulations of the optimally localized averages method for helioseismic inversions. *A&A*, **262**(Sep.), L33–L36.

Pijpers, F. P. and Thompson, M. J. 1994. The SOLA method for helioseismic inversion. *A&A*, **281**(Jan.), 231–40.

Rabello-Soares, M. C., Basu, S., and Christensen-Dalsgaard, J. 1999. On the choice of parameters in solar-structure inversion. *MNRAS*, **309**(Oct.), 35–47.

Robinson, F. J., Demarque, P., Li, L. H., Sofia, S., Kim, Y.-C., Chan, K. L., and Guenther, D. B. 2003. Three-dimensional convection simulations of the outer layers of the Sun using realistic physics. *MNRAS*, **340**(Apr.), 923–36.

Rogers, F. J. and Nayfonov, A. 2002. Updated and Expanded OPAL Equation-of-State Tables: Implications for Helioseismology. *ApJ*, **576**(Sep.), 1064–74.

Rogers, F. J., Swenson, F. J., and Iglesias, C. A. 1996. OPAL Equation-of-State Tables for Astrophysical Applications. *ApJ*, **456**(Jan.), 902.

Roxburgh, I. and Vorontsov, S. 2003. Diagnostics of the Internal Structure of Stars using the Differential Response Technique. *Ap&SS*, **284**, 187–91.

Roxburgh I. W., Audard, N., Basu, S., Christensen-Dalsgaard, J., and Vorontsov, S. V. 1998. Proc. IAU Symp. 181: Sounding Solar and Stellar Interiors, (poster vol.). eds. J. Provost, F.-X. Schmider, 245.

Schou, J., Kosovichev, A. G., Goode, P. R., and Dziembowski, W. A. 1997. Determination of the Sun's seismic radius from the SOHO Michelson Doppler Imager. *ApJ*, **489**(Nov.), L197.

Schou, J., Christensen-Dalsgaard, J., Howe, R., Larsen, R. M., Thompson, M. J., and Toomre, J. 1998. Slow poles and shearing flows from heliospheric observations with mdi and gong spanning a year. *Structure and Dynamics of the Interior of the Sun and Sun-like Stars*, **418**, 845.

Sekii, T. 1997. Internal solar rotation. *Proc IAU Symp. 181, Sounding Solar and Stellar Interiors.* Pages 189–202 of: J. Provost and F.-X. Schmider (ed.). Dordrecht, Holland: Kluwer.

Serenelli, A. M., Basu, S., Ferguson, J. W., and Asplund, M. 2009. New solar composition: the problem with solar models revisited. *ApJ*, **705**(Nov.), L123–L127.

Thompson, M. J. 1993. Seismic investigation of the Sun's internal structure and rotation. *GONG 1992. Seismic Investigation of the Sun and Stars. ASPCS*, **42**(Jan.), 141.

Ulrich, R. K. 1970. The five-minute oscillations on the solar surface. *ApJ*, **162**(Dec.), 993.

Unno, W., Osaki, Y., Ando, H., Saio, H., and Shibahashi, H. 1989. *Nonradial oscillations of stars, 2nd ed.* Tokyo, Japan: University of Tokyo Press.

5. A crash course on data analysis in asteroseismology

THIERRY APPOURCHAUX

"Throughout human history, as our species has faced the frightening, terrorizing fact that we do not know who we are or where we are going in this ocean of chaos, it has been the authorities – the political, the religious, the educational authorities – who have attempted to comfort us by giving us order, rules, regulations, informing – *forming* in our minds – their view of reality.

To think for yourself you must question authority and learn how to put yourself in a state of vulnerable open-mindedness, chaotic, confused vulnerability to inform yourself."

Timothy Leary in *Sound Bites from the Counter Culture* (1989)

5.1 Introduction

This chapter attempts to provide a summary of the course I gave during the XXII Canary Island Winter School of Astrophysics. In no way should this chapter be perceived as the final answer to a problem. I hope that this chapter can serve as a basis for students and fellow scientists to go beyond what is written here. As in many approaches that I have pursued, this work is a snapshot of where I am and hopefully a possible starting point from which one can expand to other paths not yet ventured.

This chapter starts with a short historical introduction on signal processing and statistics or how our forefathers started doing data analysis more than 200 years ago. The second part is related to the sampling and acquisition of continuous physical signals for subsequent analysis in a digital world. The third part contains a broad review of statistics from the so-called *frequentist* and *Bayesian* points of view. The last part is related to the applications of the previous concept to data analysis for asteroseismology, which also includes a description of the physics behind that latter term.

5.2 Historical overview

The basic principle of asteroseismic data analysis can be summarized as follows:

- acquire signal from a finite world,
- compute the Fourier transform of the discrete signal,
- extract the characteristics of the harmonic signals.

The first step is closely related to approximation of continuous function using decomposition on a base of orthogonal functions. The latter could be the sine and cosine functions used in the second step, that is, Fourier decomposition or transform. The last step is related to inference based on statistics or applications of probability to the data. Hereafter, I will try to place in a historical perspective all of these steps. The perspective will be presented in a chronological order for each subject. This historical review does not pretend to be complete, as I am not an epistemologist. The main goal of this historical review is to give the reader some keys on reflecting on some tools that we regularly use. The reader will then be free to delve into the subject or leave it aside.

5.2.1 *Spectral analysis and digital signal processing*

The father of spectral analysis is the well-known Joseph Fourier. He pioneered the decomposition of an arbitrary function in cosine and sine functions in his book on heat. In

Chapter 3 of his book (p. 257), Fourier (1822) expressed the decomposition in harmonic functions that later became, using Euler's notation, the complex Fourier transform. In the original formulation of Fourier lies explicitly the potential for a finite summation. In other words, any arbitrary function can be approximated with a finite summation over the Fourier coefficients, which is the discrete Fourier transform (DFT). In essence, this is the introduction of a representation of a continuous world using a finite set: a *digital* world. The expression provided by Fourier was already a digital description of the world. The use of sine and cosine functions was very much at the center of the mathematical world at the beginning of the nineteenth century. In 1805, Carl F. Gauss devised an algorithm for interpolation of cosine and sine functions that would later be recognized as the fast Fourier transform (FFT), a work that was posthumously published (Gauss, 1866). This work was published in Latin but a translation can be found in Goldstine (1977) and Heideman *et al.* (1985). The proposal 27 of Gauss's work is what we now call the FFT (Heideman *et al.*, 1985). The *original* algorithm was discovered (not to say rediscovered) by James Cooley and John Tuckey in 1965, while they were working at IBM and the Bell Telephone Laboratory, respectively (Cooley and Tuckey, 1965). As anticipated by Gauss, they showed that the DFT could be speeded up by using the fact that the number of points N in the transform could be expressed as products of prime numbers, thereby speeding the computation by $O(\frac{N}{\log N})$.

At the time of the publication of the paper by Cooley and Tuckey (1965), spectral analysis had already entered the *modern* age of the digital world. When working at the AT&T Bell Telephone Laboratory, Claude (Shannon, 1949) introduced the so-called sampling theorem, which is key for reducing a continuous finite-bandwidth signal to a digital sample. The frequency at which the signal should be sampled was derived by Nyquist (1924), hence bearing the name the Nyquist frequency (Harry Nyquist belonged to what later became the Bell Telephone Laboratory). The sampling theorem is at the basis of all digital audio equipment since the invention by Sony of the digital audio disc in 1972, later to become the compact disc of Philips in 1978. Digital signal processing (DSP) is used in many applications, such as speech recognition, compression, audio sampling, home cinema, and, of course, in scientific processing.

It is rather surprising to realize that the existence of the digital world dates back to the beginning of the nineteenth century. In a way, it is not so strange that we need to describe an infinite world with a finite set of data, for instance using function interpolation. Being ourselves *finite* or limited in *time* and *space*, our world was bound to sooner or later become digital.

5.2.2 *Probability, statistics, and inference*

Probability and statistics are related to one another. Probability provides the mathematical foundations for assessing the chance that a random event will occur. Statistics using probability theory provides inference on what has been *really* observed. Probability theory started with Jakob Bernoulli, who was applying combinatorial analysis for calculating probabilities related to games of chance. He published what is known as the Bernoulli distribution and also introduced the law of large numbers (Bernoulli, 1713). The same Bernoulli distribution was approximated by De Moivre (1718), which was a special case of the central limit theorem. Later on, Reverend Thomas Bayes solved a problem that was left untouched by De Moivre related to the probability of occurrence of unrelated events. Proposition 5 of Bayes (1763) is what is known today as the *Bayes theorem.* This theorem was also found independently by Laplace (1774), who was working on the same subject. Unfortunately, this view on probability quite advanced at the time of Bayes and Laplace and was not used until it was *rediscovered* by Jeffreys (1939).

Inference is related to how one can deduce from data a theoretical model of the world being observed (with error bars on this model). The origin of the first inference can be traced back to the work of Laplace, related to the use of the arithmetic mean (Laplace, 1774). Gauss demonstrated, using the *maximum likelihood principle*, that an estimate of a parameter measured many times can indeed be expressed as an arithmetic mean of these observations (Gauss, 1809). A typical inference called *least squares*[1] *minimization* was used by Legendre (1805) for deriving the orbit of comets, and for verifying the length of the meter through the measurement of the Earth's circumference. This technique was also found before Legendre by Gauss but was not published until later (Gauss, 1809). Gauss also derived what was called the law of errors: the so-called Gaussian distribution of errors (Gauss, 1809). The use of this ubiquitous error distribution is a simple consequence of the principle of the *maximum entropy distribution* (Jaynes, 2009), the distribution simply reflecting the state of our knowledge (or lack of) by knowing the mean value μ and the root means square deviation σ of a set of observations. The principles behind *maximization* are at the very heart of inference. While Gauss introduced the concept for likelihood, Fisher set the proper mathematical background and theory behind the use of *maximum likelihood estimators*, notably their asymptotic properties and their information content (Fisher, 1912, 1925). Around that time, a controversy between Ronald Fisher and Harold Jeffreys marked the start of different view on probability and statistics: *frequentist* versus *Bayesian*. Jeffreys (1939) was key in reviving the approach touched upon by Bayes and Laplace, an approach that was not the main stream of statistical thinking at the time of Fisher. At that point in time, the way was open for Bayesian probability and statistics to be applied in various fields of physics and astrophysics. Jaynes (2009) provides many possible applications of Bayesian approaches such as one used for Fourier analysis.

Now, in the twenty-first century, it would be naive to believe that inference can only be based on Bayesian approaches, it is surely not a *panacea*. It should be kept in mind that as much as we evolve, any field of science evolves accordingly. Even in statistics, the evolution is not finished. The current stream is to try to reconcile the various approaches promoted by Fisher and Jeffreys (Berger, 1997). Therefore, it is the responsibility of any physicist or astrophysicist to follow this evolution. It is perhaps superfluous to remind the reader that all inferences on the world as we see it come from instrumentation, observation, phenomena that are neither perfect nor deterministic. Let me remind you of a quote by Poincaré (1914): "From this point of view all the sciences would only be unconscious applications of the calculus of probabilities. And if this calculus be condemned, then the whole of the sciences must also be condemned."

5.3 Digital signal processing and spectral analysis

5.3.1 *Time series sampling*

Shannon (1949) provided the theorem that allows us to sample a continuous signal whose frequencies are contained in a finite bandwidth. Using the decomposition in Fourier series, Shannon (1949) wrote that: "If a function $x(t)$ contains no frequencies higher than $\Delta\nu$, it is completely determined by giving its ordinates at a series of points spaced $1/(2\Delta\nu)$ apart"; he then wrote:

$$x(t_n) = \int_{-\Delta\nu}^{+\Delta\nu} X(\nu)\mathrm{e}^{\mathrm{i}2\pi\nu t_n}\,\mathrm{d}\nu \tag{5.1}$$

[1] Translated literally from *Moindres quarrés.*

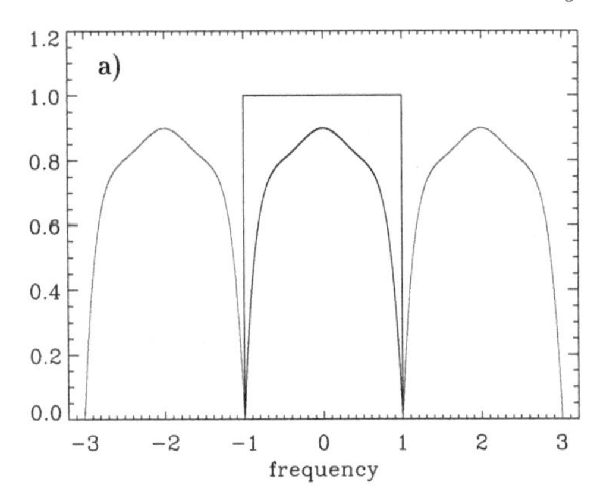

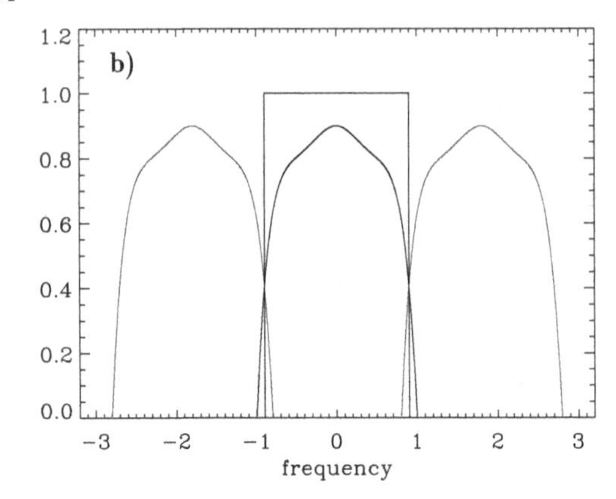

FIG. 5.1. *(a)* The frequency response of a finite bandwidth signal with the boxcar on top, properly sampled according to Shannon's theorem. *(b)* Undersampled and aliased.

where $t_n = n\Delta t$ (with $\Delta t = 1/2\Delta\nu$), $x(t)$ is the function to be sampled, and $X(\nu)$ is the Fourier transform of $x(t)$. From this theorem, and the use of Fourier decomposition, one can then write:

$$X(\nu) = \Pi(2\Delta\nu)\left[\Delta t \sum_{n=-\infty}^{n=+\infty} x(t_n)\mathrm{e}^{\mathrm{i}2\pi\nu t_n}\right] \tag{5.2}$$

where Π is the boxcar. The term between brackets is simply the original Fourier decomposition, which is implicitly periodic, hence the use of the boxcar for delimiting the frequency space. This is the case represented by the left hand side of Fig. 5.1. Using the inverse Fourier transform of Equation 5.2, one can shows that we can fully reconstruct $x(t)$ by writing:

$$x(t) = \sum_{n=-\infty}^{n=+\infty} x(t_n)\mathrm{sinc}\left(\frac{t-t_n}{\Delta t}\right) \tag{5.3}$$

where sinc ($=\sin x/x$) is the sinus cardinal function. This equation shows that one can recover *perfectly* a continuous function using samples of that function at regular spacing, whose cadence is provided by the spectral content of that function. In practice, the recovery can only be approximated, as the summation over ∞ is impractical.

5.3.2 *Aliasing*

There are two cases that provide unwanted frequency leaking into the original spectrum:

- undersampling of a band-limited signal, and
- nonband-limited signal.

This is what is called *aliasing*, the effect to which is shown on the right hand side of Fig. 5.1. In this case, high frequency signal leaks back, or aliases, at frequency $\nu_{\mathrm{alias}} = \nu_{\mathrm{true}} - \Delta\nu$. Undersampling occurs when the sampling time is larger than Shannon's sampling time ($1/2\Delta\nu$). Undersampling of a band-limited signal can be easily resolved by applying Shannon's theorem. The case of signal having nonlimited frequency content is clearly not covered by Shannon's theorem. In that case, there are two techniques that can provide a reduction of the aliasing power: integration and weights.

I shall focus on the effect of integration that is often encountered either when observing time series or making images with discrete arrays such as charge coupled devices (CCD). When one integrates a signal x over some time (Δt), one can write:

$$x_{\mathrm{obs}}(t) = \sum_n \left[\frac{1}{\Delta t}\int_{t_n}^{t_n+\Delta t} x(t)\mathrm{d}t\right] \tag{5.4}$$

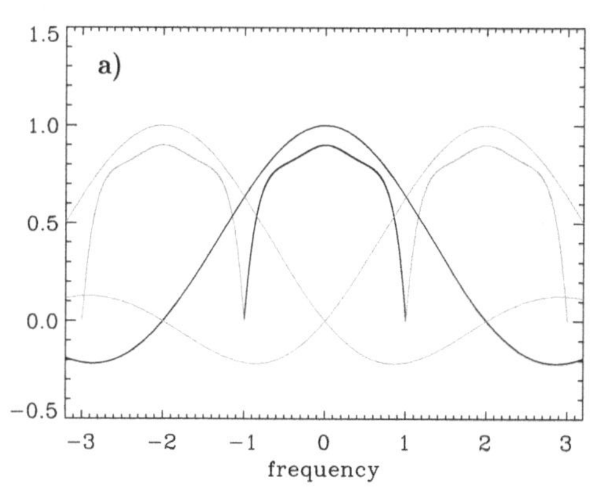

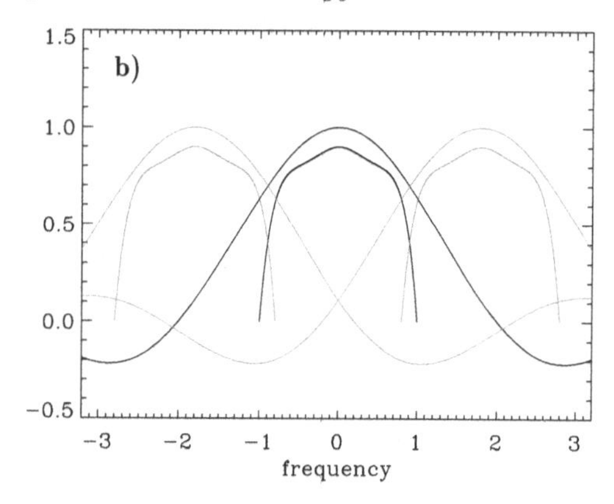

FIG. 5.2. *(a)* The finite bandwidth signal and the sinc function related to the integration over 100% of the sampling time (in black at $\nu = 0$), (in gray at $\nu = \pm 1/2\Delta\nu$); when the signal is sampled according to Shannon's theorem. *(b)* When the signal is undersampled.

where x is regularly sampled at $\Delta t_s = t_{n+1} - t_n$. This equation can be rewritten using convolution as:

$$x_{\rm obs}(t) = \text{III}(\Delta t_s) * \left[\frac{1}{\Delta t}(x * \Pi(\Delta t))(t)\right] \tag{5.5}$$

where III is the Dirac comb. Since the Fourier transform of a Dirac comb is also a Dirac comb, the Fourier spectrum of the signal can then be written as:

$$X_{\rm obs}(\nu) = \frac{1}{\Delta t_s}\text{III}\left(\frac{1}{\Delta t_s}\right)\left[X(\nu)\text{sinc}(\Delta t \nu)\text{e}^{\text{i}\pi\nu\Delta t}\right] \tag{5.6}$$

Figure 5.2 shows the resulting effect of the integration when $\Delta t = \Delta t_s$. In practice, when the signal is band limited, integration will introduce a filtering effect of the high frequencies. This effect can only be reduced by having a very short integration time with respect to the sampling or $\Delta t \ll \Delta t_s$. The effect of aliasing is also shown on Fig. 5.2. When the signal is nonband limited, there are two possible solutions for reducing the effect of the high-frequency signal:

- introduction of a window function in frequency (the Π function);
- integration at 100% duty cycle ($\Delta t = \Delta t_s$).

The first solution requires the combination of the time sample by using the Fourier transform of the Π function, which is a sinc in time. This solution can only be partially implemented as it would require a sum over ∞. This solution is naturally implemented when making images through a telescope using discrete arrays. In that case, the highest spatial frequencies are cut off by having the telescope diameter providing a cut-off D/λ half that of the spatial pixel sampling; this is illustrated by Fig. 5.2a. In other terms, there is no aliasing in a telescope when the pixel resolution is half of the telescope resolution, that is, two samples per resolution element. De facto, in a telescope, the second solution is also used in combination with the first solution. The second solution, although not perfect, will reduce the amplitude of the high frequency noise in time series.

5.3.3 *Filtering*

The effect of time series filters can be studied simply by understanding the effect of smoothing on time series. Smoothing can be understood as being the application of a weighting function sliding in time: a convolution. This can be written as:

$$x_{\rm sm}(t) = (x * w)(t) \tag{5.7}$$

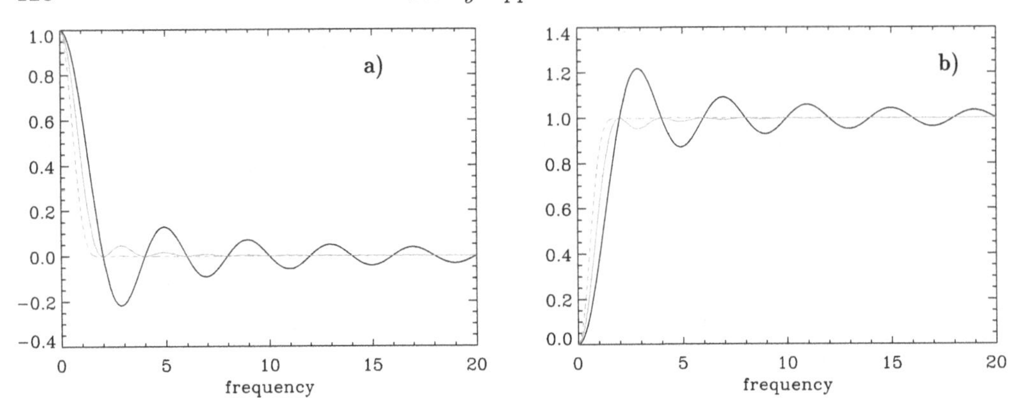

FIG. 5.3. *(a)* Frequency response of several low-pass filters: boxcar (black), triangle or two times the boxcar (gray), bell shape of four times the boxcar (dashed gray). *(b)* Frequency response of several high-pass filters resulting from the previous low-pass filters corresponding to the same color coding.

where w is the weighting function, which can be complex. Using the Fourier transform, this becomes:

$$X_{\mathrm{sm}}(\nu) = X(\nu)W(\nu) \tag{5.8}$$

where W is the Fourier transform of w. The smoothing filter typically provides a low-pass filter that can be used to derive a high-pass filter of the original function x by computing:

$$x_{\mathrm{fil}}(t) = x(t) - x_{\mathrm{sm}}(t). \tag{5.9}$$

Using the Fourier transform, I get:

$$X_{\mathrm{fil}}(\nu) = X(\nu)(1 - W(\nu)) \tag{5.10}$$

Figure 5.3 shows the results of applying one time, two times, and four times the boxcar on a time series. Applying twice the boxcar is equivalent to a triangular weighting function (convolution of two boxcar functions), while applying four times the boxcar is equivalent to a bell shape weighting function (convolution of two triangle functions). It is rather clear from Fig. 5.3 that boxcar smoothing should be avoided because it provides too much *ringing* effect of the type Gibbs (1898, 1899) discovered. The introduction of a less sharp transition for all derivatives by multiple boxcar smoothing provides a neat solution to this Gibbs effect.

Another form of filtering is to use the original series shifted in time by t_0 and then subtract the shifted series from the unshifted. In that case, the weighting function is simply the Dirac distribution $w(t) = \delta(t - t_0)$. Then, using Equation 5.10, the modulus of the filter is simply

$$|1 - W(\nu)| = 2\sin(\pi\nu t_0). \tag{5.11}$$

The frequency at which the transmission is down by half is given by $\nu_{\mathrm{cuton}} = 1/3t_0$. This kind of smoothing filter has been used for the data of the Global Oscillation Network Group (GONG) using the first difference obtained with $t_0 = \Delta t_s$ (Appourchaux *et al.*, 2000).

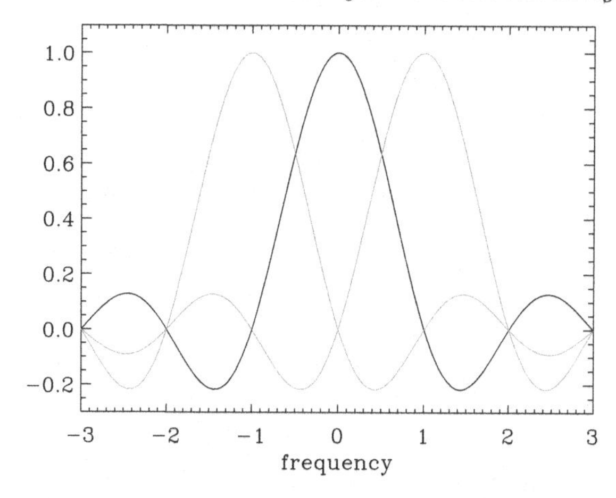

FIG. 5.4. The sinc function as a function of frequency normalized to the resolution $(1/T)$, for $\nu = 0$ (black), for $\nu = \pm 1$ (gray).

5.3.4 *Time limits and the discrete Fourier transform*

The Fourier transform is not applicable when doing *real* data analysis. There is no way that we can observe an infinite strings of data. We usually observe during a finite time T for which we can compute the following Fourier transform:

$$X_T(\nu) = \int_{-T/2}^{+T/2} x(t) \mathrm{e}^{\mathrm{i}2\pi\nu t} \mathrm{d}t, \tag{5.12}$$

which can be rewritten as:

$$X_T(\nu) = [X * \mathrm{sinc}(T)](\nu). \tag{5.13}$$

The convolution function of the term in brackets is the sinc function. Figure 5.4 shows the sinc function for adjacent frequency spaced at $\pm\frac{1}{T}$. It is obvious that the spectrum is correlated between the various frequency bins. I will show later that the correlation is in fact null for frequency bins separated by integer values of $\frac{1}{T}$, and for slowly varying power spectra. If I combine the finite observation with a finite bandwidth signal, I then have the DFT:

$$X_{\mathrm{DFT}}(\nu_p) = \Delta t \sum_{n=1}^{n=N} x(t_n) \mathrm{e}^{\mathrm{i}2\pi\nu_p t_n} \tag{5.14}$$

with $\nu_p = \frac{p}{T}$, $t_n = n\Delta t_s$, $T = N\Delta t_s$. Equation 5.14 is simply the truncated original Fourier series, which is then by definition periodic with a period of $\Delta\nu = \frac{1}{t_s}$. This property directly gives that N is also the maximum value of p. The DFT can be computed using the given definition but it is rather time consuming, as the time for computing the summation scales as N^2. As mentioned in Section 5.2, Cooley and Tuckey (1965) provided a faster way based on the factorization properties of the Fourier transform; in that case, the time for computing the summations scales as $N \log N$.

5.3.5 *Fourier transform of stationary processes*

The application of Fourier transforms to nondeterministic or random functions is key in doing data analysis in astrophysics. For such a process, I define for the random variable x, the *power spectral density*, as:

$$S_T(\nu) = \frac{1}{T} |X_T(\nu)|^2 \tag{5.15}$$

where $X_T(\nu)$ is given by Equation 5.12. I can then write $S_T(\nu)$ as:

$$S_T(\nu) = \frac{1}{T} \int_{-T/2}^{+T/2} \int_{-T/2}^{+T/2} x(t)x(t')\mathrm{e}^{\mathrm{i}2\pi\nu t}\mathrm{e}^{-\mathrm{i}2\pi\nu t'}\mathrm{d}t\mathrm{d}t'. \tag{5.16}$$

Because the process is stationary, I can make the change of variable $\tau = t - t'$, and then I have:

$$S_T(\nu) = \int_{-T/2}^{+T/2} \left[\frac{1}{T} \int_{-T/2}^{+T/2} x(t')x(t'+\tau)\mathrm{d}t' \right] \mathrm{e}^{\mathrm{i}2\pi\nu\tau}\mathrm{d}\tau. \tag{5.17}$$

The term in brackets is by definition the autocorrelation function ($C_T(\tau)$) of the process taken over a finite window T. Then, I have:

$$S_T(\nu) = \int_{-T/2}^{+T/2} C_T(\tau)\mathrm{e}^{\mathrm{i}2\pi\nu\tau}\mathrm{d}\tau. \tag{5.18}$$

Equation 5.18 introduces the so-called Wiener-Khinchin theorem, which is essential for understanding the spectral analysis of random processes. As a matter of fact, if I also assume ergodicity of the process (i.e., the temporal average can be exchanged with the spatial average), then I have:

$$C(\tau) = \mathrm{E}(x(t)x(t+\tau)) = \lim_{T\to\infty} C_T(\tau). \tag{5.19}$$

I can then write:

$$S_x(\nu) = \lim_{T\to\infty} \mathrm{E}\left[S_T(\nu)\right] = \int_{-\infty}^{+\infty} C(\tau)\mathrm{e}^{\mathrm{i}2\pi\nu\tau}\mathrm{d}\tau, \tag{5.20}$$

which is the proper Wiener-Khinchin theorem (Wiener, 1930; Khinchin, 1934). Using the inverse Fourier transform, I can then also write:

$$C(\tau) = \int_{-\infty}^{+\infty} S_x(\nu)\mathrm{e}^{-\mathrm{i}2\pi\nu\tau}\mathrm{d}\nu \tag{5.21}$$

This relation between the autocorrelation function and the power spectral density is absolutely key in understanding the Fourier analysis of stationary processes.

5.3.6 *Parseval's theorem*

For $\tau = 0$, I also have

$$C(0) = \int_{-\infty}^{+\infty} S_x(\nu)\mathrm{d}\nu = \mathrm{E}(x^2) \tag{5.22}$$

This equation is also an alternate formulation of the energy conservation principle known as Parseval's theorem (Parseval des Chênes, 1806). This important property provided by Equation 5.22 can be adapted for normalizing the DFT. For the DFT given by Equation 5.14, I can write:

$$\alpha^2 \sum_{p=1}^{p=N} |X_{\mathrm{DFT}}(\nu_p)|^2 = \frac{1}{N} \sum_{n=1}^{n=N} x(t_n)^2 \tag{5.23}$$

where α is the required normalization factor used for the spectral density. For Equation 5.14, it is easy to show that the normalization factor α is $1/N$, which is the usual factor used when computing the DFT. If other definitions of the DFT are to be used, the proper normalization factor α can be derived with Equation 5.23. This equation is used for calibrating the spectra coming from different routines, all having different normalization factors.

5.3.7 *Statistics of the discrete Fourier transform*

Equation 5.14 is *our bread and butter* for extracting the frequencies associated with periodic phenomena observed, for instance, in stars. Whenever one carries out observations, one should not forget that the observations come with noise either associated with the observed phenomenon or with the instrument providing the data. Therefore, it is essential to understand the statistics of the Fourier transform of random variables. Let us start with a simple and commonly used example: a random variable x with an unknown distribution (with $\mathrm{E}(x) = 0$ and $\mathrm{E}(x^2) = \sigma^2$) for which all time samples $x(t_n)$ are independent from each other; in addition, the process is assumed to be stationary. Using Equation 5.14, I can then write the DFT of the $x(t_n)$ as:

$$X^{\mathrm{r}}_{\mathrm{DFT}}(\nu_p) + iX^{\mathrm{i}}_{\mathrm{DFT}}(\nu_p) = \sum_{n=1}^{n=N} x(t_n)\cos(2\pi\nu_p t_n) + \mathrm{i}\sum_{n=1}^{n=N} x(t_n)\sin(2\pi\nu_p t_n) \tag{5.24}$$

where $X^{\mathrm{r}}_{\mathrm{DFT}}$ and $X^{\mathrm{i}}_{\mathrm{DFT}}$ are the real and imaginary part of the Fourier spectrum. Since the $x(t_n)$ are independent and identically distributed (i.i.d.), by virtue of the central limit theorem, the statistical distribution of $X^{\mathrm{r}}_{\mathrm{DFT}}$ and $X^{\mathrm{i}}_{\mathrm{DFT}}$ is a normal distribution for $N \gg 1$ with:

$$\mathrm{E}(X^{\mathrm{r}}_{\mathrm{DFT}}(\nu_p)) = \mathrm{E}(X^{\mathrm{i}}_{\mathrm{DFT}}(\nu_p)) = 0 \tag{5.25}$$

$$\mathrm{E}([X^{\mathrm{r}}_{\mathrm{DFT}}(\nu_p)]^2) = \mathrm{E}([X^{\mathrm{i}}_{\mathrm{DFT}}(\nu_p)]^2) = \frac{N}{2}\sigma^2. \tag{5.26}$$

Then, because $X^{\mathrm{r}}_{\mathrm{DFT}}$ and $X^{\mathrm{i}}_{\mathrm{DFT}}$ are independent and have the same normal distribution, the statistics of the power spectrum are then by definition a χ^2 with two degrees of freedom (d.o.f.).

Unfortunately (or fortunately), none of the processes that we observe have the properties of being i.i.d. The processes are usually stationary processes but are not i.i.d. because these processes have usually a memory such that the correlation of $x(t_n)$ and $x(t_m)$ are different from zero when $t_n \neq t_m$. Nevertheless, in that case, it can be demonstrated that the components of the Fourier transform are *also* both normally distributed with the same mean of zero and the same variance, which depends on frequency (Peligrad and Wu, 2010). It is amusing to quote Peligrad and Wu on their finding: *In this sense, Theorem 2.1* [of Peligrad and Wu (2010)] "justifies the folklore in the spectral domain analysis of time series: the Fourier transforms of stationary processes are asymptotically independent Gaussian."

5.3.8 *Time series sampled at uneven times*

It is quite common in astrophysics to have samples that are not equally spaced in time. This lack of uniformity could be due to a variable number of photon per seconds or due to variable detector read-out time. Usually, this can be avoided by carefully designing the electronics; see, for instance, Fröhlich *et al.* (1997). Nevertheless, if the time series are unevenly sampled, there are ways and means to find solutions. Equation 5.14 can still be applied, obviously by dropping Δt_s. The problem with this equation is that for uneven times, the statistics of the Fourier spectrum is no longer χ^2 with two d.o.f. (Scargle, 1982). Equation 5.24 can be adapted such that this statistical property is kept by writing:

$$X^{\mathrm{r}}_{\mathrm{LS}}(\nu_p) = \frac{1}{w(\tau)}\sum_{n=1}^{n=N} x(t_n)\cos(2\pi\nu_p(t_n - \tau)) \tag{5.27}$$

$$X^{\mathrm{i}}_{\mathrm{LS}}(\nu_p) = \frac{1}{v(\tau)}\sum_{n=1}^{n=N} x(t_n)\sin(2\pi\nu_p(t_n - \tau)) \tag{5.28}$$

where w and v are given by:

$$w(\tau) = \sum_{n=1}^{n=N} \cos^2(2\pi\nu_p(t_n - \tau)) \tag{5.29}$$

$$v(\tau) = \sum_{n=1}^{n=N} \sin^2(2\pi\nu_p(t_n - \tau)) \tag{5.30}$$

and τ is introduced for keeping the invariance in time of the transform given by Equations 5.27 and 5.28. τ is given by:

$$\tan(2\pi\nu\tau) = \frac{\sum_{n=1}^{n=N} \sin(2\pi\nu t_n)}{\sum_{n=1}^{n=N} \cos(2\pi\nu t_n)} \tag{5.31}$$

The definition provided by Equations 5.27 and 5.28 has the benefit of giving a power spectrum or Lomb-Scargle (LS) periodogram ($[X^{\rm r}_{\rm LS}(\nu_p)]^2 + [X^{\rm i}_{\rm LS}(\nu_p)]^2$), which is χ^2 with two d.o.f. (Scargle, 1982). It must also be pointed out that Equations 5.27 and 5.28 are also the solution obtained when applying *least squares* minimization to

$$\sum_{n=1}^{n=N} \left[x(t_n) - a_{\rm c}\cos(2\pi\nu t) - a_{\rm s}\sin(2\pi\nu t)\right]^2 \tag{5.32}$$

where $a_{\rm c}$ and $a_{\rm s}$ are given by Equations 5.27 and 5.28, respectively. For speed, the LS periodogram is usually computed using the implementation prescribed by Press and Rybicki (1989), which is an approximation of the LS periodogram based upon *extirpolation*[2] on a regular mesh and the use of the FFT. The prescription is then very close to interpolating onto a regular mesh. It is worth noting that most users of the LS periodogram for unevenly sampled data are in fact computing the FFT of the original data resampled onto a regular mesh, but with a proper normalization as given by Scargle (1982). It must be noted that the LS periodogram does not provide a better solution to coping with the presence of gaps. The reason is that although the Fourier transform *explicitly* includes gaps as zeros, adding zeros is also *implicitly* performed with the LS periodogram. As a consequence, correlations between frequency bins also exist with the LS periodogram, but these are generally ignored. The correlations in the Fourier transform in the presence of gaps are addressed in the next section.

5.3.9 *The influence of gaps in the time series*

The impact of the gaps on the Fourier spectrum has been described by Gabriel (1994). The gaps introduce correlation between frequency bins that need to be taken into account when one wants, for example, to fit the power spectrum (see, for applications, Stahn and Gizon, 2008). Hereafter, I will provide the result regarding the correlation of the Fourier spectrum between the frequency bins. Assuming that I observe a random variable x through a window W, the Fourier transform can be written as:

$$\tilde{\mathcal{X}}(\nu) = \int_{-\infty}^{+\infty} x(t)W(t)\mathrm{e}^{\mathrm{i}2\pi\nu t}\mathrm{d}t \tag{5.33}$$

[2] Reverse interpolation, or extirpolation, replaces a function value at any arbitrary point by several function values on a regular mesh.

The mean correlation between two frequency bins ν_1 and ν_2 is given by:

$$\mathrm{E}[\tilde{\mathcal{X}}(\nu_1)\tilde{\mathcal{X}}^*(\nu_2)] = \mathrm{E}\left[\int_{-\infty}^{+\infty}\int_{-\infty}^{+\infty} x(t)x(t')W(t)W(t')\mathrm{e}^{\mathrm{i}2\pi\nu_1 t}\mathrm{e}^{-i2\pi\nu_2 t'}\mathrm{d}t\mathrm{d}t'\right] \quad (5.34)$$

where * denotes the complex conjugate. Following Gabriel (1993), I have the following properties for the real and imaginary parts of $\mathcal{X}$:

$$\mathrm{E}[\tilde{\mathcal{X}}_{\mathrm{r}}(\nu_1)\tilde{\mathcal{X}}_{\mathrm{r}}^*(\nu_2)] = \mathrm{E}[\tilde{\mathcal{X}}_{\mathrm{i}}(\nu_1)\tilde{\mathcal{X}}_{\mathrm{i}}^*(\nu_2)] \quad (5.35)$$

$$\mathrm{E}[\tilde{\mathcal{X}}_{\mathrm{r}}(\nu_1)\tilde{\mathcal{X}}_{\mathrm{i}}^*(\nu_2)] = -\mathrm{E}[\tilde{\mathcal{X}}_{\mathrm{i}}(\nu_1)\tilde{\mathcal{X}}_{\mathrm{r}}^*(\nu_2)] \quad (5.36)$$

Using the Fourier transform of $x(t)$, I can rewrite Equation 5.34:

$$\begin{aligned}&\mathrm{E}[\tilde{\mathcal{X}}(\nu_1)\tilde{\mathcal{X}}^*(\nu_2)] = \\ &\int\int\int\int \mathrm{E}[X(\nu)X(\nu')]W(t)W(t')\mathrm{e}^{\mathrm{i}2\pi[(\nu_1-\nu')t-(\nu_2-\nu)t']}\mathrm{d}t\mathrm{d}t'\mathrm{d}\nu\mathrm{d}\nu'\end{aligned} \quad (5.37)$$

By construction, I assume that there is no correlation between the real and imaginary parts of the original spectrum X, that their variances have the same value $\mathrm{E}(X_{\mathrm{r}}^2(\nu)]$ and the frequency bins of the original spectrum are not correlated (see also Gabriel, 1993), such that I have:

$$\mathrm{E}[X_{\mathrm{r}}(\nu)X_{\mathrm{r}}(\nu')] = \mathrm{E}[X_{\mathrm{r}}^2(\nu)]\delta(\nu'-\nu) \quad (5.38)$$

$$\mathrm{E}[X_{\mathrm{r}}(\nu)X_{\mathrm{r}}(\nu')] = \mathrm{E}[X_{\mathrm{i}}(\nu)X_{\mathrm{i}}(\nu')] \quad (5.39)$$

where δ is the Dirac distribution. Then, I can rewrite:

$$\mathrm{E}[\tilde{\mathcal{X}}(\nu_1)\tilde{\mathcal{X}}^*(\nu_2)] = 2\int_{-\infty}^{+\infty} \mathrm{E}[X_{\mathrm{r}}^2(\nu)][W(\nu_1-\nu)W(\nu-\nu_2)]\mathrm{d}\nu \quad (5.40)$$

where W is the Fourier transform of the window function w. For understanding the impact of Equation 5.40, let us assume that we observe white noise of mean 0 and of variance σ_0 in frequency. In that case, I have:

$$\mathrm{E}[\tilde{\mathcal{X}}(\nu_1)\tilde{\mathcal{X}}^*(\nu_2)] = 2\sigma_0^2\int_{-\infty}^{+\infty} W(\nu_1-\nu)W(\nu-\nu_2)\mathrm{d}\nu. \quad (5.41)$$

Using the properties of convolution and the inverse Fourier transform, it can be shown[3] that I have:

$$\mathrm{E}[\tilde{\mathcal{X}}(\nu_1)\tilde{\mathcal{X}}^*(\nu_2)] = 2\sigma_0^2\int_{-\infty}^{+\infty} w^2(t)\mathrm{e}^{\mathrm{i}2\pi(\nu_1-\nu_2)t}\mathrm{d}\nu. \quad (5.42)$$

The integral in this equation is simply the Fourier transform of the square of the window function, or W_{sq}. For a window function such as the one provided by an observing window of length T (see Equation 5.12), the correlation is then given by the sinc function. This justifies *a posteriori* the sampling at frequency interval of $1/T$, for which the correlation is null. If the variations of $\mathrm{E}[X_{\mathrm{r}}^2(\nu)]$ are slow with respect to $W(\nu)$, I can also rewrite Equation 5.40 as:

$$\mathrm{E}[\tilde{\mathcal{X}}(\nu_1)\tilde{\mathcal{X}}^*(\nu_2)] \approx 2\mathrm{E}[X_{\mathrm{r}}^2(\nu_1)]W_{\mathrm{sq}}(\nu_1-\nu_2) \quad (5.43)$$

where W_{sq} is the Fourier transform of w^2. Of course, this approximation does not hold when the variations in frequency are over scales of $1/T$. Nevertheless, Equation 5.43

[3] I leave the demonstration to the reader.

gives an interesting solution for understanding the correlation between frequency bins in a Fourier spectrum.

It is also useful to understand the correlation between the real and imaginary parts of the Fourier transform. Using Equations 5.35 and 5.36, I can derive the very useful formulation:

$$\mathrm{E}[\tilde{\mathcal{X}}_{\mathrm{r}}(\nu_1)\tilde{\mathcal{X}}_{\mathrm{r}}^{*}(\nu_2)] = \int_{-\infty}^{+\infty} \mathrm{E}[X_{\mathrm{r}}^2(\nu)]\mathcal{R}\left[W(\nu_1-\nu)W(\nu-\nu_2)\right]\mathrm{d}\nu \tag{5.44}$$

$$\mathrm{E}[\tilde{\mathcal{X}}_{\mathrm{r}}(\nu_1)\tilde{\mathcal{X}}_{\mathrm{i}}^{*}(\nu_2)] = \int_{-\infty}^{+\infty} \mathrm{E}[X_{\mathrm{r}}^2(\nu)]\mathcal{I}\left[W(\nu_1-\nu)W(\nu-\nu_2)\right]\mathrm{d}\nu \tag{5.45}$$

where $\mathcal{R}$ and $\mathcal{I}$ denote the real and imaginary operators, respectively. These two relations can be used when a specific window that is different from the boxcar is used. For example, the use of weights across the observing window will then introduce correlation provided by Equations 5.44 and 5.45. The introduction of these weights or tapers is described in the next section.

5.3.10 *Taper estimates of Fourier power spectrum*

Fourier spectrum estimation is well adapted for periodic signals (pure sine waves or stochastic waves) but not necessarily well suited for estimating the spectral density of frequency-dependent noise (pink or red noise). For that purpose, one can:

- average the power spectrum over an ensemble of n sub-series,
- smooth the power spectra over n frequency bins, or
- use multitapered spectra using the full time series for deriving a similar average.

Fourier spectrum estimation can be replaced by multitapered spectra that are widely used in geophysics (for a review, see Thomson, 1982). Multitapered spectra are generated by applying a set of tapers to a single time series, and an estimate of the mean power spectrum is derived from an average of these spectra. Using tapers due to Slepian (1978), the multitapered spectra are statistically independent from one another, and the statistics of the mean spectrum follow a χ^2 distribution with $2n$ degrees of freedom (where n is the number of tapers; Thomson, 1982). The statistics of the average power spectrum (or smoothed power spectrum) also follow a χ^2 distribution with $2n$ degrees of freedom: the resolution of the average spectrum is n times lower than that of the multitapered spectrum. In helioseismology, the use of these Slepian tapers has been replaced by more practical (but less accurate) sine tapers (Komm *et al.*, 1999). Unfortunately, for sine waves, tapers tend to broaden the peaks, as shown by Thomson (1982). Tapers as such provide more benefit for broader peaks than for narrower peaks.

5.4 Data analysis and statistics

5.4.1 *Hypothesis testing*

Statistical testing is essential when one wants to decide: *have we found a signal* or *not*? This is related to *decision theory*, which can be summarized as: *how do we choose between one hypothesis versus another in the presence of uncertainties?* In this area, there are two schools of thought: the frequentist school and the Bayesian school.

The difference between a Bayesian and a frequentist relates to their views of *subjective* versus *objective* probabilities. A frequentist thinks that the laws of physics are *deterministic*, while a Bayesian subscribes to a belief that the laws of physics are true or *operational*. The *subjective* approach to probability was first coined by De Finetti (1937).

For the rest of us, the difference in views between frequentists and Bayesians can be outlined by taking an example from the Six Nations rugby tournament. For a frequentist,

France has been winning over England in their direct confrontation only 39% of these matches since 1906. Based on this result, for a frequentist, France has only 39% chance of winning any future game against England. For a Bayesian, this is a completely different story. A Bayesian may attribute higher chance or lower chance to France to win a given game based on the current physical and technical skills of each player, on the current ability of the players to play as a team, and on the psychological and mental health of the players and of the team as a whole. Based on this assessment, a Bayesian might have attributed a 70% chance to win the 2010 game, or a 20% chance to win the 2011 game. This is an *a posteriori* evaluation of the chances, as both games took place while writing this chapter. It is used as an example.

In short, frequentists assign probability to measurable events that can be measured an infinite number of times, while Bayesians assign probability to events that cannot be measured, such as the survival time of the human race (Gott, 1994) or the future outcome of sporting matches.

In what follows, I will try to give an overview of what *I believe* I know on a subject that is rapidly evolving; what I write is certainly *not* gospel.

5.4.1.1 Frequentist hypothesis testing

For a frequentist, statistical testing is related to hypothesis testing. In short, we have two types of hypotheses:

- H_0 hypothesis or null hypothesis: what has been observed is pure noise.
- H_1 hypothesis or alternative hypothesis: what has been observed is a signal.

For the H_0 hypothesis, I assume known statistics for the random variable Y observed as y and assumed to be pure noise; and then set a *false alarm probability* that defines the acceptance or rejection of the hypothesis. The so-called *detection significance* (or *p-value*, terms not widely used in astrophysics) is the probability of having a value as extreme as *the one actually observed.* There is an ongoing confusion because statisticians call *the significance level* what astronomers call the *false alarm probability*; and statisticians call the *p-value* what is set in astronomy as the *detection significance* (which is *not* the significance level). Here, I shall use the current vocabulary understood in astronomy. For example, the *false alarm probability* p for the H_0 hypothesis is defined as:

$$p = P_0(\tilde{T}(Y) \geq \tilde{T}(y_c)), \tag{5.46}$$

where $\tilde{T}$ is the statistical test, P_0 is the probability of having $\tilde{T}(Y) \geq \tilde{T}(y_c)$ when H_0 is true, and y_c is the cutoff threshold derived from the test $\tilde{T}$ and the value p. For example, take the case of a random variable Y distributed with χ^2, two d.o.f. statistics, having a mean of σ. If I further assume that $\tilde{T}(Y) = Y$, I then have that:

$$p = P_0(Y \geq y_c) = e^{-\frac{y_c}{\sigma}}. \tag{5.47}$$

If one observes a value $\tilde{y}$ of the random variable Y that is larger than y_c, the H_0 hypothesis is rejected. The value that is quoted in this case is the *detection significance* $\mathcal{D}$, that is,

$$\mathcal{D} = e^{-\frac{\tilde{y}}{\sigma}} \tag{5.48}$$

The H_0 hypothesis was used by Scargle (1982) for setting a false alarm probability, and by Appourchaux *et al.* (2000) to impose an upper limit on g-mode amplitudes. The method was based on the knowledge of the statistical distribution of the power spectrum of full-disc asteroseimic instruments, namely the χ^2 distribution with 2 d.o.f. For the H_1 hypothesis, I assume given statistics both for the noise and for the signal that we wish to detect, and set a level that defines the acceptance or rejection of that hypothesis.

In this example, I took as given that the test $\tilde{T}$ was known ($\tilde{T}(Y) = Y$). As a matter of fact, such a test is not obtained in an *ad hoc* manner but can be rationally derived using the Neyman-Pearson lemma. An example of the application of this lemma for deriving the test and level provided by Equation 5.47 is explained in the next section.

5.4.1.2 The Neyman-Pearson lemma

The derivation of the best test and of the levels associated with the H_0 and H_1 hypotheses is provided by the Neyman-Pearson lemma (Neyman and Pearson, 1933). This lemma is very useful for designing tests that will maximize signal detection while minimizing noise effects.

Lemma 1 $\exists \eta > 0$ such that $\Lambda(y) = \frac{L(y|H_0)}{L(y|H_1)} \leq \eta$ where $P(\Lambda(y) \leq \eta | H_0) = \alpha$

where $L(y|H_0)$ and $L(y|H_1)$ are the likelihood for each hypothesis and α is called the *power* of the test. I will show later how one can use such a lemma for a specific case that is often encountered in astronomy: the detection of a single frequency peak in a power spectrum of a star having eigenmodes with very long lifetimes. Let us assume that we observe pure noise in a power spectrum, as the statistics are χ^2 with two d.o.f. I can write the likelihood of observing a value $\tilde{y}$ as:

$$L(\tilde{y}|H_0) = \frac{1}{B} e^{-\tilde{y}/B} \tag{5.49}$$

where B is the mean noise level in the power spectrum. Next, I assume that the peak is not deterministic but that its amplitude is stochastic with an amplitude A. The likelihood for H_1 is then:

$$L(\tilde{y}|H_1) = \frac{1}{B+A} e^{-\tilde{y}/(B+A)}. \tag{5.50}$$

The likelihood ratio then can be written as:

$$\Lambda(\tilde{y}) = (1+H) e^{-\frac{\tilde{y}H}{1+H}} \tag{5.51}$$

with $H = A/B$. Then, applying the Neyman-Pearson lemma leads to:

$$\Lambda(\tilde{y}) \leq \eta \Rightarrow \tilde{y} \geqslant \eta' \tag{5.52}$$

where η' is given by solving

$$P(\tilde{y} \geqslant \eta' | H_0) = e^{-\frac{\eta'}{B}} = \alpha, \tag{5.53}$$

which justifies *a posteriori* the use of Equation 5.47. Then, I can also write the detection probability of a sine wave as:

$$P(\tilde{y} \geqslant \eta' | H_1) = e^{-\frac{\eta'}{A+B}} = \alpha^{\frac{1}{1+H}} \tag{5.54}$$

Figure 5.5 shows the result for the detection probability for sine waves stochastically excited. Such a diagram is also called the *receiver operating characteristic* (roc). It provides a very efficient way of assessing the performance of the statistical test used. In summary, the Neyman-Pearson lemma can be used for deriving in a nonarbitrary fashion the best test for accepting/rejecting H_0. With this lemma, the design of a test is therefore more systematic and less prone to improvization.

5.4.1.3 Is the world dichotomic?

Taking a decision based on the result given by a single test, for either hypothesis, could lead to errors in the decision process. For instance, the null hypothesis could be wrongly rejected when it is true (*false positive* or wrong detection), but could also be wrongly accepted while it is false (*false negative* or no detection in presence of a signal). The *false positive* results in a *Type I* error, while *the false negative* results in a *Type II* error (see Table 5.1). The ideal case would be, using the Neyman-Pearson lemma, to set a test that would minimize the occurrence of both types of errors.

TABLE 5.1. Types of error obtained for different decisions, based upon the statistical test performed, and how the error relates to the status of the H_0 hypothesis.

		Status of H_0	
		True	*False*
Decision	*Reject*	Type I	Correct
	Accept	Correct	Type II

It has been customary when applying the H_0 hypothesis to set the decision level arbitrarily at 10% (Appourchaux *et al.*, 2000). From the frequentist viewpoint, there is nothing wrong in setting *a priori* the decision level before the test is applied. There are three types of results we might obtain from applying the test:

(i) H_0 always rejected.
(ii) H_0 rejected or accepted at a level very *close* to 10%.
(iii) H_0 always accepted.

Decision (i) will lead to the mention of a *detection being statistically significant* at a level provided by the *detection significance* (e.g., from Equation 5.48). The question is to know what was detected. The next step would then be the application of a test for the H_1 hypothesis, taking into account assumptions about the detected signal, which may very likely result in the detection of signal. Decision (iii) seems straightforward, that is, noise dominates, but might one then be tempted to lower, *a posteriori*, the decision level? Decision (ii) is the more difficult borderline case, forcing us to either accept or reject H_0. Here, we might ask: are things really that clear cut? What are the chances that if we accept H_0 it is actually wrong (Type II error), or truly right if rejected (Type I error)?

These potential actions result from the application of a frequentist test trying to answer the following question: what is the likelihood of the observed data set $\tilde{y}$, given that H_0 is true or $p(\tilde{y}|H_0)$? The *detection significance* mentioned when the test rejects the H_0 hypothesis is nothing but $p(\tilde{y}|H_0)$, when actually what we want to know is the likelihood that H_0 is true given the data, that is, $p(H_0|\tilde{y})$ ($\neq p(\tilde{y}|H_0)$). The frequentist view does provide a useful answer when one can repeat the observations ad infinitum. But when

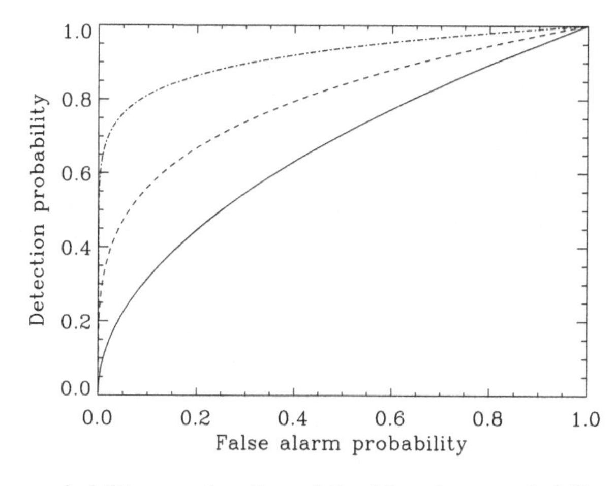

FIG. 5.5. Detection probability as a function of the false alarm probability for pure sine waves stochastically excited for various signal-to-noise ratio: 1 (continuous line), 2 (dashed line), 10 (dot-dashed line).

we have only one universe, one observation, another approach must be used based on Bayes' theorem; an approach that in principle gives access directly to $p(\mathrm{H_0}|\tilde{y})$.

5.4.2 *Bayesian hypothesis testing*

5.4.2.1 On the posterior probability

We should never forget the *two sides of the coin*: if probability (likelihood) can justify *alone* the rejection or acceptance of a hypothesis, this probability *is not* the significance that the hypothesis is rejected or accepted. The decision levels that have been discussed are related directly to a well-known controversy in the medical field concerning improper use of Fisher's p values as measures of the probability of effectiveness of a medicine or drug (Sellke *et al.*, 2001). The *detection significance* (or p value) is improperly used as the significance of the evidence against the null hypothesis. It is far from trivial at first sight to understand what is wrong with the *detection significance.* Let us recall the example I gave for a random variable Y having a χ^2 with two d.o.f. statistics. In that case, the *detection significance* is given as:

$$\mathcal{D} = \mathrm{e}^{-\frac{\tilde{y}}{\sigma}} \not\equiv P_0(Y \geq \tilde{y}). \tag{5.55}$$

The latter statement ($\not\equiv$) is fundamental. The observation is performed only once, providing a value of $\tilde{y}$ and hence the detection significance $\mathcal{D}$. But in no way does it provide the probability that the random variable is *always* above $\tilde{y}$ (or $P_0(Y \geq \tilde{y})$). It is not correct to assume that if the observation were repeated it would provide the same level $\tilde{y}$. The mistake is to ascribe a significance to a measurement performed only once, that is, not repeated, and spanning just a very small volume of the parameter space (e.g. $Y \in [\tilde{y}, \tilde{y} + \delta y]$). If one makes a measurement $\tilde{y}$ of the random variable Y that is above y_{c}, the significance of that measurement is *not* $\mathrm{e}^{-\tilde{y}/\sigma}$. In the framework of Bayesian statistics, we are not interested in the *detection significance* but in the posterior probability of the hypothesis $p(\mathrm{H_0}|\tilde{y})$, in other words, as already stated, $p(\mathrm{H_0}|\tilde{y}) \neq p(\tilde{y}|\mathrm{H_0})$. A similar description of this misunderstanding has been presented by Sturrock and Scargle (2009).

In order to derive the posterior probability $p(\mathrm{H_0}|y)$, let us first recall Bayes' theorem. The theorem of Bayes (1763) relates the probability of an event A given the occurrence of an event B to the probability of the event B given the occurrence of the event A, and the probability of occurrence of the events A and B alone.

$$P(\mathrm{A}|\mathrm{B}) = \frac{P(\mathrm{B}|\mathrm{A})P(\mathrm{A})}{P(\mathrm{B})} \tag{5.56}$$

For example, the probability of having rain given the presence of clouds is related to the probability of having clouds given the presence of rain by Equation 5.56. The term *prior probability* is given to $P(\mathrm{A})$ (probability of having rain, in general). The term *likelihood* is given to $P(\mathrm{B}|\mathrm{A})$ (probability of having clouds given the presence of rain). The term *posterior probability* is given to $P(\mathrm{A}|\mathrm{B})$ (probability of having rain given the presence of clouds). The term *normalization constant* is given to $P(\mathrm{B})$ (probability of having clouds in general).

The posterior probability of a hypothesis H, given the data D and all other prior information I, is stated as:

$$P(\mathrm{H}|\mathrm{D},\mathrm{I}) = \frac{P(\mathrm{H}|\mathrm{I})P(\mathrm{D}|\mathrm{H},\mathrm{I})}{P(\mathrm{D}|\mathrm{I})}. \tag{5.57}$$

where $P(\mathrm{H}|\mathrm{I})$ is the prior probability of H given I, otherwise known as the prior; $P(\mathrm{D}|\mathrm{I})$ is the probability of the data given I, which is usually taken as a normalizing constant; $P(\mathrm{D}|\mathrm{H},\mathrm{I})$ is the direct probability (or likelihood) of obtaining the data given H and I. Berger and Sellke (1987) obtained, using Bayes' theorem, $p(\mathrm{H_0}|\tilde{y})$ with respect to $p(\tilde{y}|\mathrm{H_0})$

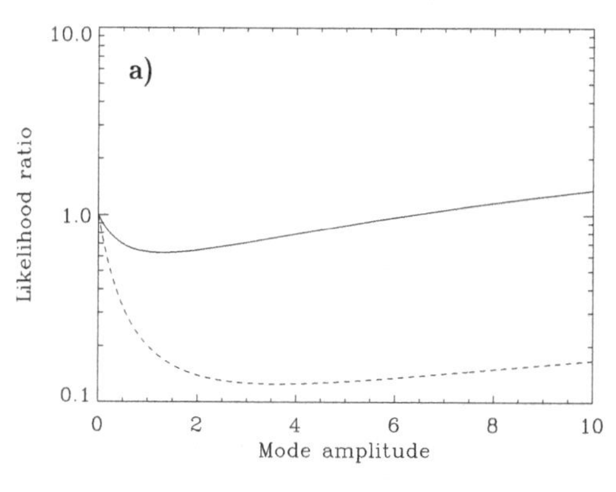

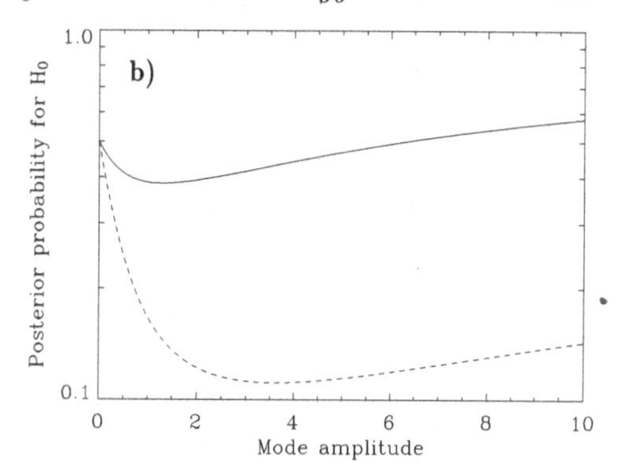

FIG. 5.6. *(a)* Likelihood ratio $\mathcal{L}$ as a function of the mode amplitude for *detection significances* of 10 % (solid line), and of 1 % (dashed line); the noise is set to unity. *(b)* The posterior probability of $\mathrm{H_0}$ as a function of mode amplitude for *detection significances* of 10 % (solid line), and of 1 % (dashed line) (from Equation 5.78).

and $p(\tilde{y}|\mathrm{H_1})$, where $\mathrm{H_1}$ is the alternative hypothesis:

$$p(\mathrm{H_0}|\tilde{y}) = \frac{p(\mathrm{H_0})p(\tilde{y}|\mathrm{H_0})}{p(\mathrm{H_0})p(\tilde{y}|\mathrm{H_0}) + p(\mathrm{H_1})p(\tilde{y}|\mathrm{H_1})}. \tag{5.58}$$

I set $p_0 = p(\mathrm{H_0})$, and since we have $p(\mathrm{H_1}) = 1 - p_0$, they finally obtained:

$$p(\mathrm{H_0}|\tilde{y}) = \left(1 + \frac{(1-p_0)}{p_0}\mathcal{L}\right)^{-1}, \tag{5.59}$$

with $\mathcal{L}$ being the likelihood ratio defined as:

$$\mathcal{L} = \frac{p(\tilde{y}|\mathrm{H_1})}{p(\tilde{y}|\mathrm{H_0})}. \tag{5.60}$$

Here, $p(\mathrm{H_0}|\tilde{y})$ is the so-called posterior probability of $\mathrm{H_0}$ given the observed data $\tilde{y}$. Naturally, there is no way to favor $\mathrm{H_0}$ over $\mathrm{H_1}$, or vice versa, otherwise our own prejudice would most likely be confirmed by the test, that is, $p_0 = 0.5$. Subsequently, Berger *et al.* (1997) recommended to report the following when performing hypothesis testing:

$$\text{if } \mathcal{L} > 1, \text{ reject } \mathrm{H_0} \text{ and report } p(\mathrm{H_0}|\tilde{y}) = \frac{1}{1+\mathcal{L}}, \tag{5.61}$$

$$\text{if } \mathcal{L} \leq 1, \text{ accept } \mathrm{H_0} \text{ and report } p(\mathrm{H_1}|\tilde{y}) = \frac{1}{1+\mathcal{L}^{-1}}. \tag{5.62}$$

The advantage of such a presentation is that even for a borderline case, say, when the ratios above are close to unity, it is clear that there is only a 50 % chance that the $\mathrm{H_0}$ hypothesis is wrongly accepted, or wrongly rejected. This presentation is more honest and better encapsulates human judgement and prejudice.

5.4.2.2 Example of posterior probability

Using the example given in Section 5.4.1.1 for the detection of sine waves, we can derive $p(\mathrm{H_0}|\tilde{y})$ using Equation 5.51 as:

$$p(\mathrm{H_0}|\tilde{y}) = \left(1 + \frac{1}{1+H}p^{-H/(1+H)}\right)^{-1} \tag{5.63}$$

where $p = \mathrm{e}^{-\tilde{y}/B}$ is the *detection significance.* Figure 5.6 shows the results for two different *detection significances.* When the *detection significance* is 10 %, the likelihood ratio can be greater than unity for large values of the mode amplitude, leading to the acceptance of the null hypothesis. This is rather paradoxical, that is, that large mode amplitude can lead to the rejection of the alternative hypothesis. To resolve the paradox, we note that the posterior probability of H_0 is in any case never lower than 40%, or the posterior probability of H_1 is never higher than 60%. This implies that both hypotheses are equally likely when the *detection significance* is as low as 10 %. In other words, when we set, *a priori*, a large mode amplitude and get a low *detection significance*, the alternative hypothesis is as likely as the null hypothesis. In other words, the assumption about a large mode amplitude is *not supported by the data.*

The main conclusion to be drawn from this calculation is that the *detection significance* should be set much lower than 10 % in order to avoid misinterpretation of the result. For example, with a *detection significance* of 1 %, the posterior probability for H_0 can fall to 10 % when the signal-to-noise ratio is above unity. Sellke *et al.* (2001) showed that the posterior probability can never be lower than the lower bound:

$$p(\mathrm{H}_0|x) \geq \left(1 - \frac{1}{ep\ln p}\right)^{-1}. \tag{5.64}$$

The reader may verify for him- or herself that this lower bound is effectively reached for Equation 5.63. In the case, when the amplitude of the mode A is not known, one needs to set, *a priori*, the value for the likely range of amplitudes. In the case of a uniform prior, the posterior probability $p(\mathrm{H}_0|\tilde{x})$ then does reach a minimum that is higher than the lower bound of Equation 5.64 (Appourchaux *et al.*, 2009).

In summary, the significance level should not be used for justifying a detection (or a nondetection). Instead, I recommend using the prescription of Berger *et al.* (1997), as given by Equations 5.61 and 5.62 and to specify the alternative hypothesis H_1.

5.4.2.3 On the choice of the prior probability

One important question when applying Bayesian statistics is what value should the prior probability of the hypothesis H_0, that is, p_0, take? We define the prior probability as the probability that the H_0 hypothesis is correct. The probability that the alternative H_1 hypothesis is correct can then be defined as $p(\mathrm{H}_1) = 1 - p_0$. It is common to set $p_0 = 0.5$ so as to avoid prejudicing one hypothesis over the other. Would we expect the probability that H_1 and H_0 are true to be the same in all instances? Since Bayesian statistics requires *a priori* knowledge, it is possible to use our knowledge of physics/astrophysics to tell us which hypothesis is more likely to be true in a given circumstance.

5.4.3 *Parameter estimation*

The previous section on *hypothesis testing* is really the prerequisite when one wants to assess if there is a signal sought in the observation. Unfortunately, knowing that a signal is present does not provide any pertinent information for doing physics or astrophysics. This is the goal of *parameter estimation.*

Parameter estimation is a vast subject in statistics. In this section, I will introduce the estimations that are the most commonly used in astrophysics. The estimations described hereafter are also related to the frequentist and Bayesian world. As we will see, these estimations are not so foreign from each other. The frequentist estimation can be used when the signal-to-noise ratio is high, while the Bayesian estimation is more useful (but more time consuming) when the the signal-to-noise ratio is low.

5.4.4 *Maximum likelihood estimation*

As shown in the *Historical Overview* section, Gauss (1809) introduced the concept of *maximum likelihood estimation*, or MLE. The aim of MLE is to find the set of parameters that maximize the likelihood of the observed event; this is a *point-like* estimation. Having observed a random variable x with a probability distribution $p(x, \boldsymbol{\lambda})$, where $\boldsymbol{\lambda}$ is a vector of p parameters describing the model behind the random variable x, the likelihood L of N observations of x is given by:

$$L(x, N, \boldsymbol{\lambda}) = \prod_{k=1}^{N} f(x_k, \boldsymbol{\lambda}) \tag{5.65}$$

where the product implicitly expresses the fact that all x_k are *independent* of each other. Usually, we define the logarithmic likelihood function ℓ as

$$\ell(x, N, \boldsymbol{\lambda}) = \ln L(x, \boldsymbol{\lambda}) = -\sum_{k=1}^{N} \ln f(x_k, \boldsymbol{\lambda}). \tag{5.66}$$

The estimate of $\boldsymbol{\lambda}$ is derived from the maximization of the likelihood as given by Equation 5.66 such that we have

$$\tilde{\boldsymbol{\lambda}} = \max_{\boldsymbol{\lambda}} \ell(x, N, \boldsymbol{\lambda}). \tag{5.67}$$

Such an estimator has several interesting properties in the limit of very large sample ($N \to \infty$) which are:

- MLE are asymptotically unbiased,
- MLE are of minimum variance,
- MLE are asymptotically normal.

The first property implies that:

$$\lim_{N \to \infty} E(\tilde{\boldsymbol{\lambda}}) = \boldsymbol{\lambda_0} \tag{5.68}$$

where $\boldsymbol{\lambda_0}$ is the true value of the model. The second property implies that no other asymptotically unbiased estimator has lower variance. Using the Cramer-Rao theorem (Cramer, 1946; Rao, 1945), this can be rewritten as:

$$\mathrm{cov}\left[\tilde{\boldsymbol{\lambda}}\right] \geq \frac{1}{I(\boldsymbol{\lambda_0})} \tag{5.69}$$

where $I(\boldsymbol{\lambda})$ is the Fisher information matrix whose elements are given by:

$$I_{ij} = \mathrm{E}\left[\frac{\partial^2 \ell(x, \boldsymbol{\lambda})}{\partial \lambda_i \partial \lambda_j}\right]. \tag{5.70}$$

Asymptotically, we also have:

$$\lim_{N \to \infty} \mathrm{cov}\left[\tilde{\boldsymbol{\lambda}}\right] = \frac{1}{I(\boldsymbol{\lambda_0})}. \tag{5.71}$$

Finally, the third property, regarding the estimator being asymptotically normal, can be expressed as:

$$p(\boldsymbol{\lambda}) = \mathcal{N}\left(\boldsymbol{\lambda_0}, \frac{1}{I}\right) \tag{5.72}$$

where $\mathcal{N}$ is the normal distribution. This equation is usually used for providing the statistical distribution of $\tilde{\boldsymbol{\lambda}}$ as:

$$p(\boldsymbol{\lambda}) \approx \mathcal{N}\left(\tilde{\boldsymbol{\lambda}}, \frac{1}{H(\tilde{\boldsymbol{\lambda}})}\right) \tag{5.73}$$

where H is the so-called Hessian matrix whose elements are derived from:

$$H_{ij} = \left[\frac{\partial^2 \ell(x, \tilde{\boldsymbol{\lambda}})}{\partial \lambda_i \partial \lambda_j}\right] \tag{5.74}$$

with the property given by the Cramer-Rao theorem as:

$$\frac{1}{H(\tilde{\boldsymbol{\lambda}})} \geq \frac{1}{I(\boldsymbol{\lambda}_0)}. \tag{5.75}$$

Equation 5.74 is used when computing the so-called formal error bars on $\tilde{\boldsymbol{\lambda}}$; as a matter of fact, according to the Cramer-Rao theorem, Equation 5.74 gives only a lower bound to the error bars.

5.4.4.1 Significance of estimates

When one uses least squares for fitting data, one can test the significance of its fitted parameters using the so-called R test (Frieden, 1983). For MLE, a useful test can be used: the likelihood ratio test. This method requires maximizing first the likelihood $\mathrm{e}^{-\ell(\omega_p)}$ of a given event where p parameters are used to describe the statistical model of the event. Then, if one wants to describe the same event with n additional parameters, the likelihood $\mathrm{e}^{-\ell(\Omega_{p+n})}$ will be maximized. The likelihood ratio test consists in determining the ratio of the two likelihoods. Using the logarithmic likelihood, we can define the ratio Λ as:

$$\ln(\Lambda) = \ell(\Omega_{p+n}) - \ell(\omega_p). \tag{5.76}$$

If Λ is close to 1, it means that there is no improvement in the maximized likelihood and that the additional parameters are not significant. On the other hand, if $\Lambda \ll 1$, it means that $\ell(\Omega_{p+n}) \ll \ell(\omega_p)$ and that the additional parameters are very significant. In order to define a significance for the n additional parameters, we need to know the statistics of $\ln(\Lambda)$ under the null hypothesis, that is, when the n additional parameters are not needed to describe the model. For this null hypothesis, Wilks (1938) showed that for large sample size the distribution of $-2\ln\Lambda$ tends to the $\chi^2(n)$ distribution.

5.4.4.2 Calibration of error bars

Before applying MLE to real data, it is always advisable to test the power of this approach on synthetic data, that is, to perform Monte-Carlo simulations. They are not merely for playing games; these simulations are real tools for understanding what we fit and how we fit it. Assuming that the statistics of the data are known, performing Monte-Carlo is useful for the following reasons:

- Assessing the model of the data.
- Assessing the statistical distribution of the parameters of the model.
- Assessing the precision on the fitted parameters of the models.

The parameters derived by the MLE should have the desirable properties of having a normal distribution (as described earlier); if not, we advise applying a change of variable on the fitted parameters ($\log x$, for instance). A normal distribution is necessary to derive meaningful error bars; this is the assumption behind Equation 5.74. In order to be able to derive a good estimate of the error bars using one realization, the standard deviation of a large sample of fitted parameters should be equal to the mean of formal errors return by the fit, that is, this is the approximation of Equation 5.72 by Equation 5.73. In other words, we should have the following approximation for the inverse of the covariance of the parameters:

$$H(\tilde{\boldsymbol{\lambda}}) \approx I(\boldsymbol{\lambda}_0) \tag{5.77}$$

where H is related to the formal error bars and I is related to the asymptotic error bars. This calibration as expressed in this equation is key to deriving meaningful error bars. I advise the reader to use such a calibration procedure for checking the formal error bars derived from software codes fitting function using least squares. The formal error bars derived by a single noise realization are only *a lower limit* to the *real* error bars. This lower limit is never reached when, for instance, the signal-to-noise ratio is too low. In this case, we are very far from the asymptotic behavior. This case is covered by the Bayesian approach to parameter estimation.

5.4.5 *Bayesian parameter estimation*

Parameter estimation can also be done using Bayes' theorem. In this case, this is the so-called *Bayesian inference*. Using the same notation as before, I can express using Bayes' theorem the probability distribution as:

$$p(\boldsymbol{\lambda}|x,\mathrm{I}) = \frac{p(\boldsymbol{\lambda}|\mathrm{I})p(x|\boldsymbol{\lambda},\mathrm{I})}{p(x|\mathrm{I})} \tag{5.78}$$

where $\boldsymbol{\lambda}$ are the observables for which I seek the *posterior probability*, x is the observed data set, and I is the information. The *prior probability* of the observables is given by $p(\boldsymbol{\lambda}|\mathrm{I})$: this is the way to quantify our *belief* about what I seek. The likelihood is given by $p(x|\boldsymbol{\lambda},\mathrm{I})$, which is exactly the $L(x,\boldsymbol{\lambda})$ of the previous section. Therefore the *frequentist* approach is related to the *Bayesian* approach simply by the *frequentist* likelihood, the *prior* probability, and the normalization factor $p(x|\mathrm{I})$. The main advantage of the Bayesian approach is that the posterior probability $p(\boldsymbol{\lambda}|x,\mathrm{I})$ is directly accessible, whereas for the frequentist approach, only the location of the maximum of the likelihood is known. In this approach, there is no direct visibility of the parameter probability distribution but only an approximation provided by Equation 5.73. This is why the frequentist approach is a point-like estimation, whereas the Bayesian approach is more global. For instance, the power of the Bayesian approach is such that it provides the full posterior probability, which may not be necessarily a normal distribution but could be the sum of many normal distributions due to many local minima. In that case, only the posterior probability can provide a correct assessment of the statistics of the derived parameters.

5.4.5.1 Posterior probability estimation

The main difficulty in Bayesian inference is to derive the *posterior* probability. If the derivation of Equation 5.78 is analytical, then parameter estimation can easily be done (see an example in the Application to asteroseismology section). When this is not possible, the easiest method is to compute the posterior probability by using a random walk algorithm that will provide the posterior probability from a representative samples of the $\boldsymbol{\lambda}$. A famous example of such a procedure is derived from the so-called Metropolis-Hastings algorithm (MH) (Metropolis *et al.*, 1953; Hastings, 1970). Let us see how this algorithm works in practice for a probability distribution $p(\boldsymbol{\lambda})$ for which we want to have a representative sample. We start from a given point $\boldsymbol{\lambda}^{(t)}$ and from a known probability distribution Q. We draw at random from the probability distribution a value $\boldsymbol{\lambda}'$ knowing $\boldsymbol{\lambda}^{(t)}$ and we compute the following ratio:

$$r = \frac{p(\boldsymbol{\lambda}')Q(\boldsymbol{\lambda}^{(t)}|\boldsymbol{\lambda}')}{p(\boldsymbol{\lambda}^{(t)})Q(\boldsymbol{\lambda}'|\boldsymbol{\lambda}^{(t)})} \tag{5.79}$$

then the new proposed value $\boldsymbol{\lambda}'$ is accepted or rejected following this scheme:

$$\begin{array}{lll} \text{If} & r \geq 1 & \text{then } \boldsymbol{\lambda}^{(t+1)} = \boldsymbol{\lambda}' \\ \text{If} & r < 1 & \text{then} \\ & \boldsymbol{\lambda}^{(t+1)} = \boldsymbol{\lambda}' & \text{if } r < \alpha \\ & \boldsymbol{\lambda}^{(t+1)} = \boldsymbol{\lambda}^{(t)} & \text{if } r \geq \alpha \end{array} \tag{5.80}$$

where α is a random number drawn from a uniform distribution. Asymptotically, the $\boldsymbol{\lambda}^{(t)}$ will then tend to have the probability distribution $p(\boldsymbol{\lambda})$. The benefit of this algorithm is that the computation of the normalization factor in the denominator of Equation 5.78 is not needed. Another obvious benefit is that very complex probability distributions can be derived, thereby providing the potential correlations between the various parameters λ_i.

The difficulty in the use of the MH is not in the algorithm itself, which is quite easy to implement, but in the proper choice of the input distribution Q. This is a vast subject that goes far beyond this course. The reader will find in Gregory (2005) what is required for delving into the subject of obtaining the posterior probability distribution using various techniques: Gibbs sampling, thermal annealing, convergence, and so forth.

5.4.5.2 Mean, rms, and moment estimation

As soon as the *posterior probability* is known, we can derive an estimate of the moment k of the parameter λ_i by deriving first the posterior probability distribution of λ_i only or $p(\lambda_i|x, \mathrm{I})$. This is done by integrating (or marginalizing) over the so-called *nuisance* parameters as follows:

$$p(\lambda_i|x, \mathrm{I}) = \int_{\Omega_i} p(\boldsymbol{\lambda}|x, \mathrm{I}) \mathrm{d}\lambda_n \tag{5.81}$$

where $\lambda_n \in \Omega_i$ with $n \neq i$. Then, we can compute the moments of the posterior probability by writing:

$$< \lambda_i^k > = \int_{\lambda_i} \lambda_i^k p(\lambda_i|x, \mathrm{I}) \mathrm{d}\lambda_i. \tag{5.82}$$

The first and second moments provide the mean value and rms deviations of λ_i as

$$\tilde{\lambda}_i = < \lambda_i^1 > \tag{5.83}$$

$$\sigma_{\lambda_i} = \sqrt{< \lambda_i^2 > - \tilde{\lambda}_i^2}. \tag{5.84}$$

The use of these two moments is enough to describe a normal distribution, as all the other moments can be derived from these two. Here, I note that the Cramer-Rao criteria is also relevant. It means that the rms value just derived is bounded as follows:

$$\sigma_{\lambda_i}^2 \geq H^{-1}(\tilde{\lambda}_i). \tag{5.85}$$

As a result, the error bars derived from the Bayesian approach are larger than those returned using MLE. It means that, contrary to popular belief, the Bayesian approach is more conservative than MLE.

If the posterior probability is not normal, higher moments could be quoted. It is rather impractical to quote all the moments higher than 2. Instead, we can use the value of the median and of the percentiles. We can define, for a random variable x with a probability distribution p, two values x_1 and x_2 such that for a given percentile q we have:

$$q = \int_{-\infty}^{x_1} p(x) \mathrm{d}x = \int_{x_2}^{+\infty} p(x) \mathrm{d}x. \tag{5.86}$$

When $q = 50\%$, we have $x_1 = x_2$, thereby providing the median. Usually, for a Gaussian distribution, the 1-σ and 2-σ values provide a percentile of 15.9% and 2.3%, respectively, that is, 68.2% and 95.4% of the values are in the range defined by x_1 and x_2. The use of the four percentiles and the median can be shown in a so-called box plot. The main advantage of using percentiles is that they are invariant under a change of variable. In other words, when making a change of variable from x to $g(x)$ in Equation 5.86, the x_1 and x_2 returned are unaffected by the transform (provided that g is a monotonic function of x).

In practice, when the analytical integration cannot be done, Equation 5.82 is computed using the MH algorithm we have mentioned, such that we have:

$$< \lambda_i^k > = \frac{1}{N_t} \sum_t \left(\lambda_i^{(t)}\right)^k \tag{5.87}$$

where N_t is the total number of samples computed to be returned by the MH algorithm. Then, the median and the percentiles are computed by sorting the values of $\lambda_i^{(t)}$. Examples of such results can be found in Benomar *et al.* (2009b).

5.4.5.3 Role of the prior

The *prior probability* expresses what we believe we know (or not) about the parameters λ_i. The choice of the prior is related to the amount of information at our disposal. The most obvious *prior probability* of the parameters λ_i of interest is the one that is uniformly distributed over some range; this is an *uninformative* prior. The role of the prior and its impact on the *posterior probability* should ideally be as small as possible. Objective uninformative priors are derived using the procedure described by Jeffreys (1946), related to the calculation of the determinant of the Fisher matrix (Fisher, 1925). An informative prior could be, for example, a Gaussian distribution of a given parameter λ_i. Priors are not always *proper* in the sense that they are not always related to a *proper* probability distribution, that is, providing finite moments. The $1/\sigma$ prior for the unknown rms value of a parameter is a specific example of an improper prior. An extensive discussion on the impact of the prior in a Bayesian framework has been well developed by Jaynes (1987).

5.4.5.4 Significance and model comparison

The power of the Bayesian approach is also to be able to compare different models. The approach used by frequentists using the likelihood ratio test outlined in the previous section is quite similar to what is called the Bayesian odd ratio. Let us assume that one wants to compare between different model M_n. The odd ratios between any two sets of models are:

$$\mathrm{O}_{n/m} = \frac{p(\mathrm{M}_n|x, I)}{p(\mathrm{M}_m|x, I)} = \frac{p(\mathrm{M}_n|I)}{p(\mathrm{M}_m|I)} \frac{p(x|\mathrm{M}_n, I)}{p(x|\mathrm{M}_m, I)}. \tag{5.88}$$

The second part of the equation was derived from Bayes' theorem. The second fraction closely resembles the likelihood ratio presented earlier, but it is the product of the ratio of the *prior* model probabilities by the ratio of the *global* likelihood. It is termed *global* because these probabilities are not a point-like estimate (as in the frequentist approach) but an integration over all the possible values of the estimated parameters λ_i. The global likelihood $p(x|\mathrm{M}_m, I)$ is written as:

$$p(x|\mathrm{M}_m, I) = \int_\Omega p(\boldsymbol{\lambda}|\mathrm{M}_m, I) p(x|\boldsymbol{\lambda}, \mathrm{M}_m, I) \mathrm{d}\boldsymbol{\lambda} \tag{5.89}$$

where $p(\boldsymbol{\lambda}|\mathrm{M}_m, I)$ is the prior probability, and $p(x|\boldsymbol{\lambda}, \mathrm{M}_m, I)$ is the likelihood. The computation of the global likelihood is rather difficult but can be done using the MH algorithm under parallel tempering. The integration of Equation 5.89 can then be done using the so-called thermodynamic integration (Gelman and Meng, 1998). Applications of this kind of integration can be found in Gregory (2005).

It is also useful to express the posterior probability of each model $p(\mathrm{M}_n|x, I)$ as follows:

$$p(\mathrm{M}_n|x, I) = \frac{p(\mathrm{M}_n|I) p(x|\mathrm{M}_n, I)}{\sum p(\mathrm{M}_m|I) p(x|\mathrm{M}_m, I)}. \tag{5.90}$$

Usually, if the model comparison is an *objective* Bayesian analysis, then all prior model probabilities are equal $(p(x|\mathrm{M}_m, I) = p(x|\mathrm{M}_k, I), \forall\ m, k)$. This assumption can be used when the models are strictly different and are not *nested*. Here, *nested* means that a child model relies on a parent model, the former having more parameters describing the model than the latter. Most of the models we used are indeed *nested*; they differ from each other by a few parameters. In this case, it results in the so-called model *multiplicity* that must be taken into account under the *subjective* Bayesian approach. Under this approach, it is possible to have model probabilities that differ from each other ($p(\mathrm{M}_m|I) \neq p(\mathrm{M}_k|I)$). A byproduct of the *subjective* Bayesian approach is also to provide an estimate of these $p(\mathrm{M}_m|I)$ based on the data. Model multiplicity has just started to be taken into account in model comparison (Scott and Berger, 2010). Because this is very recent, I advise the reader to inform him- or herself on whether this approach can be useful for model comparison.

5.5 Application to asteroseismology

In the two previous sections, I laid down the foundations for applying harmonic analysis and statistics to astrophysics. In particular, the field of asteroseismology is extremely relevant for these applications. Stars have been known to oscillate since, at least, the sixteenth century when David Fabricius found that o Ceti (Mira) was variable. Since then, many others stars such as β Cephei, δ Scuti, Cepheids, γ Dor, and solar-like stars have been found to oscillate (see Christensen-Dalsgaard, 2004, and references therein). Stellar oscillations are mainly excited by an opacity-driven mechanism (κ mechanism) and by turbulence occurring in convection zones (Gautschy and Saio, 1995, 1996). Broadly speaking, the world of stellar oscillations can be divided in two categories:

- periodic pulsations having a weakly time-dependent amplitude, and
- oscillatory eigenmodes stochastically excited.

The first type results from overstable oscillations in stars being driven by the κ mechanism, whose amplitudes are limited by a nonlinear mechanism. The functional form of the variation of the luminosity with time is then periodic but not necessarily sinusoidal. In that case, the amplitude of the oscillations varies more slowly that the periods of the oscillations.

The second type results from modes being randomly excited in solar-like stars whose amplitudes are damped by various mechanisms (Houdek *et al.*, 1999). In that case, the amplitude of the oscillations varies on time scales shorter than the periods of oscillations. There are stars for which the two types of oscillations coexist (Belkacem *et al.*, 2009, 2010).

When applying various tools for obtaining the frequencies of the oscillation (frequencies that describe the internal structure of the star), then one should ask oneself which type of functional form the stellar oscillation will have. Typically, there are three types of functional forms:

- periodic non-sinusoidal,
- sinusoidal, and
- harmonic oscillator stochastically excited.

These functional forms being periodic, the Fourier transform is the obvious choice for time series analysis. As for the last functional form, the random nature of the excitation imposes the application of a proper statistical treatment of the time series and its associated power spectrum. Hereafter, I will treat the two most common cases encountered in asteroseismology: classical pulsators (periodic non-sinusoidal function) and solar-like oscillators (harmonic oscillators).

5.5.1 *Classical pulsators*

For stars having luminosity variations whose functional forms are sinusoidal, the common practice is to use the Fourier transform described in the previous sections. The first application of statistics and Fourier transform was for finding periodicities in earthquakes due to Schuster (1897), which has been termed the *Schuster periodogram.*

For variable stars (or classical pulsators), the first application of the periodogram is attributed to Wehlau and Leung (1964). Since that date, the analysis of time series has been evolving to take into account various aspects related to the presence of gaps and the large dynamic range in the amplitude of the periodicities. One of the problems encountered when observing stars from the ground is that the periodic gaps in the observation (due to the day–night cycle) introduce aliases. The presence of these gaps produces then spurious peaks located on either side of the main peak located at multiple of $\pm$ 11.57 μHz (1/24 hr). One solution for taking into account such a frequency response is to apply the CLEAN algorithm (Roberts *et al.*, 1987), which is used in radioastronomy for aperture synthesis (Högbom, 1974). Another approach is to use a combinaison of the CLEAN algorithm and of prewhitening, which consists in removing the signal of largest amplitude in the time series (after having being detected) and then recomputing the periodogram; the procedure iterates until there is no large signal detected in the periodogram (Belmonte *et al.*, 1991). This technique is now the most commonly used for classical pulsators.

5.5.2 *Spectral analysis revisited*

5.5.2.1 Single sine wave

I already touched on the Fourier analysis of pure sine waves using the frequentist framework (application of least squares). Here, I shall briefly revisit what can be done when using a Bayesian approach. Bretthorst (1988), using a Bayesian approach to the analysis of the time series of a pure sine wave sampled regularly and embedded in noise having a Gaussian distribution, demonstrated that the *posterior probability* for the frequency ν of the sine wave can be written as:

$$\ln P(\nu|x,\sigma,\mathrm{I}) = \frac{C(\nu)}{\sigma^2} + \cdots \tag{5.91}$$

where x are the data (x_i taken at time t_i), σ is the rms value of the noise assumed to be known, and $C(\nu)$ is the Schuster periodogram given by:

$$C(\nu) = \frac{1}{N}\left|\sum_{i=1}^{N} d_i \mathrm{e}^{\mathrm{i}2\pi\nu t_i}\right|^2. \tag{5.92}$$

This is nothing less than the power spectrum. For a pure sine wave with frequency ν_0, the periodogram has a maximum at that frequency. Using the posterior probability for ν, we then have the first moment of the frequency as:

$$\tilde{\nu} = \int P(\nu|x,\sigma,\mathrm{I})\mathrm{d}\nu = \nu_0 \tag{5.93}$$

Following this approach, Jaynes (1987) demonstrated using Equation 5.92 that for a pure sine wave of amplitude A, the rms error on the frequency ν_0 is given by:

$$\delta\nu = \frac{\sqrt{6}}{\pi}\frac{\sigma}{A}\frac{\sqrt{\Delta t}}{T^{\frac{3}{2}}} \tag{5.94}$$

where T is the observing time and Δt is the sampling time. This formula is the same as given by Cuypers (1987) and derived by Koen (1999). The frequency precision is inversely proportional to the signal-to-noise ratio computed in the time domain. This equation also shows that the Rayleigh criterion ($1/T$) for frequency resolution is very pessimistic compared with the precision with which frequencies can be measured.

5.5.2.2 *Many sine waves*

The approach outlined above linking the posterior probability to the Fourier transform justifies *a posteriori* the use of that transform: "The highest peak in the discrete Fourier transform is an optimal frequency estimator for a data set which contains a single harmonic frequency in the presence of Gaussian white noise" (Bretthorst, 1988). Fortunately, for several harmonic signals whose frequencies (ν_i) are far from each other, the *naive* use of Fourier can be broken down in the application of the preceding section as many times as required or:

$$\ln P(\nu_1, \ldots \nu_r|x, \sigma, \mathrm{I}) = \left(\sum_{j=1}^{r} \frac{C(\nu_j)}{\sigma^2}\right) + \cdots \tag{5.95}$$

This justifies the use of the periodogram for finding harmonic signal in a time series (Bretthorst, 1988).

5.5.2.3 *Periodic signals*

When the signal is periodic but not sinusoidal, the decomposition using the periodogram will provide the fundamental ν_f of the period and spurious frequencies at $n\nu_f$. The presence of these spurious frequencies can become difficult to handle when there are a lot of frequencies such as in δ Scuti (Poretti *et al.*, 2009). The periodic signal can also be the result of nonlinear saturation, which cannot be easily described in terms of harmonic signals (Yoachim *et al.*, 2009). As mentioned above, prewhitening together with the use of the CLEAN algorithm is a possible solution for analyzing such periodic signals. The analysis can also be performed using other techniques such as principal components analysis (Tanvir *et al.*, 2005).

I showed in a previous section how one could analyze data that are not equally sampled in time using the Lomb-Scargle periodogram. The LS periodogram has been extended not only to a decaying sinusoid (Bretthorst, 2001a) but also to periodic functions in general (Bretthorst, 2001b). This revisitation of spectral analysis by Bretthorst is extremely useful when one wants to understand how to apply a Bayesian analysis to time series. Based on the work of Bretthorst (2001b), a possible application would be to model the amplitude limitation δ Scuti to obtain the functional forms of the periodic signal. A different functional form from a sinusoid could be used as an advantage for disentangling the various possible combinations of frequencies (Poretti *et al.*, 2009), either due to the gaps, the periodic signal, or the physics of the stars. This is an avenue yet to be explored.

5.5.3 *Solar-like oscillations*

The analysis of solar-like oscillations is slightly more complicated because of the random excitation of the modes. In order to apply the Fourier transform and use statistical tools for extracting mode frequencies and mode parameters for such stars, one has to understand how the functional form of the mode amplitude is related to the harmonic oscillator being randomly excited.

5.5.4 *Randomly excited harmonic oscillations*

It is well known that pressure modes (or p modes) are stochastically excited oscillators (Kumar *et al.*, 1988). The source of excitation lies in the many granules covering the star (turbulent convection; Houdek *et al.*, 1999; Samadi and Goupil, 2001). The modes are assumed to be independently excited (Kumar *et al.*, 1988) because the eigenfunction of the modes is primarily radial and nearly independent of degree in the upper convection zone, where the modes are excited. The eigenmodes are harmonic oscillators being intrinsically damped, and excited through a forcing function F. The differential equation of such a damped harmonic oscillator is written as:

$$\frac{\mathrm{d}^2 x_{\mathrm{osc}}}{\mathrm{d}t^2} + 2\pi\gamma \frac{\mathrm{d}x_{\mathrm{osc}}}{\mathrm{d}t} + (2\pi)^2 \nu_0^2 x_{\mathrm{osc}} = F(t) \tag{5.96}$$

where t is the time, x_{osc} is the displacement, γ is the damping term related to the mode linewidth, ν_0 is the frequency of the mode, and $F(t)$ is the forcing function. Equation 5.96 is also the expression of an auto-regressive (AR) process which in this case is a stationary process. From this equation, the Fourier transform of x can be written as:

$$\tilde{x}_{\mathrm{osc}}(\nu) = \frac{\tilde{F}(\nu)}{(2\pi)^2(\nu_0^2 - \nu^2 + i\gamma\nu)} \tag{5.97}$$

where $\tilde{x}_{\mathrm{osc}}(\nu)$ and $\tilde{F}(\nu)$ are the Fourier transform of $x_{\mathrm{osc}}(t)$ and $F(t)$. From the large number of granules, it can be derived that the forcing function is normally distributed. Therefore, the two components (the real and imaginary parts) of the Fourier transform of the forcing function are also normally distributed (see Section 5.3.7). Therefore, for the harmonic oscillator, each component of $\tilde{x}_{\mathrm{osc}}(\nu)$ is normally distributed with a mean of zero, and the same variance. The variance is related to the spectral density given by:

$$S_{\mathrm{osc}}(\nu) = \frac{S_F(\nu)}{(2\pi)^4\left[\left(\nu_0^2 - \nu^2\right)^2 + \nu^2\gamma^2\right]}. \tag{5.98}$$

The spectral density of $\tilde{x}_{\mathrm{osc}}(\nu)$ has then a χ^2 with two degrees of freedom statistics. For such statistics, I outline that the variance is the same as the mean. Equation 5.98 is the p-mode profile that is usually approximated by a Lorentzian profile when $\nu \approx \nu_0$ as:

$$S_{\mathrm{osc}}(\nu) \approx \frac{S_F(\nu)}{\gamma^2(2\pi)^4\nu_0^2} \frac{1}{1+x^2} \tag{5.99}$$

with $x = 2(\nu - \nu_0)/\gamma$. An asymmetry effect can also be introduced in the profile of Equation 5.99. The asymmetry effect was first detected with instruments making an image of the Sun (Duvall *et al.*, 1993). The asymmetry is due to the location of the excitation of the modes with respect to the resonant cavity. There is a direct analogy with a source being placed inside a Fabry-Perot cavity (Duvall *et al.*, 1993). The asymmetry was then detected by Toutain *et al.* (1997) with instruments observing the Sun as a star, the sign of the asymmetry depending upon the observables (intensity of velocity). The empirical theoretical explanation for the different asymmetry in intensity with respect to velocity was given by Nigam *et al.* (1998). It is related to the different correlation between the background and the modes in velocity and in intensity, which were studied by Severino *et al.* (2001). The theoretical framework provided then a simple solution for the expression of the mode profile with asymmetry (Nigam and Kosovichev, 1998), given by:

$$S_{\mathrm{osc}}(\nu) = H\frac{1+2Bx}{1+x^2} + B^2 \tag{5.100}$$

where B is the asymmetry effect and H is the mode height. The effect of the linear slope is to change the sign of asymmetry when the sign of B changes. When B is positive (negative), there is more power at high (low) frequency. Therefore, if the asymmetry is not taken into account, the sign of B will affect the true location of the mode frequency differently. This effect, if not taken into account, may provide large systematic errors when comparing different observables thereby providing potential problems for the inferred model of the star.

5.5.5 *Random nonharmonic field*

Another source contributing to the observed power spectrum is the noise generated by the star itself. It is quite customary to have the stellar noise increasing at low frequencies.

As a matter of fact, the noise observed is not limited to stars but more related to an intrinsic behavior encountered in many physical phenomena. This so-called $1/f$ noise is so ubiquitous that it is found in almost any physical measurements and also in stars. The $1/f$ noise appears in many electrical applications and other applications (Keshner, 1982). Although white noise is known not to have any memory (autocovariance is the Dirac distribution), the $1/f$ noise possesses some memory. Mathematically, AR models have also these properties. Random processes following AR models have the desired properties of having memory, thereby producing $1/f$-like power spectra. This fact was used by Harvey (1985) in deriving the solar background noise spectrum for instruments observing solar radial velocities. The spectrum is derived from a superposition of four components related to solar activity and to three different types of granulation having different lifetimes. In Harvey (1985), the power law used had a -2 slope; it was later refined to be arbitrary b (Harvey *et al.*, 1993) such that we have for the spectral density of background noise $x_n(t)$:

$$S_{\rm n}(\nu) = \sum_i \frac{H_i}{1 + (2\pi\tau_i\nu)^{b_i}} \tag{5.101}$$

where τ_i is the characteristic time of the decaying autocorrelation function of the process, H_i is the height in the power spectrum at $\nu = 0$, and b_i is the power law exponent (with the b_i ranging typically from 2 to 6 for intensity measurements; Appourchaux *et al.*, 2002; Aigrain *et al.*, 2004). It is clear that Equation 5.101 also has a Lorentzian-like shape as much as the harmonic oscillator of the previous section. Therefore, I also outline that modes stochastically excited because they also follow an AR model have the same mathematical property as the random nonharmonic field.

5.5.6 *Power spectrum model*

5.5.6.1 *Stationary processes*

I showed in Section 5.3.7 that stationary processes have indeed a χ^2 with two d.o.f. statistics. The question is: what is the mean value when different stationary processes are in play? Let us assume that two time series for two different processes coexist: $x_{\rm osc}(t)$ related to the excitation of the harmonic oscillators (the eigenmodes), and $x_{\rm n}(t)$ related to the nonharmonic noise (the background stellar noise). The total time series is then given by:

$$x(t) = x_{\rm osc}(t) + x_{\rm n}(t) \tag{5.102}$$

and the autocorrelation is given by:

$$\mathrm{E}[x(t_1)x(t_2)] = C_{\rm osc}(\tau) + C_{\rm n}(\tau) + [\mathrm{E}[x_{\rm osc}(t_1)x_{\rm n}(t_2)] + \mathrm{E}[x_{\rm n}(t_1)x_{\rm osc}(t_2)]] \tag{5.103}$$

with $\tau = t_2 - t_1$ and where $C_{\rm osc}$, $C_{\rm n}$ are the autocorrelation for the harmonic and nonharmonic signals. The last term in brackets is related to the potential correlation between the two types of stationary processes. Usually, we assume that there is no correlation between these processes such that the term in brackets is zero. Using the Wiener-Khinchine theorem, we have for the spectral density:

$$S_x(\nu) = S_{\rm osc}(\nu) + S_{\rm n}(\nu). \tag{5.104}$$

So, the spectral density of the sum of two independent stationary process is the sum of the spectral density of each stationary process. This implicit assumption, not frequently mentioned, is usually a good approximation for solar-like oscillations. If we assume that the processes are correlated but that their correlation is stationary, we can then write:

$$S_x(\nu) = S_{\rm osc}(\nu) + S_{\rm n}(\nu) + \tilde{I}(\nu) \tag{5.105}$$

where $\tilde{I}(\nu)$ is simply the Fourier transform of the expression in brackets of Equation 5.103. Such correlations between the background and the harmonic oscillators have been studied by Severino *et al.* (2001). The term $\tilde{I}$ is responsible for the mode profile asymmetry, as explained earlier (Nigam and Kosovichev, 1998).

5.5.6.2 Model of the spectrum

The complete model of the power spectrum is given by the superposition of *all* the harmonic signals and of the nonharmonic signals. Apart from the Sun, so far we have only observed disk-integrated oscillations in stars. Instruments integrating over the stellar surface observe the velocity or the intensity signal as a superposition of various modes of different degrees. The impact of this integration is to filter out the higher degree modes, keeping typically only the degrees below 4. The resulting mode sensitivity (V_l) was first computed for velocity measurements for nonrotating stars by Christensen-Dalsgaard and Gough (1982), then by Christensen-Dalsgaard (1989) taking into account Doppler imaging due to rotation; for intensity, it was first computed by Toutain and Gouttebroze (1993). The impact of the inclination angle of the star upon the mode sensitivity ($c_{l,m}$) was also computed by Toutain and Gouttebroze (1993), calculation which was later rediscovered by Gizon and Solanki (2003). The mode sensitivity V_l has also been extensively computed by different authors; for example, Bedding *et al.* (1996), Appourchaux *et al.* (2000), and Ballot *et al.* (2006).

Using the equations given earlier and taking into account the geometrical mode sensitivity, the full spectral density is therefore given by:

$$S_x(\nu) = \sum_{n,l,m} H_n V_l^2 c_{l,m} \tilde{L}\left(\frac{\nu - \nu_{nlm}}{\Gamma_{nlm}}\right) + \sum_i \frac{H_i}{1 + (2\pi\tau_i\nu)^{b_i}} \tag{5.106}$$

where H_n is the mode height for radial order n, $\tilde{L}$ is the mode profile, Γ_{nlm} is the inverse mode lifetime, ν_{nlm} is the mode frequency, and the H_i, τ_i, and b_i represent the stellar and instrumental background noise (including also the photon noise). The mode profile is here assumed to be Lorentzian. It can be replaced by an asymmetrical profile but at the expense of making an incorrect approximation of the correlation effect $\tilde{I}$. If one wants to properly take into account this correlation effect, one should refer to Severino *et al.* (2001).

Several effects will lift the degeneracy of the mode frequency ν_{nl}, such as stellar rotation or stellar activity, thereby splitting the frequency of the mode in different components (Zeeman-like effect). In this case, the mode frequency can be decomposed into the Clebsch-Gordan coefficient as:

$$\nu_{nlm} = \nu_{nl} + \sum_{i=1}^{i_{\max}} a_i(n,l) l \mathcal{P}_l^{(i)}(m/l) \tag{5.107}$$

where $i_{\max} \leq 2l$, and $a_i(n,l)$ are the splitting coefficients and $\mathcal{P}_l^{(i)}$ are polynomials given by Ritzwoller and Lavely (1991) or by Chaplin *et al.* (2004). The advantage of the $\mathcal{P}_l^{(i)}$ is that they are by construction orthogonal to each other. With this property, the max i for the decomposition can be different from $2l$; in other words, removing or adding polynomials will not affect the value of the $a_i(n,l)$. To second order, this property may not be completely true as there could be slight variations in the mode height due to stochastic excitation or other effects not modeled. For example, the observation of the integrated signal over the stellar surface with a star observed perpendicular to its rotation axis ($i = 90^{\circ}$) will *remove* modes for which $l + m$ is odd. If this averaging effect is not taken into account, the decomposition will affect the value of the $a_i(n,l)$ (Chaplin *et al.*, 2004).

Depending on the model used, different assumptions concerning the mode and background parameters are possible. For instance, the mode linewidth can be assumed to be independent of frequency, or to vary with frequency according to some polynomials, or to be independent for each order n, or to be independent for each degree l, and so forth. It is sometimes desirable to reduce the number of fitted parameters using such assumptions. Examples of such assumptions are found in the references given in the next two sections.

5.5.7 *Frequentist parameter estimation*

Now, having observed a star, we have a time series of intensity or velocity fluctuations which is properly sampled, and for which the filtering effect of the data reduction is understood. We compute the Fourier transform and then obtain the power spectral density properly scaled using Parseval's theorem. The amount of gaps is negligible such that the frequency bins are independent of each other. The statistics of the power spectrum are χ^2 with two d.o.f. Assuming that all the stationary processes are nearly independent, we have a complete model of the power spectrum including harmonic and non harmonic signals.

Now that I have set the scene, all the actors are in place for the play. The method applied for deriving all the parameters of Equation 5.106 is usually to use MLE. The use of MLE for fitting solar power spectrum was first mentioned by Duvall and Harvey (1986), then applied and tested by Anderson *et al.* (1990), in which they also mentioned the use of Monte-Carlo simulations for validating the method. Their pioneering work led in helioseismology to the understanding of the derivation of error bar for mode frequencies (Libbrecht, 1992; Toutain and Appourchaux, 1994), which is given by:

$$\delta\nu = \sqrt{\frac{\Gamma}{4\pi T}(\beta+1)^{\frac{1}{4}}\left(\sqrt{\beta+1}+\sqrt{\beta}\right)^{\frac{3}{2}}}. \qquad (5.108)$$

This equation shows that the mode precision is proportional to the square root of the mode linewidth and is proportional to the square root of the frequency resolution. This results greatly contrasts with that of a pure sine wave given by Equation 5.94. Similarly, other error bars were also derived for linewidth, mode height, white noise (Toutain and Appourchaux, 1994); rotational splitting (Toutain and Appourchaux, 1994; Gizon and Solanki, 2003; Ballot *et al.*, 2008); inclination angle (Gizon and Solanki, 2003; Ballot *et al.*, 2008); the correlation between the mode height, the linewidth, and background noise (Toutain and Appourchaux, 1994) (see Appendix); and between the rotational splitting and the inclination angle (Ballot *et al.*, 2008).

Monte-Carlo simulations have also been used for verifying the veracity of using the inverse of the Hessian for estimating error bars on the parameters of the model (Anderson *et al.*, 1990; Toutain and Appourchaux, 1994; Schou and Brown, 1994; Appourchaux *et al.*, 1998; Gizon and Solanki, 2003) (see Section 5.4.4). When doing so, it was patent that the distribution for mode height, linewidth, and background noise was not normal but log-normal. There are two main reasons for this state of affairs: first, these simulations are far from the asymptotic behavior, otherwise the distribution would be normal; second, the mathematical transform $\log(x)$ allows us to have a distribution closer to a normal distribution than x alone. In other words, the asymptotic behavior is reached faster with such a transform. The same would apply to any transform reducing the amount of value larger than the median with respect to value smaller than the median.[4]

Although the MLE has been in use for more than 20 years by helioseismologists, it is not yet completely adopted by our fellow asteroseismologists. The body of evidence provided by helioseismology is rather large; hereafter I will necessarily refer to a few

[4] The reader may convince him- or herself by simulating a χ^2 with n d.o.f., for example, together with a $\log x$ or $\sqrt{x}$ transform.

examples: IPHIR (InterPlanetary Helioseismology by IRradiance measurements) photometers aboard the Russian mission Mars94 (Sun as a star; Toutain and Fröhlich, 1992), VIRGO (Variability of solar IRradiance and Gravity Oscillations) photometers aboard SoHO (Solar and Heliospheric Observatory) (Sun as a star and low resolution; Fröhlich *et al.*, 1997; Appourchaux *et al.*, 1997), GOLF (Global Oscillations at Low Frequency) spectrometer aboard SOHO (Sun as a star; Gelly *et al.*, 2002), BiSON (Birmingham Solar Oscillations Network) instrument (Sun as a star; Chaplin *et al.*, 1996), LOWL (Low-l) instrument (imager; Schou, 1992; Schou and Brown, 1994), GONG (Global Oscillation Network Group) instrument (imager; Hill *et al.*, 1996; Appourchaux *et al.*, 1998), SOI/MDI (Solar Oscillations Investigation/Michelson Doppler Imager) instrument aboard SoHO (imager; Kosovichev *et al.*, 1997).

Following these successful applications for solar data, MLE was preliminarily tested on synthetic stellar timeseries for use on the CNES CoRoT mission (Appourchaux *et al.*, 2006a,b). The recipe laid down in this publication was found later not be applicable to the real CoRoT data. Appourchaux *et al.* (2008) introduced a scheme using not fit on narrow frequency window (*local* fit as in Appourchaux *et al.*, 2006a) but using fit on large frequency window (*global* fit). This new scheme has now been applied to solar-like stars and red giants observed by CoRoT (Barban *et al.*, 2009; García *et al.*, 2009; Mosser *et al.*, 2009; Deheuvels *et al.*, 2010; Baudin *et al.*, 2011); and to solar-like stars observed by the NASA (National and Aeronautics Space Administration) Kepler mission (Mathur *et al.*, 2011).

One of the recent controversies in asteroseismology is related to the degree identification in the power spectra of HD49333 (observed by CoRoT). In Appourchaux *et al.* (2008), the degree of the modes could not easily be identified, leaving two possible solutions: the ridges in the echelle diagram were either $l = 0 - 1$ or $l = 1 - 0$. The reason for such a difficult identification, already anticipated by Appourchaux *et al.* (2006a), is related to a mode linewidth larger than in the Sun combined with a narrower small spacing than in the Sun. As a result, the two ridges $l = 0 - 2$ and $l = 1$ were indistinguishable from each other. One of the solutions found by Appourchaux *et al.* (2008) was to base the choice of the ridge upon the likelihood ratio test. Here, I would like to cite a warning in Appourchaux *et al.* (2008) as to the use of this test: "However, it is important to remember that a higher likelihood does not mean that a given model is physically more meaningful, rather, it means the model is statistically more likely." It was clear at this stage that in order to take into account prejudices and *a priori* knowledge that could have helped to get the physically meaningful model, a Bayesian approach to the problem was required. This is the subject of the next section.

5.5.8 *Bayesian parameter estimation*

In asteroseismology, the first attempt at applying a Bayesian approach was due to Brewer *et al.* (2007). They used a model based on pure sine waves that did not correctly model stochastic excitation of harmonic oscillations. Their model can be used for classical pulsators or when the modes have a lifetime longer than the observation time. Following the attempt by Brewer *et al.* (2007), I realized that the approach and framework provided by Bretthorst (1988) is not generally applicable in asteroseismology. The separation in time between the deterministic signal and the noise cannot be done when oscillators are stochastically excited. Since the separation is not possible, the statistics of the stellar signal suggested by Brewer *et al.* (2007) need to be replaced by the likelihood usually used by frequentist Appourchaux (2008). At the time, it was not clear that the replacement was so *obvious*.

Following the early work of Appourchaux (2008), the application of Bayesian analysis to solar-like oscillators has been quite extensively developed. The difficult data analysis of HD49333 led to the development of several Bayesian data analyses in asteroseismology

(Benomar *et al.*, 2009b; Gruberbauer *et al.*, 2009). A simplified Bayesian approach called Maximum A Posteriori (MAP) has also been used for putting constraints on the mode height (Gaulme *et al.*, 2009). Recently, Handberg and Campante (2011) produced a helpful guide for anyone wanting to perform Bayesian data analysis in asteroseismology. A more extensive guide in French is also available in Benomar (2010). These two guides covers many aspects of the application of Bayes' theorem such as power spectrum models, statistics, priors, parallel tempering, and global likelihood. Here, I should add that the Bayesian framework has been used not only for parameter estimation but also for making decisions based on posterior probability estimates: mode detection (Appourchaux *et al.*, 2009; Broomhall *et al.*, 2010), presence of $l = 3$ modes, and mixed modes (Deheuvels *et al.*, 2010).

As anticipated by Appourchaux *et al.* (2008), the HD49333 controversy led to the first useful application of the Bayesian approach. Benomar *et al.* (2009b) using the same time series as Appourchaux *et al.* (2008) applied a Bayesian analysis and derived the global likelihood of the various models previously used. In doing so, Benomar *et al.* (2009b) showed that amongst the four models used, the identification of Appourchaux *et al.* (2008) was the most probable one *but* for a model having *no* $l = 2$ modes! When these are included (as in Appourchaux *et al.*, 2008), Benomar *et al.* (2009b) found that the identification of Appourchaux *et al.* (2008) was less probable (14%) than the other identification (23%); thereby showing that both models were nearly equiprobable. The absence of $l = 2$ modes in HD49333 was also the conclusion of Gruberbauer *et al.* (2009) who also used a Bayesian approach. The main differences of the work of Gruberbauer *et al.* (2009) from that of Benomar *et al.* (2009b) are that they do not use global likelihoods for decision and that their model has no splitting (no degree identification possible). The controversy was finally closed when 180 days of data were available for HD49333. Using these data, Benomar *et al.* (2009a) showed that both frequentist and Bayesian approaches favor a model having $l = 2$ modes with a degree identification different from that of Appourchaux *et al.* (2008).

What lessons can we draw from such a controversy? The process of the advancement of science cannot be done without making *errors*. There can be several kinds:

- done on purpose for hiding the facts (a lie),
- done by ignorance of the facts, and
- done in knowledge of the facts.

The first kind of error has led to several infamous publications, such as a cold fusion (Fleischmann and Pons, 1989) finding that was never reproduced. The second kind of error is due to lack of information or knowledge; in this world of information flow, this can easily happen to any one of us. The third kind can either be perceived as an error from an outsider or as a decision done in full conscience of what has been decided. The *error* of Appourchaux *et al.* (2008) is obviously akin to the last kind. If we would have had the choice, we had certainly opted for applying the Bayesian approach in Appourchaux (2008), but time prevented us to do so. This is another lesson to be learned: even in this fast-paced world of information flow and publications, one should never trade quality of publication and scientific integrity for fast publication. Science is like pasta, a slow sugar.

5.6 Conclusion

If you have been able to survive until now, I would like to thank you. It would be pretentious to believe that this *conclusion* is going to conclude anything for good. As a matter of fact, this conclusion aims at giving the reader the incentive to go beyond this *chapter* but still use this *chapter* as a foundation for the future. Data analysis clearly starts from the instrumentation itself; this aspect was not treated here. It is clear that the measurements leading to the analysis of time series are done through an instrument. From this point of view, the aspects related to the filtering due to the instrument are implicitly laid out. Apart from these mere details, the reader should find all that is needed for a proper understanding of data analysis in asteroseismology. Needless to say, this field is evolving very rapidly, thanks to the CoRoT and Kepler missions. In the last couple of years, the Bayesian data analysis has developed quickly. Despite what is commonly thought, the Bayesian approach, although using prejudices, is in the end more conservative than the frequentist approach. Actually, the main drawback of this approach is not in the methodology itself but the speed at which the computation of the results can be performed.

With the advent of the new PLATO mission, it can be anticipated that both Bayesian and frequentist data analysis will be used for deriving the mode parameters of more than 20,000 stars. I can think of a hybrid scheme where both approaches will be used depending on the signal-to-noise ratio in the power spectrum. Another aspect that was partly discussed here is how one can assess and ensure that the fitted mode parameters are of usable quality for doing stellar models. This is certainly to be one of the future challenges of asteroseismology: the automatic assessment of stellar mode parameters. I can also anticipate that we will apply quality assessments that are regularly used in factories doing mass production. For instance, sampling at random some of our output products may be useful for *qualifying* the lot as useable by the scientific community. We are really in the infant stage for such massive quality assessment.

The future of asteroseismology will reach another milestone when the surface of stars is imaged. At that point in time, we will then be able to probe the internal and dynamics structure of stars in the very same manner as we did with the Sun. The Stellar Imager could be the next asteroseismic challenge whose achievement may be not so far in the future (Christensen-Dalsgaard *et al.*, 2011).

As for any new endeavor, we need to learn from the past for going ahead while keeping in mind Timothy Leary's motto: "Think for yourself, question authority."

Acknowledgments

I would like to thank Pere Pallé for inviting me to give these lectures in the beautiful island of Tenerife. I would also like to express my thanks for the many discussions I had with my long-standing colleagues from which this course benefited: Frédéric Baudin, Othman Benomar, Patrick Boumier, William Chaplin, Patrick Gaulme, Laurent Gizon, and Takashi Sekii. This course also benefited from the acute reading of John Leibacher. Last but not least, many thanks to my wife, Maryse, and my sons, Kévin and Thibault, for their indefectible support; without forgetting the mention of the purring cat, Myrtille. Thanks to the volcano of El Teide for momentary time support, going up there was not an easy ride!

Appendix

Correlation between mode height and linewidth

Toutain and Appourchaux (1994) showed that some fitted mode parameters were intrinsically correlated. When one wants to derive the mode amplitude A (proportional to the square root of ΓH), one should take into account the correlation between the errors on Γ and H, which are the mode linewidth the mode height, respectively. Because we have $A \propto \Gamma H$, using the logarithm of these quantities, we have:

$$\log A = \frac{1}{2}(\log \Gamma + \log H) + \cdots \tag{5.109}$$

or simply

$$a = \frac{1}{2}(\gamma + h) + \cdots \tag{5.110}$$

The error bar on a is derived from:

$$\sigma_a^2 = \frac{1}{4}\left(\sigma_\gamma^2 + \sigma_h^2 + 2\mathrm{E}(ah)\right), \tag{5.111}$$

which can be rewritten using the correlation coefficient as:

$$\sigma_a^2 = \frac{1}{4}\left(\sigma_\gamma^2 + \sigma_h^2 + 2\rho\sigma_\gamma\sigma_h\right). \tag{5.112}$$

Using the work of Toutain and Appourchaux (1994), and after some mathematical manipulation, I can derive for a single mode the correlation as:

$$\rho = \frac{\mathrm{E}(\gamma h)}{\sigma_h\sigma_\gamma} = -\sqrt{\frac{\sqrt{\beta+1}}{\sqrt{\beta+1}+\sqrt{\beta}}} \tag{5.113}$$

where β is the noise to the mode height ratio in the power spectrum. This correlation is negative, and its absolute value always greater than 76.5% when $\beta < 1$. It means that even when the modes are easily detected the correlation is always very large and never negligible. Finally, the error bar on a is then given by:

$$\sigma_a = \frac{1}{2\sqrt{T\Gamma\pi}}\sqrt{\left(\sqrt{\beta+1}+\sqrt{\beta}\right)\left(\left(\sqrt{\beta+1}+\sqrt{\beta}\right)^3 - 4(1+\beta)\sqrt{\beta}\right)} \tag{5.114}$$

where T is the observation time. From this equation, I can deduce that the precision on the measurement of the mode amplitude increases with the observation time and the mode linewidth.

REFERENCES

Aigrain, S., Favata, F., and Gilmore, G. 2004. Characterising stellar micro-variability for planetary transit searches. *A&A*, **414**, 1139–52.

Anderson, E. R., Duvall, T. L., Jr., and Jefferies, S. M. 1990. Modeling of solar oscillation power spectra. *ApJ*, **364**(Dec.), 699.

Appourchaux, T. 2008. Bayesian approach for g-mode detection, or how to restrict our imagination. *Astronomische Nachrichten*, **329**(June), 485.

Appourchaux, T., Andersen, B. N., Fröhlich, C., Jiménez, A., Telljohann, U., and Wehrli, C. 1997. In-flight performance of the VIRGO Luminosity Oscillations Imager aboard SOHO. *Sol. Phys.*, **170**, 27–41.

Appourchaux, T., Gizon, L., and Rabello-Soares, M. C. 1998. The art of fitting p-mode spectra. I. Maximum Likelihood Estimation. *A&AS*, **132**(Oct.), 107–19.

Appourchaux, T., Fröhlich, C., Andersen, B. N., Berthomieu, G., Chaplin, W., Elsworth, Y., Finsterle, W., Gough, D. O., Hoeksema, J. T., Isaak, G. R., Kosovichev, A. G., Provost, J., Scherrer, P. H., Sekii, T., and Toutain, T. 2000. Observational upper limits for low-degree solar g modes. *ApJ*, **538**, 401–14.

Appourchaux, T., Andersen, B., and Sekii, T. 2002. What have we learnt with the Luminosity Oscillations Imager over the past 6 years? Pages 47–50 of: C. Fröhlich and A. Wilson (ed), *From Solar Min to Max: Half a Solar Cycle with SOHO*. ESA Special Publication, vol. 508.

Appourchaux, T., Berthomieu, G., Michel, E., Aerts, C., Ballot, J., Barban, C., Baudin, F., Boumier, P., de Ridder, J., Floquet, M., Garcia, R. A., Garrido, R., Goupil, M.-J., Lambert, P., Lochard, J., Neiner, C., Poretti, E., Provost, J., Roxburgh, I., Samadi, R., and Toutain, T. 2006a. Data analysis tools for the seismology programme. Page 377 of: M. Fridlung, A. Baglin, J. Lochard and L. Conroy (ed), *The CoRoT Mission*. ESA Publications Division, ESA Spec. Publ. 1306.

Appourchaux, T., Berthomieu, G., Michel, E., Ballot, J., Barban, C., Baudin, F., Boumier, P., de Ridder, J., Floquet, M., Garcia, R. A., Garrido, R., Goupil, M.-J., Lambert, P., Lochard, J., Mazumdar, A., Neiner, C., Poretti, E., Provost, J., Roxburgh, I., Samadi, R., and Toutain, T. 2006b. Evaluation of the scientific performances for the seismology programme. Page 429 of: M. Fridlung, A. Baglin, J. Lochard and L. Conroy (ed), *The CoRoT Mission*. ESA Publications Division, ESA Spec. Publ. 1306.

Appourchaux, T., Michel, E., Auvergne, M., Baglin, A., Toutain, T., Baudin, F., Benomar, O., Chaplin, W. J., Deheuvels, S., Samadi, R., Verner, G. A., Boumier, P., García, R. A., Mosser, B., Hulot, J.-C., Ballot, J., Barban, C., Elsworth, Y., Jiménez-Reyes, S. J., Kjeldsen, H., Régulo, C., and Roxburgh, I. W. 2008. CoRoT sounds the stars: p-mode parameters of Sun-like oscillations on HD 49933. *A&A*, **488**(Sept.), 705–14.

Appourchaux, T., Samadi, R., and Dupret, M.-A. 2009. On posterior probability and significance level: application to the power spectrum of HD 49 933 observed by CoRoT. *A&A*, **506**(Oct.), 1–5.

Ballot, J., García, R. A., and Lambert, P. 2006. Rotation speed and stellar axis inclination from p modes: how CoRoT would see other suns. *MNRAS*, **369**(July), 1281–1286.

Ballot, J., Appourchaux, T., Toutain, T., and Guittet, M. 2008. On deriving p-mode parameters for inclined solar-like stars. *A&A*, **486**(Aug.), 867–75.

Barban, C., Deheuvels, S., Baudin, F., Appourchaux, T., Auvergne, M., Ballot, J., Boumier, P., Chaplin, W. J., García, R. A., Gaulme, P., Michel, E., Mosser, B., Régulo, C., Roxburgh, I. W., Verner, G., Baglin, A., Catala, C., Samadi, R., Bruntt, H., Elsworth, Y., and Mathur, S. 2009. Solar-like oscillations in HD 181420: data analysis of 156 days of CoRoT data. *A&A*, **506**(Oct.), 51–6.

Baudin, F., Barban, C., Belkacem, K., Hekker, S., Morel, T., Samadi, R., Benomar, O., Goupil, M., Carrier, F., Ballot, J., Deheuvels, S., De Ridder, J., Hatzes, A. P., Kallinger, T., and Weiss, W. W. 2011. Amplitudes and lifetimes of solar-like oscillations observed by CoRoT. Red-giant versus main-sequence stars. *A&A*, **535**(May), A84.

Bayes, T. 1763. An Essay towards solving a Problem in the Doctrine of Chances. *Philosophical Transactions of the Royal Society of London*, **53**, 370.

Bedding, T. R., Kjeldsen, H., Reetz, J., and Barbuy, B. 1996. Measuring stellar oscillations using equivalent widths of absorption lines. *MNRAS*, **280**(June), 1155–61.

Belkacem, K., Samadi, R., Goupil, M.-J., Lefèvre, L., Baudin, F., Deheuvels, S., Dupret, M.-A., Appourchaux, T., Scuflaire, R., Auvergne, M., Catala, C., Michel, E., Miglio, A., Montalban, J., Thoul, A., Talon, S., Baglin, A., and Noels, A. 2009. Solar-like oscillations in a massive star. *Science*, **324**(June), 1540.

Belkacem, K., Dupret, M. A., and Noels, A. 2010. Solar-like oscillations in massive main-sequence stars. I. Asteroseismic signatures of the driving and damping regions. *A&A*, **510**(Jan.), A6.

Belmonte, J. A., Chevreton, M., Mangeney, A., Praderie, F., Saint-Pe, O., Puget, P., Alvarez, M., and Roca Cortes, R. 1991. Rapid photometry of the Delta Scuti variable 63 Herculis and of the F 2V star HD 155543: Observations and analysis of the time series. *A&A*, **246**(June), 71–83.

Benomar, O. 2010. *Sismologie stellaire: Méthodes statistiques appliquées aux étoiles de type solaire*. Ph.D. thesis, Université Paris-Sud XI, Orsay, France.

Benomar, O., Baudin, F., Campante, T. L., Chaplin, W. J., García, R. A., Gaulme, P., Toutain, T., Verner, G. A., Appourchaux, T., Ballot, J., Barban, C., Elsworth, Y., Mathur, S., Mosser, B., Régulo, C., Roxburgh, I. W., Auvergne, M., Baglin, A., Catala, C., Michel, E., and Samadi, R. 2009a. A fresh look at the seismic spectrum of HD49933: analysis of 180 days of CoRoT photometry. *A&A*, **507**(Nov.), L13–L16.

Benomar, O., Appourchaux, T., and Baudin, F. 2009b. The solar-like oscillations of HD 49933: a Bayesian approach. *A&A*, **506**(Oct.), 15–32.

Berger, J. O. 1997. Could Fisher, Jeffreys and Neyman have agreed upon testing? *Statistical Science*, **18**, 1–32.

Berger, J. O., and Sellke, T. 1987. Testing of a point null hypothesis: The irreconcilability of significance levels and evidence. *Journal of the American Statistical Association*, **82(397)**, 112–39.

Berger, J. O., Boukai, B., and Wang, Y. 1997. Unified frequentist and Bayesian testing of a precise hypothesis. *Statistical Science*, **12(3)**, 133–60.

Bernoulli, J. 1713. *Ars conjectandi, opus posthumum. Accedit Tractatus de seriebus infinitis, et epistola gallicé scripta de ludo pilae reticularis*. Thurneysen Brothers, Basel, Switzerland.

Bretthorst, L. G. 1988. *Bayesian Spectrum Analysis and Parameter Estimation*. Springer-Verlag, Berlin, available electronically at bayes.wustl.edu/glb/book.pdf.

Bretthorst, G. L. 2001a. Generalizing the Lomb-Scargle periodogram. Pages 241–245 of: A. Mohammad-Djafari (ed), *Bayesian Inference and Maximum Entropy Methods in Science and Engineering*. American Institute of Physics Conference Series, vol. 568.

Bretthorst, G. L. 2001b. Generalizing the Lomb-Scargle periodogram: the nonsinusoidal case. Pages 246–251 of: A. Mohammad-Djafari (ed), *Bayesian Inference and Maximum Entropy Methods in Science and Engineering*. American Institute of Physics Conference Series, vol. 568.

Brewer, B. J., Bedding, T. R., Kjeldsen, H., and Stello, D. 2007. Bayesian inference from observations of solar-like oscillations. *ApJ*, **654**(Jan.), 551–7.

Broomhall, A.-M., Chaplin, W. J., Elsworth, Y., Appourchaux, T., and New, R. 2010. A comparison of frequentist and Bayesian inference: searching for low-frequency p modes and g modes in Sun-as-a-star data. *MNRAS*, **406**(Aug.), 767–81.

Chaplin, W. J., Elsworth, Y., Howe, R., Isaak, G. R., McLeod, C. P., Miller, B. A., van der Raay, H. B., Wheeler, S. J., and New, R. 1996. BiSON Performance. *Sol. Phys.*, **168**(Sept.), 1–18.

Chaplin, W. J., Appourchaux, T., Elsworth, Y., Isaak, G. R., Miller, B. A., and New, R. 2004. On comparing estimates of low-l solar p-mode frequencies from Sun-as-a-star and resolved observations. *A&A*, **424**(Sept.), 713–17.

Christensen-Dalsgaard, J. 1989. The effect of rotation on whole-disc Doppler observations of solar oscillations. *MNRAS*, **239**(Aug.), 977–94.

Christensen-Dalsgaard, J. 2004. An Overview of helio- and asteroseismology. Page 1 of: D. Danesy (ed), *SOHO 14 Helio- and Asteroseismology: Towards a Golden Future*. ESA Special Publication, vol. 559.

Christensen-Dalsgaard, J., and Gough, D. O. 1982. On the interpretation of five-minute oscillations in solar spectrum line shifts. *MNRAS*, **198**(Jan.), 141–71.

Christensen-Dalsgaard, J., Carpenter, K. G., Schrijver, C. J., Karovska, M., and the Si Team. 2011. The stellar imager (SI): a mission to resolve stellar surfaces, interiors, and magnetic activity. *Journal of Physics Conference Series*, **271**(1), 012085.

Cooley, J. W., and Tuckey, J. W. 1965. An algorithm for the machine calculation of complex Fourier series. *Math. Comp.*, **19**, 297–301.

Cramer, H. 1946. *Mathematical Methods of Statistics*. Princeton University Press, New Jersey.

Cuypers, J. 1987. *Medelingen van de Koninklijke Academie voor Wetenschappen, Letteren en Schone Kunsten van Belgïe*, Jg. 49 Nr. 3 (Brussel: Paleis der Academiëen), 21.

De Finetti, B. 1937. La prévision: ses lois logiques, ses sources subjectives. *Annales de l'Institut Henri Poincaré*, **7**, 1.

De Moivre, A. 1718. *The doctrine of chances of a method of calculating the probability of events in play*. W. Pearson, London, United Kingdom.

Deheuvels, S., Bruntt, H., Michel, E., Barban, C., Verner, G., Régulo, C., Mosser, B., Mathur, S., Gaulme, P., Garcia, R. A., Boumier, P., Appourchaux, T., Samadi, R., Catala, C., Baudin, F., Baglin, A., Auvergne, M., Roxburgh, I. W., and Pérez Hernández, F. 2010. Seismic and spectroscopic characterization of the solar-like pulsating CoRoT target HD 49385. *A&A*, **515**(June), A87.

Duvall, T. L., Jr., and Harvey, J. W. 1986. Solar Doppler shifts: sources of continuous spectra. Page 105 of: Osaki, Y., and Shibahashi, H. (eds), *Seismology of the Sun and the Distant Stars*. Lecture Notes in Physics, Springer Verlag, Berlin.

Duvall, Jr., T. L., Jefferies, S. M., Harvey, J. W., Osaki, Y., and Pomerantz, M. A. 1993. Asymmetries of solar oscillation line profiles. *ApJ*, **410**(June), 829–36.

Fisher, R. A. 1912. On an absolute criterion for fitting frequency curves. *Messenger of Mathematics*, **41**, 155–60.

Fisher, R. A. 1925. Theory of statistical estimation. *Proc. Cambridge Philos. Soc.*, **22**, 155–60.

Fleischmann, M., and Pons, S. 1989. Electrochemically induced nuclear fusion of deuterium. *Journal of Electroanalytical Chemistry and Interfacial Electrochemistry*, **261**, 301–8.

Fourier, J. 1822. *Théorie analytique de la chaleur*. Firmin-Didot, Paris, France.

Frieden, B. R. 1983. *Probability, Statistical Optics, and Data Testing: A Problem Solving Approach*. Springer-Verlag, Berlin.

Fröhlich, C., Andersen, B. N., Appourchaux, T., Berthomieu, G., Crommelynck, D. A., Domingo, V., Fichot, A., Finsterle, W., Gomez, M. F., Gough, D., Jiménez, A., Leifsen, T., Lombaerts, M., Pap, J. M., Provost, J., Cortés, T. R., Romero, J., Roth, H., Sekii, T., Telljohann, U., Toutain, T., and Wehrli, C. 1997. First results from VIRGO, the experiment for helioseismology and solar irradiance monitoring on SOHO. *Sol. Phys.*, **170**, 1–25.

Gabriel, M. 1993. The probability-density function of solar p modes and the location of the excitation mechanism. *A&A*, **274**(July), 931.

Gabriel, M. 1994. The probability-density function of a Fourier line. *A&A*, **287**(July), 685–691.

García, R. A., Régulo, C., Samadi, R., Ballot, J., Barban, C., Benomar, O., Chaplin, W. J., Gaulme, P., Appourchaux, T., Mathur, S., Mosser, B., Toutain, T., Verner, G. A., Auvergne, M., Baglin, A., Baudin, F., Boumier, P., Bruntt, H., Catala, C., Deheuvels, S., Elsworth, Y., Jiménez-Reyes, S. J., Michel, E., Pérez Hernández, F., Roxburgh, I. W., and Salabert, D. 2009. Solar-like oscillations with low amplitude in the CoRoT target HD 181906. *A&A*, **506**(Oct.), 41–50.

Gaulme, P., Appourchaux, T., and Boumier, P. 2009. Mode width fitting with a simple Bayesian approach. Application to CoRoT targets HD 181420 and HD 49933. *A&A*, **506**(Oct.), 7–14.

Gauss, C. F. 1809. *Theoria motus corporum coelestium in sectionibus conicis solem ambientum.* Königliche Gesellschaft der Wissenschaften, Göttingen, Germany.

Gauss, C. F. 1866. *Nachlass: Theoria interpolationis methodo nova tractata: Werke band 3, 265–327.* Königliche Gesellschaft der Wissenschaften, Göttingen, Germany.

Gautschy, A., and Saio, H. 1995. Stellar pulsations across the HR diagram: part 1. *ARA&A*, **33**, 75–114.

Gautschy, A., and Saio, H. 1996. Stellar pulsations across the HR diagram: part 2. *ARA&A*, **34**, 551–606.

Gelly, B., Lazrek, M., Grec, G., Ayad, A., Schmider, F. X., Renaud, C., Salabert, D., and Fossat, E. 2002. Solar p-modes from 1979 days of the GOLF experiment. *A&A*, **394**(Oct.), 285.

Gelman, A., and Meng, X.-L. 1998. Simulating normalizing constants: from importance sampling to bridge sampling to path sampling. *Statistical Science*, **13(2)**(Oct.), 163–185.

Gibbs, J. W. 1898. Fourier series. *Nature*, **59**, 200.

Gibbs, J. W. 1899. Fourier series. *Nature*, **59**, 606.

Gizon, L., and Solanki, S. K. 2003. Determining the inclination of the rotation axis of a sun-like star. *ApJ*, **589**(June), 1009–19.

Goldstine, H. H. 1977. *A History of Numerical Analysis from the 16th through the 19th Century.* Springer, New York–Heidelberg–Berlin.

Gott, J. R. III. 1994. Future Prospects Discussed. *Nature*, **368**, 106–8.

Gregory, P. C. 2005. *Bayesian Logical Data Analysis for the Physical Sciences: A Comparative Approach with "Mathematica" Support.* Cambridge University Press, Cambridge.

Gruberbauer, M., Kallinger, T., Weiss, W. W., and Guenther, D. B. 2009. On the detection of Lorentzian profiles in a power spectrum: a Bayesian approach using ignorance priors. *A&A*, **506**(Nov.), 1043–53.

Handberg, R., and Campante, T. L. 2011. Bayesian peak-bagging of solar-like oscillators using MCMC: a comprehensive guide. *A&A*, **527**(Mar.), A56.

Harvey, J. 1985. High-resolution helioseismology. Pages 199–208 of: E. Rolfe, and B. Battrick (eds), *Future missions in solar, heliospheric and space plasma physics, ESA SP-235.* ESA Publications Division, Noordwijk, The Netherlands.

Harvey, J. W., Duvall, Jr., T. L., Jefferies, S. M., and Pomerantz, M. A. 1993. Chromospheric oscillations and the background spectrum. *ASP Conference Series*, **42**, 111.

Hastings, W. K. 1970. Monte Carlo sampling methods using Markov chains and their applications. *Biometrika*, **57(1)**(July), 97.

Heideman, M. T., Johnson, D. H., and Sidney Burrus, C. 1985. Gauss and the history of the fast Fourier transform. *Archive for History of Exact Sciences*, **34**, 265–77.

Hill, F., Stark, P. B., Stebbins, R. T., Anderson, E. R., Antia, H. M., Brown, T. M., Duvall, Jr., T. L., Haber, D. A., Harvey, J. W., Hathaway, D. H., Howe, R., Hubbard, R., Jones, H. P., Kennedy, J. R., Korzennik, S. G., Kosovichev, A. G., Leibacher, J. W., Libbrecht, K. G., Pintar, J. A., Rhodes, Jr., E. J., Schou, J. Thompson, M. J., Tomczyk, S., Toner, C. G., Toussaint, R., and Williams, W. E. 1996. The Solar Acoustic Spectrum and Eigenmode. *Science*, **272**, 1292.

Högbom, J. A. 1974. Aperture Synthesis with a Non-Regular Distribution of Interferometer Baselines. *A&AS*, **15**(June), 417–26.

Houdek, G., Balmforth, N. J̃., Christensen-Dalsgaard, J., and Gough, D. Õ. 1999. Amplitudes of stochastically excited oscillations in main-sequence stars. *A&A*, **351**(Nov.), 582.

Jaynes, E. T. 1987. Bayesian spectrum and chirp analysis. Page 1 of: C. R. Smith and G. J. Erickson (ed), *Maximum entropy and Bayesian spectral analysis and estimation problems.* D.Reidel, Dordrecht, the Netherlands.

Jaynes, E. T. 2009. *Probability Theory: The Logic of Science.* 6th edition. Larry Brethorst (ed). Cambridge University Press, Cambridge, United Kingdom.

Jeffreys, H. 1939. *Theory of Probability.* Oxford Universitty Press, Oxford, United Kingdom.

Jeffreys, H. 1946. An invariant form for the prior probability in estimation problems. *Proceedings of the Royal Society of London. Series A, Mathematical and Physical Sciences*, **186**, 155–160.

Keshner, M. S. 1982. 1/f noise. *IEEE Proceedings*, **70**, 212–18.

Khinchin, A. Y. 1934. Korrelationstheorie der stationaren stochastichen Prozesse. *Mathematische Annalen*, **109**, 604.

Koen, C. 1999. The analysis of indexed astronomical time series. V. Fitting sinusoids to high-speed photometry. *MNRAS*, **309**(Nov.), 769–802.

Komm, R. W., Gu, Y., Hill, F., Stark, P. B., and Fodor, I. K. 1999. Multitaper spectral analysis and wavelet denoising applied to helioseismic data. *ApJ*, **519**, 407–21.

Kosovichev, A. G., Schou, J., Scherrer, P. H., Bogart, R. S., Bush, R. I., Hoeksema, J. T., Aloise, J., Bacon, L., Burnette, A., de Forest, C., Giles, P. M., Leibrand, K., Nigam, R., Rubin, M., Scott, K., Williams, S. D., Basu, S., Christensen-Dalsgaard, J., Dappen, W., Rhodes, E. J., Duvall, T. L., Howe, R., Thompson, M. J., Gough, D. O., Sekii, T., Toomre, J., Tarbell, T. D., Title, A. M., Mathur, D., Morrison, M., Saba, J. L. R., Wolfson, C. J., Zayer, I., and Milford, P. N. 1997. Structure and rotation of the solar interior: initial results from the MDI medium-l program. *Sol. Phys.*, **170**, 43–61.

Kumar, P., Franklin, J., and Goldreich, P. 1988. Distribution functions for the time-averaged energies of stochastically excited solar p modes. *ApJ*, **328**(May), 879.

Laplace, M. 1774. Mémoire sur la probabilité des causes par les évènements. *Mémoires de Mathématique et de Physique*, **T6**, 621–56.

Legendre, A.-M. 1805. *Nouvelles méthodes pour la détermination des orbites des comètes.* Firmin Didot, Paris, France.

Libbrecht, K. G. 1992. On the ultimate accuracy of solar oscillation frequency measurements. *ApJ*, **387**(Mar.), 712.

Mathur, S., Handberg, R., Campante, T. L., Garcia, R. A., Appourchaux, T., Bedding, T. R., Mosser, B., Chaplin, W. J., Ballot, J., Benomar, O., Bonanno, A., Corsaro, E., Gaulme, P., Hekker, S., Regulo, C., Salabert, D., Verner, G., White, T. R., Brandao, I. M., Creevey, O. L., Dogan, G., Elsworth, Y., Huber, D., Hale, S. J., Houdek, G., Karoff, C., Metcalfe, T. S., Molenda-Zakowicz, J., Monteiro, M. J. P. F. G., Thompson, M. J., Christensen-Dalsgaard, J., Gilliland, R. L., Kawaler, S. D., Kjeldsen, H., Quintana, E. V., Sanderfer, D. T., and Seader, S. E. 2011. Solar-like oscillations in KIC11395018 and KIC11234888 from 8 months of Kepler data. *ArXiv e-prints.*

Metropolis, N., Rosenbluth, A. W., Rosenbluth, M. N., Teller, A. H., and Teller, E. 1953. Equation of state calculations by fast computing machines. *Journal of Chemical Physics*, **21(6)**(July), 1087.

Mosser, B., Michel, E., Appourchaux, T., Barban, C., Baudin, F., Boumier, P., Bruntt, H., Catala, C., Deheuvels, S., García, R. A., Gaulme, P., Regulo, C., Roxburgh, I., Samadi, R., Verner, G., Auvergne, M., Baglin, A., Ballot, J., Benomar, O., and Mathur, S. 2009. The CoRoT target HD 175726: an active star with weak solar-like oscillations. *A&A*, **506**(Oct.), 33–40.

Neyman, J., and Pearson, E. 1933. On the problem of the most efficient tests of statistical hypotheses. *Philosophical Transactions of the Royal Society of London. Series A*, **231**(July), 289.

Nigam, R., and Kosovichev, A. G. 1998. Measuring the Sun's eigenfrequencies from velocity and intensity helioseismic spectra: asymmetrical line profile-fitting formula. *ApJ*, **505**(Sept.), L51.

Nigam, R., Kosovichev, A. G., Scherrer, P. H., and Schou, J. 1998. Asymmetry in velocity and intensity helioseismic spectra: a solution to a long-standing puzzle. *ApJ*, **495**(Mar.), L115.

Nyquist, H. 1924. Certain factors affecting telegraph speed. *Bell Syst. Tech. Journal*, **3**, 324–52.

Parseval des Chênes, M.-A. 1806. Mémoire sur les séries et sur l'intégration complete d'une équation aux différences partielles linéaire du second ordre, a coefficients constants. *Mémoires présentés à l'Institut des Sciences, Lettres et Arts, par divers savans, et lus dans ses assemblées. Sciences, mathématiques et physiques*, **1**, 638.

Peligrad, M., and Wu, W. B. 2010. Central limit theorem for Fourier transforms of stationary processes. *The Annals of Probability*, **38**, 2009.

Poincaré, H. 1914. *La Science et l'Hypothèse*. Editions Flammarion, Paris, France.

Poretti, E., Michel, E., Garrido, R., Lefèvre, L., Mantegazza, L., Rainer, M., Rodríguez, E., Uytterhoeven, K., Amado, P. J., Martín-Ruiz, S., Moya, A., Niemczura, E., Suárez, J. C., Zima, W., Baglin, A., Auvergne, M., Baudin, F., Catala, C., Samadi, R., Alvarez, M., Mathias, P., Paparò, M., Pápics, P., and Plachy, E. 2009. HD 50844: a new look at delta Scuti stars from CoRoT space photometry. *A&A*, **506**(Oct.), 85–93.

Press, W. H., and Rybicki, G. B. 1989. Fast algorithm for spectral analysis of unevenly sampled data. *ApJ*, **338**(Mar.), 277–80.

Rao, C. R. 1945. Information and the accuracy attainable in the estimation of statistical parameters. *Bulletin of the Calcutta Mathematical Society*, **37**, 81–9.

Ritzwoller, M. H̃., and Lavely, E. M̃. 1991. A unified approach to the helioseismic forward and inverse problems of differential rotation. *ApJ*, **369**(Mar.), 557.

Roberts, D. H., Lehar, J., and Dreher, J. W. 1987. Time Series Analysis with clean. Part one: derivation of a spectrum. *AJ*, **93**(Apr.), 968–89.

Samadi, R., and Goupil, M.-J. 2001. Excitation of stellar p-modes by turbulent convection. I. Theoretical formulation. *A&A*, **370**, 136–46.

Scargle, J. D. 1982. Studies in astronomical time series analysis. II. Statistical aspects of spectral analysis of unevenly spaced data. *ApJ*, **263**(Dec.), 835–53.

Schou, J. 1992. *On the analysis of helioseismic data*. Ph.D. thesis, Århus Universitet, Denmark.

Schou, J., and Brown, T. M. 1994. Generation of artificial helioseismic time-series. *A&AS*, **107**(Nov.), 541.

Schuster, A. 1897. On lunar and solar periodicities of earthquakes. *Proc. Royal Soc. of London*, **61**, 455.

Scott, J. G., and Berger, J. O. 2010. Bayes and empirical-Bayes multiplicity adjustment in the variable-selection problem. *The Annals of Statistics*, **38(5)**, 2587–619.

Sellke, T., Bayarri, M. J., and Berger, J. 2001. Calibration of p-values for testing precise null hypotheses. *The American Statistician*, **55**(Mar.), 62–71.

Severino, G., Magrì, M., Oliviero, M., Straus, T., and Jefferies, S. M. 2001. The Solar intensity-velocity cross spectrum: a powerful diagnostic for helioseismology. *ApJ*, **561**(Nov.), 444–9.

Shannon, C. E. 1949. Communication in the presence of noise. *Proceedings of the IRE*, **37**, 10–21.

Slepian, D. 1978. Prolate spheroidal wave functions, Fourier analysis, and uncertainty. V. The discrete case. *Bell System Technical Journal*, **57**, 1371–430.

Stahn, T., and Gizon, L. 2008. Fourier analysis of gapped time series: improved estimates of solar and stellar oscillation parameters. *Sol. Phys.*, **251**(Sept.), 31–52.

Sturrock, P. A., and Scargle, J. D. 2009. A Bayesian assessment of p-values for significance estimation of power spectra and an alternative procedure, with application to solar neutrino data. *ApJ*, **706**(Nov.), 393–8.

Tanvir, N. R., Hendry, M. A., Watkins, A., Kanbur, S. M., Berdnikov, L. N., and Ngeow, C. C. 2005. Determination of Cepheid parameters by light-curve template fitting. *MNRAS*, **363**(Nov.), 749–62.

Thomson, D. J. 1982. Spectrum estimation and harmonic analysis. Pages 1055–1096 of: *Proc. IEEE*, vol. **70**.

Toutain, T., and Appourchaux, T. 1994. Maximum likelihood estimators: an application to the estimation of the precision of helioseismic measurements. *A&A*, **289**(Sept.), 649.

Toutain, T., and Fröhlich, C. 1992. Characteristics of solar p-modes: results from the IPHIR experiment. *A&A*, **257**(Apr.), 287–97.

Toutain, T., and Gouttebroze, P. 1993. Visibility of solar p-modes. *A&A*, **268**(Feb.), 309–18.

Toutain, T., Appourchaux, T., Baudin, F., Fröhlich, C., Gabriel, A. H., Scherrer, P., Andersen, B. N., Bogart, R., Bush, R., Finsterle, W., García, R. A., Grec, G., Henney, C. J., Hoeksema,

J. T., Jiménez, A., Kosovichev, A., Roca Cortés, T., Turck-Chièze, S., Ulrich, R., and Wehrli, C. 1997. Tri-phonic helioseismology: comparison of solar p modes observed by the helioseismology instruments aboard SOHO. *Sol. Phys.*, **175**(Oct.), 311–28.

Wehlau, W., and Leung, K.-C. 1964. The multiple periodicity of Delta Delphini. *ApJ*, **139**(Apr.), 843.

Wiener, N. 1930. Generalized harmonic analysis. *Acta Mathematica*, **55**, 117.

Wilks, S. S. 1938. The large-sample distribution of the likelihood ratio for testing com?posite hypotheses. *Annals of Mathematical Statistics*, **9**(July), 60.

Yoachim, P., McCommas, L. P., Dalcanton, J. J., and Williams, B. F. 2009. A panoply of Cepheid light curve templates. *AJ*, **137**(June), 4697–706.

6. An observer's views and tools

DONALD W. KURTZ

6.1 Introduction

The four lectures that I presented at the XXII Winter School of Astrophysics were an eclectic mix of topics loosely bound under the title of this chapter: an observer's views and tools. The presentations given by all of the lecturers at the Winter School are available at the time of this writing on the IAC Web site[1] (see "about the school" and "lecturers and topics" on that web site). This chapter can be read in conjunction with the lecture presentations on that web site, but that is not required. This chapter does not completely follow the order of the presentations.

6.2 Chemically peculiar and pulsating stars of the upper main sequence[2]

On and near the main sequence for $T_{\rm eff} > 6,600$ K, there is a plethora of spectrally peculiar stars and photometric variable stars with a bewildering confusion of names. There are Ap, Bp, CP, and Am stars; there are classical Am stars, marginal Am stars, and hot Am stars; there are roAp stars and noAp stars; there are magnetic peculiar stars and nonmagnetic peculiar stars; He-strong stars, He-weak stars; Si stars, SrTi stars, SrEuCr stars, HgMn stars, PGa stars; λ Boo stars; stars with strong metals, stars with weak metals; pulsating peculiar stars, nonpulsating peculiar stars; pulsating normal stars; nonpulsating normal stars; δ Sct stars, δ Del stars, and ρ Pup stars; γ Dor stars, SPB stars, β Cep stars; γ Cas stars, λ Eri stars, α Cyg stars; sharp-lined and broad-lined stars, some of which are peculiar and some of which are not. There are pre-main sequence Ae and Be stars, collectively called HAeBe stars; there are Oe and Be stars, which are not pre-main sequence stars.

What a mess! And some of the mess is partially the result of people having previously said, "What a mess!" then trying to clean up the mess by "simplifying" the nomenclature with the introduction of new names that not everyone adopted, thus adding to the mess. No cleanup is attempted here. This is a partial guide through the morass.

Table 6.1 shows some of the subgroups of chemically peculiar (CP) stars as a function of temperature. All of them are on or near the main sequence. All show spectral peculiarities. Members of the magnetic group are known to have global magnetic fields that are roughly dipolar with strengths of hundreds to tens of thousands of G. In comparison, the global magnetic fields of the Earth and Sun are $\sim\frac{1}{2}$ G; the magnetic field in sunspots is about 1.5 kG. It is possible that a few members of the "nonmagnetic" group may have magnetic fields, but it is clear that the vast majority do not. Mathys and Lanz (1990) discussed the detection of a 2 kG field in the hot Am star o Peg. Recently, with high-precision magnetic field determinations, magnetic fields have been found in spectroscopically normal O and B stars that are members of the ζ Oph, β Cep, and SPB classes of pulsating stars (see, e.g., Hubrig *et al.*, 2011a,b).

The magnetic stars are mostly known as Ap stars (for A peculiar; "Ap" is pronounced "A-pee"), but many of the "Ap" stars are B stars and F stars, often called Bp and Fp stars. However, note that there is a group, the HgMn stars, which are called Ap stars, even though they lack strong magnetic fields and they are B stars! There are A stars that have peculiar spectra, but not the same peculiarities as the Ap stars, and

[1] http://www.iac.es/winschool2010/.
[2] Based in part on Kurtz and Martinez (2000), with kind permission of Baltic Astronomy.

TABLE 6.1. Magnetic and nonmagnetic peculiar stars of the upper main sequence

T_{eff} K	Magnetic stars	Nonmagnetic stars
7000–10000	Ap SrCrEu A3–F4	Am; λ Boo A0–F2; A0–F0
10000–14000	Ap Si B8–A2	Ap HgMn B6–B9
13000–18000	He-weak Si, SrTi B3–B7	He-weak PGa B4–B5
18000–22000	He-strong B1–B2	

are nonmagnetic; these are the Am stars (for A metallic-lined stars; "Am" is always pronounced "A-em").

Am stars are given three spectral classifications: one based on the Balmer lines, which give a good measure of the effective temperature; one based on the Ca II K-line, which, because of its weakness relative to normal stars, gives an earlier spectral type; and one based on the metal lines, which, because of their enhanced strength, gives a later spectral type. "Classical Am" stars have K-line and metal line spectral types that differ by five or more spectral subclasses; this difference in "marginal Am" stars is less than five subtypes. Marginal Am stars are designated "Am," a potentially confusing notation in written text because of the use of a colon as part of the classification; "Am:" is always inelegantly pronounced "A-em-colon"; if you do not like that, then call them "marginal Am stars." Because the classification criteria are harder to diagnose for the hotter Am stars, the classical Am stars have H-line types that lie between A3 and F1. It was later recognized at higher dispersion that the Am phenomenon continues to A0 (Sirius is an Am star), so Am stars with H-line spectral types between A0 and A3 are known as "hot Am" stars.

Evolved Am stars with luminosity classes IV and III are classified as δ Del stars in the Michigan Spectral Catalogues (Houk and Cowley, 1975; Houk, 1978, 1982; Houk and Smith-Moore, 1988) and many other publications. However, Kurtz (1976) and Gray and Garrison (1989a) found the δ Del class to be highly inhomogeneous. Because of this, Gray and Garrison (1989a,b) re-classified the luminous, late-A to mid-F, evolved Am stars as ρ Pup stars, after the prototype. They recommended that the δ Del classification be dropped, but not everyone knows about or chooses to follow that recommendation. Note that all of the spectral types for the chemically peculiar stars are defined by spectral standard stars and do not depend on the physical characteristics, such as the magnetic field, that have been pointed out. Their classifications are phenomenologically defined by their spectra alone.

To rationalize this situation, Preston (1974) introduced the following nomenclature:

- CP $\equiv$ chemically peculiar star of the upper main sequence,
- CP1 $\equiv$ Am stars,
- CP2 $\equiv$ magnetic Ap, Bp stars,
- CP3 $\equiv$ HgMn stars, and
- CP4 $\equiv$ He-weak B stars.

This system is widely used along with the older Ap terminology. A good reference on this subject is Sydney Wolff's monograph on the A stars (Wolff 1983). See particularly her Table 1, which lists the observed properties of the CP stars. More recent reviews and contributions covering the field can be found in the conference proceedings of IAU Symposium 224 (edited by Zverko *et al.*, 2004).

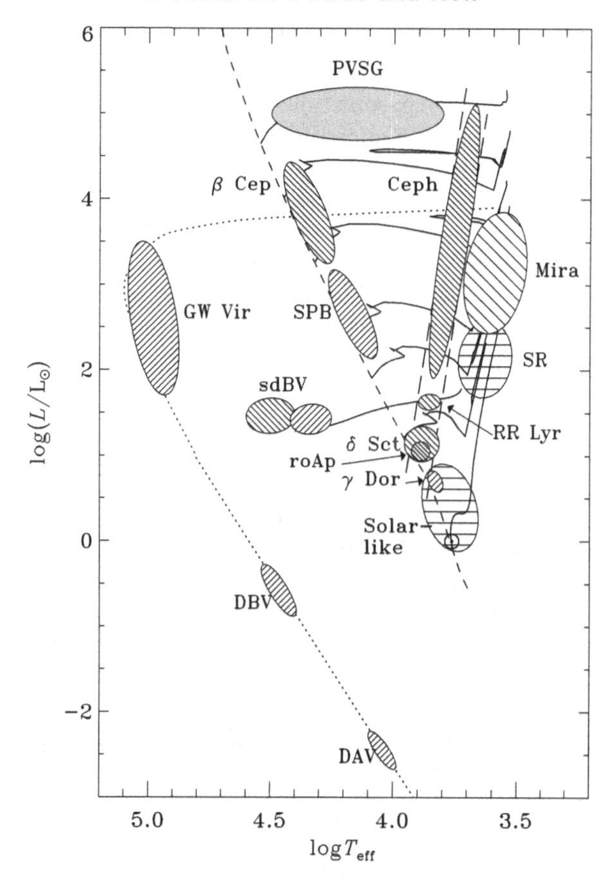

FIG. 6.1. A pulsation HR Diagram based on Fig. 1.11 of Aerts *et al.* (2010); courtesy of J. Christensen-Dalsgaard.

This list is not exhaustive for chemically peculiar stars of the upper main sequence. There are also the λ Boo stars, which have H-line types between A0 and F0, a Ca II K-line type of A0, or slightly later, and *weak* metallic lines, particularly Mg II 4481 Å. Abundance analyses show marked *underabundances* of the Fe-peak elements, with lighter elements essentially normal. Many λ Boo stars are pulsating δ Sct stars, and γ Dor stars. See, for example, the discussion of Moya *et al.* (2010) for the fascinating planet-hosting star HR 8799, which is also a γ Dor star and a λ Boo star.

6.2.1 *Pulsating stars of the upper main sequence*

It is useful here to review the nomenclature associated with the different types of pulsating stars found along the upper main sequence stars. A much more extensive, thorough discussion of these various types of pulsating stars can be found in Aerts *et al.* (2010). See particularly Tables A1 and A2 in the appendices to that book for basic parameters of the various classes of pulsating stars across the HR Diagram. Figure 6.1 shows a pulsation HR Diagram based on Fig. 1.11 of Aerts *et al.* (2010), putting the pulsating stars discussed here into context.

- The δ Sct stars are H-core-burning, main sequence and post-main sequence stars that lie in a p-mode instability strip that ranges from A2 to F0 on the main sequence and from A3 to F5 at luminosity class III. They pulsate primarily in low-overtone radial and non-radial p modes with periods between about 18 min and 8 h. With new *Kepler Mission* data, it is clear that many, if not most, δ Sct stars also pulsate in g modes (Grigahcène *et al.*, 2010), and with the μmag precision Kepler data, higher degree, ℓ, modes are also visible.

- The rapidly oscillating Ap (roAp; sometimes pronounced "r-o-A-p", sometimes "row-ap") stars are mid-A to early-F main sequence Ap SrCrEu, CP2 stars that pulsate in high-overtone p modes with periods in the range of 5–21 min and photometric peak-to-peak amplitudes ≤ 0.016 mag (generally, much less). Radial velocity amplitudes for individual spectral lines range up to 5 km s^{-1} (Freyhammer *et al.*, 2009). The roAp stars mostly lie within the δ Sct instability strip, but a few of them are cooler. Many CP2 Ap stars do not show roAp pulsations. Mathys *et al.* (1996) dubbed these stars noAp (nonoscillating Ap) stars. It is not yet clear whether the stars are truly constant, or if their oscillations are below current detection limits.
- The Herbig Ae and Be stars, known collectively as HAeBe stars, are pre-main sequence A and B stars that show emission lines from strong stellar winds and from cocoons of remnant gas from which they collapsed. These are massive counterparts of the T Tau stars. Some Herbig Ae stars are known to be δ Sct stars. Detection and study of pulsation in these interesting stars is hindered by the large photometric variations caused by their variable circumstellar envelopes.
- The γ Dor stars are multiperiodic, non-radial g-mode main sequence pulsators with periods in the range of 0.3 to 3 d. A study of 70 γ Dor candidates from the Hipparcos catalogue shows that they are confined to an instability strip that partially overlaps the δ Sct stars in the HR Diagram; they range in temperature from 7,200 to 7,700 K on the main sequence, and 6,900 to 7,500 K for $\log g \approx 4$ (Handler, 1999). Kepler mission data show that there are pure g mode γ Dor stars, but that many stars are γ Dor–δ Sct hybrids, so the distinction between these two classes is now blurred (Grigahcène *et al.*, 2010).
- The slowly pulsating B (SPB) stars are multiperiodic, non-radial g-mode main sequence pulsators with periods in the range of 0.6 to 3 days and photometric amplitudes typically less than a few hundredths of a magnitude. They lie in a narrow instability box in the HR Diagram that ranges from B2 to B9, thus they do not overlap with the δ Sct stars.
- The β Cep stars are giant and subgiant, p-mode pulsators with periods in the range of 2 to 7 h. Some are singly periodic, some are multiperiodic; they pulsate in both radial and non-radial modes of low degree, $\ell \leq 3$, and low overtone, n. Their photometric amplitudes are ≤ 0.1 mag, except for the high amplitude star BW Vul, about which there is extensive literature.
- Even hotter than the β Cep stars there are (possibly) pulsating O stars known as α Cyg variables, which we now call periodically variable supergiant (PVSG) stars. Their periods are typically of the order of 10 to 100 days, and their amplitudes are less than a few tenths of a magnitude. They are thought to be pulsating in g modes, or so-called "strange modes" – modes associated with a sound-speed inversion, caused by a density inversion, caused by an opacity bump, most likely from Fe, H, and/or He.
- There was one report of rapid pulsations with a period of 627 s driven by the ϵ-mechanism (where ϵ represents energy generation in the core) in the Wolf-Rayet star HD 96548 (WR40), but this could not be confirmed. Discovery of such pulsations in 30–40 M$_\odot$ Wolf-Rayet stars would be tremendously exciting, since they are almost-bare helium cores of massive stars that shed their H in 10^{-5} M$_\odot$ yr^{-1} winds as they try to evolve to giants, but run up against the Eddington limit instead. This causes them to return repeatedly to a new, higher mean molecular weight (μ) main sequence where CNO cycle products show in the spectra of WN stars, and 3α products show in the spectra of WC stars. Finding pulsations in these stars would provide important constraints on their internal structure.

- Amongst the B stars, there are emission-line Be stars that have periodic variations with periods between of 0.3 to 3 days and amplitudes between 0.01 and 0.3 mag. These are known as λ Eri stars. It is contentious whether their variability is caused by g mode pulsation, or by some rotational effect.

To round out this discussion of the nomenclature of the pulsators of the upper main sequence, it is useful to be aware of the nonpulsating variables, too. They include the following kinds of stars:

- Eruptive luminous blue variables (LBVs), also called S Dor stars, with variations of about 0.2 mag on time scales of weeks to months. If there is thought to be periodicity in the variations, then these are called PVSG stars as has been discussed. LBVs also have eruptions of 0.5 to 2 mag on a time scale of years to decades, and giant eruptions on the time scale of a millennium, η Car being the best known, most spectacular example of this. Thus, finding pulsation amid all the other, higher amplitude, nonperiodic variations is a challenge similar to that in the HAeBe stars.
- Wolf-Rayet stars that show nonperiodic variations with amplitudes of the order of 0.02 mag caused by the variable winds and possibly by rotation.
- Nonperiodic, long-time-scale light variable Oe and Be stars known as γ Cas stars, which are in addition to the periodic Be stars (λ Eri stars) mentioned previously. They range from spectral type O6 to B9 and through luminosity classes V to III. Their light variations are caused by variable winds.

6.2.2 *The roAp stars*

The rapidly oscillating Ap (roAp) stars are cool Ap SrCrEu (CP2) stars that exhibit many unusual characteristics:[3] strong abundance anomalies, particularly of rare earth elements; strong, global magnetic fields with polar field strengths typically of several kG, but up to 24.5 kG (Hubrig *et al.*, 2005; Kurtz *et al.*, 2006a); nonuniform abundances, both horizontally (spots) and vertically (stratification); high radial overtone pulsation with periods in the range of 5.65 to 21.2 min and amplitudes of individual pulsation modes up to 6 mmag in broadband photometry and up to 5 km s^{-1} in radial velocity (see, e.g., Freyhammer *et al.*, 2009 and Table 1 of Kurtz *et al.*, 2006a).

Theoretical interpretation of the pulsations in roAp stars is complex: The modes appear to be more or less aligned with the magnetic field (see, e.g., Bigot and Dziembowski, 2002; Bigot and Kurtz, 2011). They are low degree, $\ell \leq 3$, modes and preferentially dipole (or distorted dipole) modes, where this can be determined. Both the large and small asteroseismic separations are significantly perturbed by the magnetic field and, at least in some cases, by the rotation. The modes are magneto-acoustic – generally, magnetic pressure dominates in the observable atmosphere, although for some models with ~kG magnetic fields the magnetic and gas pressure are comparable around $\tau_{5000} \sim 1$; at higher observable atmospheric levels magnetic pressure dominates in all models (Saio, 2005; Sousa and Cunha, 2011). Atomic diffusion (the competition between radiative levitation and gravitational settling; see Section 6.3) not only alters the vertical and horizontal abundances, but also the atmospheric temperature structure – manifestations of the latter may be the corewing anomaly in the hydrogen lines (Cowley *et al.*, 2001; Kochukhov *et al.*, 2002) and the wing-nib anomaly in the Ca K line (Cowley *et al.*, 2006). The frequencies in many roAp stars exceed the acoustic cutoff frequency of models with standard stellar atmospheres, thus also indicating abnormal atmospheric structure (see, e.g., Cunha, 2006; Audard *et al.*, 1998).

The magnetic field directly alters the eigenfrequencies and eigenfunctions, it indirectly contributes to driving the pulsations by suppression of convection, and its geometry

[3] Based in part on Kurtz *et al.* (2007), with kind permission of MNRAS.

contributes to, or defines, the mode selection. For detailed discussion of the theory of these stars, see, for example, Dziembowski and Goode (1985); Shibahashi and Takata (1993); Takata and Shibahashi (1995); Dziembowski and Goode (1996); Gautschy *et al.* (1998); Bigot *et al.* (2000); Cunha and Gough (2000); Balmforth *et al.* (2001); Bigot and Dziembowski (2002); Saio and Gautschy (2004); Saio (2005); Cunha (2005, 2006, 2007); Bigot and Kurtz (2011); and Sousa and Cunha (2011). For more introduction to the observations, see Kurtz *et al.* (2006b). A list of 35 of the more than 40 known roAp stars, with some photometric and spectroscopic amplitudes, is given by Kurtz *et al.* (2006a).

All of these complexities make the roAp stars a challenge to understand, but the rewards for doing so are great. Here are some examples:

(i) As mentioned earlier, abundances of many elements and ions are nonuniformly distributed over the stellar surfaces of the roAp stars, and they are also vertically stratified. Both of these are a consequence of atomic diffusion, that is, radiative levitation and gravitational settling. For some ions, overabundances can be up to five orders of magnitude greater than that of the Sun and other (relatively) chemically normal stars; in the spots this may rise to seven orders of magnitude (see Fig. 6.3). The horizontal nonuniformity allows Doppler imaging maps such as the one in Fig. 6.3 to be made of the surface abundance distributions and pulsation geometry (see, e.g., Kochukhov *et al.*, 2004; Kochukhov, 2006; Freyhammer *et al.*, 2009) and the vertical stratification provides depth information. As yet, the vertical and horizontal components of the abundances and pulsation geometries have not been disentangled, but several groups are actively working on this complex and potentially highly rewarding problem (see Kurtz *et al.*, 2006b; Sousa and Cunha, 2011, for more extended introductions to the subject and references).

(ii) Fe ranges from slightly overabundant to somewhat depleted in the atmospheres of roAp stars as a consequence of atomic diffusion; lines of Fe I and Fe II typically form around continuum optical depth $\log \tau_{5000} \sim -0.5$. The narrow cores of the Hα lines are a probable consequence of abnormal $T - \tau$ that is, as yet, not fully modeled; these Hα line cores form between optical depths $-4 \leq \log \tau_{5000} \leq -2$, levels that are well into the chromosphere of the Sun (evidence for chromospheres in Ap stars is not strong, so we do not refer to these upper atmospheric levels as "chromospheric" here). The first and second ionization states of Nd and Pr and some other rare earth elements form at, or above, optical depths $\log \tau_{5000} \leq -4$ (see, e.g., Mashonkina *et al.*, 2005), as they have been radiatively levitated. Thus, by studying lines of different ions, and by studying the line profile variations of individual lines, it is possible to probe the abundance distributions and pulsation behavior as a function of atmospheric depth in some detail. This in itself has potentially great rewards, as the geometry of the pulsation modes provides new, independent constraints on the stratification of the atmospheres, and hence on the atomic diffusion that gives rise to this.

Figure 6.3 shows a Doppler imaging map of the distribution of Nd III over the surface of the roAp star HD 99563 (Freyhammer *et al.*, 2009) where the extreme variation of the abundance of this ion over the stellar surface can be seen. Figure 6.2 shows schematically the different levels of the stellar atmosphere that are observable in roAp stars as a consequence of vertical stratification from atomic diffusion and strong opacity in the hydrogen lines. These two diagrams thus illustrate how it is possible to observe the atmospheres and pulsation modes of the roAp stars in 3D, and how it is possible to see high in the atmosphere – at levels that would be in the chromosphere in the Sun – as a consequence of the radiative levitation of, in particular, rare earth elements.

(iii) Theoretical models (see, e.g., Saio and Gautschy, 2004; Saio, 2005; Cunha, 2006; Sousa and Cunha, 2011) indicate that the Alfvén speed and sound speed are comparable at optical depths of the order of 1 for magnetic field strengths of the order

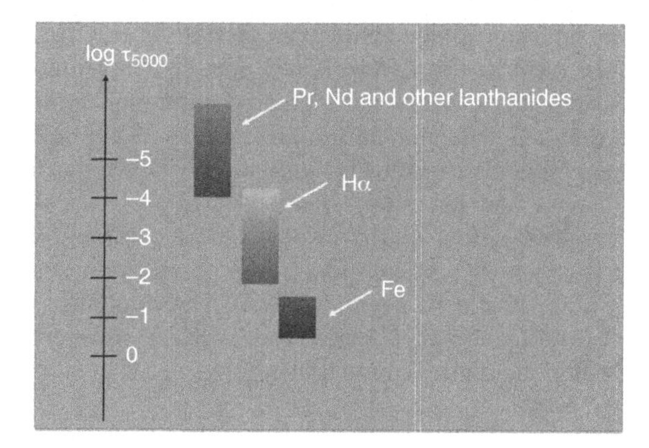

FIG. 6.2. A schematic diagram showing how the combination of vertical stratification and high opacity of hydrogen allows vertical resolution of atmospheric structure in roAp stars. Courtesy of Lars Freyhammer.

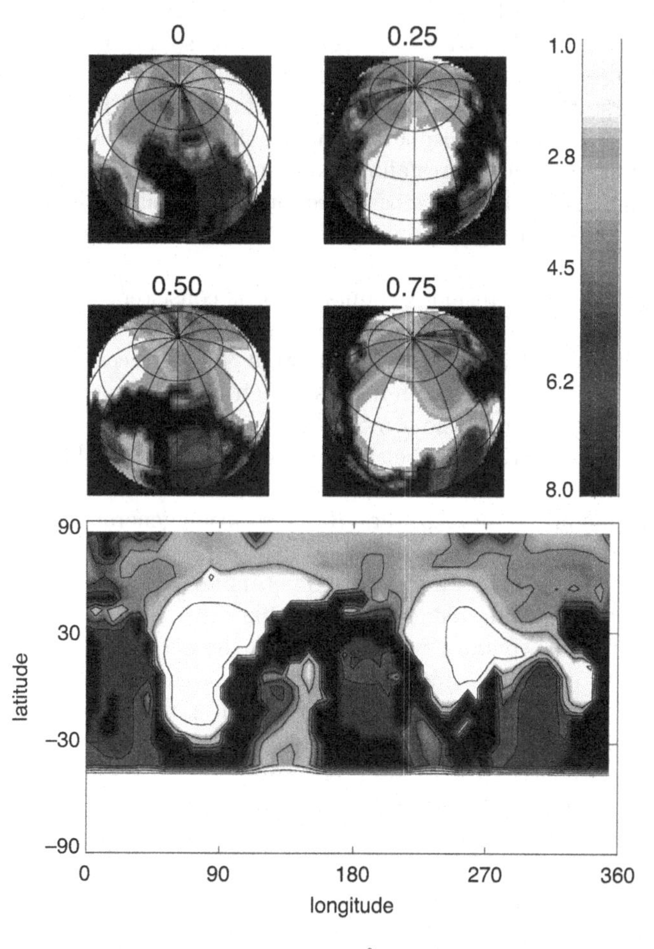

FIG. 6.3. Doppler imaging map for Nd III 6145.07 Å for HD 99563. The abundances are on a scale of $\log N_H = 12$; the solar abundance of Nd is 1.41 on this scale (Asplund *et al.*, 2009), thus the highest overabundances of Nd III in the spots rise to $10^{6.5}$ times solar at the centers of the two spots associated with the magnetic poles. These are evident at rotation phases 0.0 and 0.5, and they lie near the stellar rotational equator (latitude 0°) as a consequence of the magnetic obliquity being near 90°. The white color below latitude −44° represents the part of the star that is never visible because of the $i = 44°$ rotational inclination. From Freyhammer *et al.* (2009).

of 1 kG; higher in the atmosphere the Alfvén speed exceeds the sound speed significantly. For stronger fields, the Alfvén speed exceeds the sound speed throughout the observable atmosphere. Thus, the observed modes in roAp stars are magnetoacoustic. The detailed behavior of the interaction of the pulsation modes with kG-strength magnetic fields can thus be compared with similar interactions of solar p modes and sunspots that have been studied with local helioseismology. Whereas there is a good understanding of the interaction of p modes with sunspots, with basic agreement between theoretical models and observations, the variety of behavior for the roAp stars is still to be understood. Disentangling the complexities of the line profile variations into their horizontal, vertical, convective, and pulsational mode geometry components is highly challenging.

6.3 Atomic diffusion

The combination of gravitational settling and radiative levitation of ions, atomic diffusion, is widely applicable in stellar astrophysics; it is an important process that is part of the standard solar model (at least for helium settling and the radiative levitation of a few atomic species), and has to be taken into account in any solution to the current disagreement between helioseismic sound speeds and solar abundances. It is also an important factor in our understanding of pulsational driving in subdwarf B (sdB) stars, the stratification of white dwarf star atmospheres, and globular cluster ages. In the roAp stars, the pulsation amplitudes and phases allow us to specify which ions share horizontal and vertical distributions in the roAp atmospheres. Since these stars show the most extreme evidence of atomic diffusion, they provide the strongest constraints on atomic diffusion theory.

In the CP stars, atomic diffusion accounts for or is consistent with:

- Overabundances of the Fe-peak and rare earth elements.
- Underabundances of Ca, Sc, C, He in Am stars. This is a consequence of their noble gas electronic configuration, which results in tightly bound electrons and few spectral lines near flux maximum in the atmospheres of A stars.
- The ages of CP stars. They are relatively young. The atmospheric anomalies disappear when the stars evolve to be red giants.
- The slow rotation in Am and Ap stars. Rapid rotation generates turbulent meridional circulation that inhibits diffusion.
- The binary fraction of Am stars. Most Am stars are found in short-period binaries. The resulting tidal locking reduces the rotation rate.
- Magnetic braking in Ap stars, which results in the necessary slow rotation.
- The spots in Ap stars, which are a consequence of the effects of the magnetic field on atmospheric stability, and of the magnetic field guiding diffusion.
- The stratification of Ap stars' atmospheres.
- The pulsationpeculiarity relationship: Ap stars may be roAp stars pulsating in high radial overtone p modes, but not δ Sct stars pulsating in low radial overtones. This is a consequence of the suppression of the low overtones by the magnetic field and the driving (Saio, 2005).
- The Am–δ Sct star relation. It was previously thought that an exclusion existed between Am and δ Sct stars. It is now known that Am stars do not pulsate with high amplitudes, but are low-amplitude δ Sct stars in many cases. There may be a high incidence of δ Sct–γ Dor hybrids among Am stars (Grigahcène *et al.*, 2010).

Here are some other examples, both real and imagined, of atomic diffusion to illustrate the idea and its applications: Atomic diffusion velocities are typically of the order 10^{-6} to $10^{-2}\,\mathrm{m\,s^{-1}}$, hence the significant effects only occur where there is little or no turbulent mixing. For perspective and amusement, imagine if the Sun had no mixing processes at all and time were no object. The Sun would become stratified with a uranium sphere at the center and shells of less dense atomic nuclei lying above that up to a hydrogen atmosphere. How big would the central U sphere be? Using the solar abundances of Asplund *et al.* (2009), where the abundance of U is taken to be $\log N_U = -0.54$ (on a $\log gN_H = 12.00$ scale), we can calculate that if all the uranium settled to the center of the Sun, the central U sphere would have a radius of about 50 kms. For the Sun, this is utterly unphysical, but it gives a mental picture of gravitational settling and of the relative abundance of U. Although such total stratification of elements does not happen for main sequence stars, to first order it is not a bad description of white dwarfs, a point to which we will return.

Another example of gravitational settling can be found in Jupiter. Most of the radiation emitted by Jupiter is not reflected sunlight, but the self-generated luminosity of Jupiter radiated in the infrared as a consequence of Jupiter's surface temperature. This intrinsic luminosity is about three times greater than the reflected sunlight, and amounts to $L = 4 \times 10^{17}$ W. The source of this luminosity is not nuclear reactions, of course; Jupiter is not a star. It comes from gravitational potential energy generated by He settling – atomic diffusion. For perspective, if we imagine that it is the result of overall contraction of Jupiter, it is straightforward to calculate that Jupiter would be shrinking by about $0.1\,\mathrm{mm\,yr^{-1}}$. Yet that drives the tremendous weather we see on the planet, the belts and storms, and even the Great Red Spot.

Helium settling is included in the standard solar model, where it affects surface luminosity of the Sun at about the 1 mmag level. It is a much more important effect in red giant stars, where its inclusion in models may reduce the ages of globular clusters by as much as 1–2 Gyr (Chaboyer *et al.*, 1996). The source of this age reduction is easy to see. Helium burning is extremely sensitive to temperature: $\epsilon(3\alpha) \sim T^{30}$. Gravitational settling in the He core of a red giant star releases a small amount of gravitational potential energy that heats the core slightly, but the consequence of that is a large increase in the energy generation rate, hence a reduction in the lifetime of the star.

Lithium fuses at 2–2.4×10^6 K for its two isotopes, Li^6 and Li^7. The convective layers in cooler main sequence stars drag Li down to depths that are hot enough to fuse the Li, hence deplete it from the observable atmosphere. Thus for main sequence stars in the range $7{,}500 \geq T_{\rm eff} \geq 5{,}000$ K, Boesgaard and Tripicco (1986) found a steady decrease in Li abundance with decreasing surface temperature, as expected as a result of the growth of the depth of the surface convection zone with decreasing temperature. However, between about 7,000 K and 6,300 K in Hyades stars, Boesgaard and Tripicco (1986) found Li to be unusually depleted; they suggested this is a consequence of gravitational settling. More recently, a suggestion of enhanced depletion of Li in an exoplanet-hosting star has been refuted by Baumann *et al.* (2010).

Another manifestation of radiative levitation is found in Moon rocks. The Moon has been collecting the solar wind for 4.5 Gyr. Baking Moon rocks to release the trapped helium reveals a He^3/He^4 ratio that is much higher than in the solar atmosphere. This is a consequence of He^3 having spectral lines slightly shifted in wavelength to those of He^4. Both isotopes are partially driven from the surface of the Sun by radiation pressure. Because He^3 is less abundant, its spectral lines are less saturated and individual atoms are thus able to absorb more photons; they do not share them with so many other He^3 atoms as in the case of He^4, hence He^3 is preferentially driven in the solar wind. A similar effect leads to the extreme isotopic anomalies of Hg seen in the late B (CP3) HgMn stars.

The most extreme cases of gravitational settling are observed in white dwarf stars. We illustrate this in the next section with the interesting problem of the "DB gap."

6.3.1 *White dwarfs and the DB gap*

White dwarf stars are spectroscopically classified into six major subtypes with additional classifications indicating crossover spectra.[4] In addition, there are further subtypes that

[4] Based in part on Kurtz *et al.* (2008), with kind permission of MNRAS.

specify the presence of polarization, magnetic fields, and pulsation, plus there are classes for stars that do not fit any other class! For an introduction to the white dwarf alphabet zoo, see Table 1 of McCook and Sion (1999). See Fig. 6.1 for perspective on where the white dwarf stars lie in the HR Diagram. The hottest pre-white dwarf stars that appear on the white dwarf cooling sequence in the HR diagram have effective temperatures of nearly 200,000 K, ranging down to about 80,000 K. Some of these hottest white dwarf stars are central stars of planetary nebulae, some are not; some pulsate, some do not.

For white dwarfs cooler than $T_{\text{eff}} \leq 80{,}000$ K, there is a clear spectroscopic sequence into which the vast majority of known white dwarfs fit. The DO stars lie in the range $80{,}000 \geq T_{\text{eff}} \geq 45{,}000$ K and show strong spectral lines of He II. It is the presence of the He II lines that gives them their "O" subclass, in analogy with the main sequence O stars, but note that there is not a direct correlation in temperature with the main sequence O stars, which are generally cooler. Thus the DOs are helium-atmosphere white dwarfs. Much cooler than the DOs are the DB white dwarfs, with approximately $30{,}000 \geq T_{\text{eff}} \geq 12{,}000$ K. These stars are also helium-atmosphere white dwarfs, where in this case the DB classification indicates that generally only spectral lines of He I are seen with little or no H or metal lines. Note again that the "B" classification is in analogy with main sequence B stars that show lines of neutral helium in their spectra, but that the temperatures do not necessarily correspond.

The vast majority of the rest of the white dwarfs are classified as DA stars because they show only lines of the Balmer series of hydrogen in their visible wavelength spectra. They are thus subclass "A" in analogy to the main sequence A stars, which by definition show the strongest Balmer lines of all main sequence stars. But even more so than for the DOs and DBs, the temperatures of the DAs do not in general correlate with main sequence A stars. While there are DAs in the main sequence A star temperature range of $10{,}000 \geq T_{\text{eff}} \geq 7{,}400$ K, other DAs may be found with temperatures as hot as 170,000 K and as low as those of the coolest white dwarf stars known, $T_{\text{eff}} \sim 4{,}500$ K, a lower limit set by the age of the galaxy; there has not yet been time for the first white dwarfs to cool beyond this limit.

While white dwarf stars are predicted to be the end state of evolution for main sequence stars with $M \leq 8\ \text{M}_{\odot}$ and are the most common class of stars in the galaxy, the number known is not very great as a consequence of their intrinsic faintness. Thus the study of white dwarfs is plagued by the same problem as the study of the coolest main sequence dwarfs, brown dwarfs, and especially extrasolar planets: a dearth of photons.

The number of known and studied white dwarfs has increased dramatically in recent years as a consequence of the data available from the Sloan Digital Sky Survey (SDSS; York *et al.*, 2000). While the SDSS is primarily an extra-galactic project, its uniform, relatively deep (magnitude 23) wide-field data set is of significant use for stellar astronomy, in particular for the discovery of new white dwarf and hot subdwarf stars. McCook and Sion (1999) presented a catalogue of 2,249 white dwarfs with spectroscopic classifications; their data were complete through early 1996. Kleinman *et al.* (2004) increased the number of spectroscopically classified white dwarfs by 2,551 stars using the first SDSS data release, thus more than doubling the number known. A subsequent, more extensive work from the SDSS data release 4 (Eisenstein *et al.*, 2006a) presented a catalogue of 9,316 spectroscopically confirmed white dwarf stars. Thus, the majority of white dwarfs now known have been found in the SDSS. A study of these SDSS white dwarfs by Kepler *et al.* (2007) further illuminates the temperature and mass distribution for DA and DB stars roughly in the range $40{,}000 \geq T_{\text{eff}} \geq 12{,}000$ K.

Eisenstein *et al.* (2006a) confirm what was earlier known from smaller samples of white dwarfs: DA white dwarfs dominate, constituting about 86% of all white dwarfs in their sample (8,000 of 9,316 stars). DBs are the next most numerous group, comprising 8 percent of the sample (713 of 9,316 stars). All other classes comprise the remaining

6 percent of the sample. Hence hydrogen-atmosphere DA white dwarfs are ten times more common than helium-atmosphere DB white dwarfs.

While DAs can be found at all temperatures under 170,000 K down to the low temperature limit set by the age of the galaxy, it is a remarkable and intriguing fact that few helium atmosphere white dwarfs occur in the effective temperature range of 45,000 $\geq T_{\rm eff} \geq$ 30,000 K. The DOs and DBs are found on either side of this temperature range, but only a very few genuine helium atmosphere white dwarfs are found within it (see, e.g., Eisenstein *et al.*, 2006a). This exclusion is known as the "DB gap" (Liebert, 1986).

The process of atomic diffusion discussed in the last section is important in this context. Gravitational settling stratifies the structure of white dwarf stars, and in the presence of such strong gravitational fields it does so quickly. In the absence of any mixing – particularly in the absence of convection – we would expect all white dwarf stars with sufficient residual hydrogen ($>10^{-15}$ $M_\odot$ of H) to be DA stars. Some non-DA white dwarfs probably do not fulfil this condition, that is to say, they lack sufficient hydrogen to be DA stars. The amount of mass loss, and in particular the amount of hydrogen loss during stellar evolution, must depend on an individual star's circumstances. The major pathway to white dwarfdom is through single-star evolution with envelope loss, leaving behind as the white dwarf the previous stellar core. Two other pathways are: a less common one of evolution from a hot subdwarf on the extreme horizontal branch to white dwarf, and a process with unknown frequency: binary merger. It is assumed that the vast majority of white dwarfs follow the first evolutionary path.

In that context, Fontaine and Wesemael (1987) explained the DB gap as a natural consequence of the evolution of almost all white dwarfs from planetary nebulae nuclei. They supposed that a slow rise of hydrogen to the surface, as heavier nuclei sink in the strong gravitational field, eventually makes helium-rich white dwarfs appear as DA stars at the blue edge of the DB gap at about 45,000 K. Note that hydrogen is not being radiatively levitated in this case, but rises instead as a consequence of being the lightest nucleus in an environment of gravitational settling. They then explained the red edge of the DB gap as a natural consequence of the onset of a significant convection zone at the temperature where the He I/II ionization zone coincides with the upper atmosphere, thus mixing the small amount of residual hydrogen into a deeper sea of helium, so the star then appears as a DB white dwarf. The H is essentially overwhelmed by the more abundant He and becomes observationally undetectable, or, at best, difficult to detect.

Shibahashi (2005) revisited this idea and proposed a different model for the onset of the blue edge of the DB gap as white dwarfs cool: the blue edge of the DB gap occurs at the effective temperature where the He III/II ionization zone becomes deep enough that the surface convection zone of the DO stars disappears. Hence, in this model, the stars are always potentially DAs, but convection in the He III/II ionization zone mixes the atmosphere so the dominant helium appears in the spectrum for the DO stars, then similar mixing occurs again when the stars cool to the red edge where the He I/II ionization zone again generates a convection zone, as in Fontaine and Wesemael's original suggestion. The difference between the two models is in the time scale for gravitational settling, hence hydrogen floating to the surface. For Fontaine and Wesemael (1987), this happens slowly, during the cooling of a star from a PG 1159 pre-white dwarf stage; for Shibahashi (2005), it happens quickly, as soon as the convection is turned off at 45,000 K. One possible explanation for some of the small number of DB stars that do appear in the DB gap is that they are truly stars for which there is virtually no residual hydrogen to form an optically thick atmosphere. However, this is inconsistent with the conclusion of Eisenstein *et al.* (2006b) from statistical studies that some of the DB stars in the DB gap cool to be DA white dwarfs. They conclude that the DB gap is real, but clearly there are mysteries concerning this gap that are still to be solved.

Based on this scenario of the spectral evolution from DA to DB at the cool end of the DB gap, Shibahashi (2005) suggested that the photospheric level of an apparently

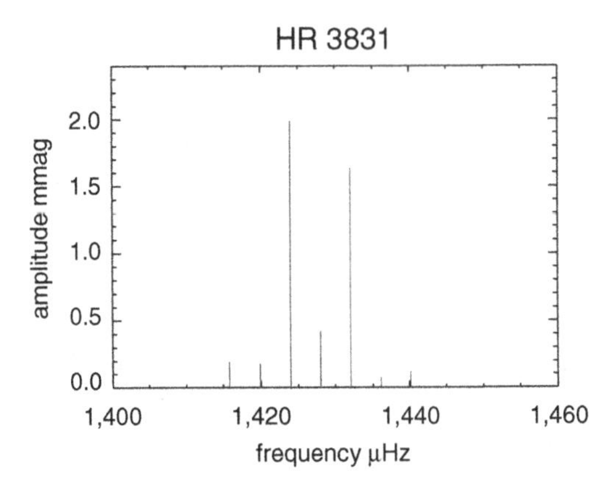

FIG. 6.4. A schematic amplitude spectrum of the principal frequency septuplet for the roAp star HR 3831. The splitting of the components is exactly equal to the rotation frequency to high precision, showing that this is not a set of frequencies arising from $\ell = 3, m = -3, -2, -1, 0, +1, +2, = 3$ modes. It is the signature of a single pulsation mode along an axis that is not the rotation axis of the star.

DA white dwarf – yet potentially helium-atmosphere DB white dwarf in the DB gap – is superadiabatic while it is convectively stabilized by a chemical composition gradient (i.e., cooler but lighter hydrogen is stratified above hotter but heavier helium). He further pointed out that under such a condition, the radiative heat exchange brings about an asymmetry in the g-mode oscillatory motion in such a way that an oscillating element overshoots its equilibrium position with increasing velocity. A local stability analysis of a white dwarf model in the DB gap by Shibahashi (2007) then suggests that g modes may be excited for stars at the cool end of the gap, that is, with temperatures around 30,000 K. He finds higher degree, l, modes preferentially excited, but some modes with $l < 3$ are excited in the models, so they may reach observable amplitudes.

There have been few searches for pulsation in DA white dwarfs of this temperature. Kurtz *et al.* (2008) surveyed stars that have DA atmospheres, but are structurally close to the known DBV (DB variable) pulsating stars, the V777 Her stars. They found some positive results that need confirmation, but suggest that there may be a new class of white dwarf pulsators for asteroseismic study.

6.4 The oblique pulsator model

6.4.1 *HR 3831*

When I discovered the roAp stars (Kurtz, 1982), the third star of this type found was the bright star HR 3831. Intriguingly, the amplitude spectrum of this star showed an equally split multiplet. With the early data this appeared to be a triplet, but with further monitoring it became clear that it is a septuplet. Figure 6.4 shows this from data obtained on 125 nights between 1993 and 1996 (Kurtz *et al.*, 1997). This appears at first to be the signature of a set of rotational split modes for $\ell = 3, m = -3, -2, -1, 0, +1, +2, +3$.

In that case, to first order, the splitting is expected to be given by the Ledoux relation (Ledoux, 1951):

$$\nu_{n\ell m} = \nu_{n\ell 0} + m(1 - C_{n\ell})\Omega/2\pi \tag{6.1}$$

where $\nu_{n\ell m}$ is the observed frequency, $\nu_{n\ell 0}$ is the rotationally unperturbed central frequency of the multiplet for $m = 0$, Ω is the rotation frequency, and $C_{n\ell}$ is a small constant that depends on the pulsation mode and the model of the star. For δ Sct stars, for example, $C_{n\ell} \sim 0.15$ and for the high radial overtones of roAp stars it is expected to be smaller, perhaps as little as $C_{n\ell} \sim 0.01$. Importantly, the value of $C_{n\ell}$ is *not* zero for rotationally perturbed m modes.

For HR 3831, the splitting of the frequencies in the multiplet shown in Fig. 6.4 is so precise that it can be shown that $C_{n\ell} < 10^{-5}$, if Equation 6.1 is applied. Because the roAp stars have abundance spots, as discussed in Section 6.2.2, their rotation periods can be measured to high precision. For HR 3831 and for some other roAp stars, it is clear that the time of pulsation maximum nearly coincides with the time of rotational light extremum and magnetic extremum. If $C_{n\ell}$ has any value other than exactly zero, this is a coincidence, because the separation of the frequencies in the multiplet would then be slightly different to the rotation frequency. This led me to propose the simple oblique pulsator model (Kurtz, 1982).

What happens when the pulsation axis of a mode is not the rotation axis? The mode is then seen from a varying aspect angle with the rotation of the star. For a normal mode, the luminosity variations then go as

$$\frac{\Delta L}{L} \propto P_\ell^m(\cos\alpha) \tag{6.2}$$

where α is the angle between the pulsation axis and the line of sight; that is, this is the co-latitude of the pulsation mode, as the angle is measured from the pulsation pole, not the equator.

It is useful to note that in the absence of limb darkening or spots, the observed amplitude is zero when the line of sight is directly over a pulsation node. Take some examples of this: (1) For an axisymmetric $\ell = 1, m = 0$ dipole pulsation mode $P_1^0(\cos\alpha) \propto \cos\alpha$, so when the mode is observed equator-on, the line of sight is over the node and the observed amplitude is zero. This is obvious, because the variations in opposite hemispheres are in antiphase and both are seen with equal visibility in this case, so they cancel each other exactly. For this same mode, maximum amplitude is observed when the mode is seen pole-on. (2) For a sectoral $\ell = 1, m = 1$ dipole mode $P_1^1(\cos\alpha) \propto \sin\alpha$, so that pole-on the observed amplitude is zero, because the line of sight is over the node (even though it is rotating, this is always true over the pulsation axis), whereas maximum amplitude is seen when the mode is equator-on. (3) The case of an axisymmetric quadrupole mode is slightly more complex, since $P_2^0(\cos\alpha) \propto \frac{3}{2}\cos\alpha - \frac{1}{2}$. Then, we see maximum amplitude when viewing the mode pole-on, there is a secondary maximum when viewing the mode equator-on that has half the amplitude of the pole-on view, and we see zero amplitude when the line of sight is over either of the nodes that occur in both hemispheres when $\alpha = \pm\cos^{-1}\sqrt{\frac{1}{3}} = \pm 54.74°$. Remembering that this is the co-latitude, the nodes for an axisymmetric quadrupole mode are at latitudes $\pm 35.26°$.

With the perspective that the observed amplitude of a mode varies with rotation when the pulsation axis is not the rotation axis, we can look at a general form of this. The angle between the line of sight and the pulsation pole, α, is given by

$$\cos\alpha = \cos i \cos\beta + \sin i \sin\beta \cos\Omega t \tag{6.3}$$

where i is the rotational inclination, β is the obliquity of the pulsation axis to the rotation axis, and Ω is the rotation frequency of the star. Taking the example of an axisymmetric dipole mode with $\ell = 1, m = 0$, a bit of trigonometric manipulation shows that

$$\frac{\Delta L}{L} = A_{-1}\cos[(\omega - \Omega)t + \phi] + A_0 \cos[\omega t + \phi] + A_{+1}\cos[(\omega + \Omega)t + \phi] \tag{6.4}$$

where ω is the pulsation frequency and ϕ is its phase. In this case

$$\frac{A_{+1} + A_{-1}}{A_0} = \tan i \tan\beta \tag{6.5}$$

contains information about the mode geometry. Thus the single dipole pulsation mode gives rise to a frequency triplet when the pulsation axis is not the rotation axis. In general,

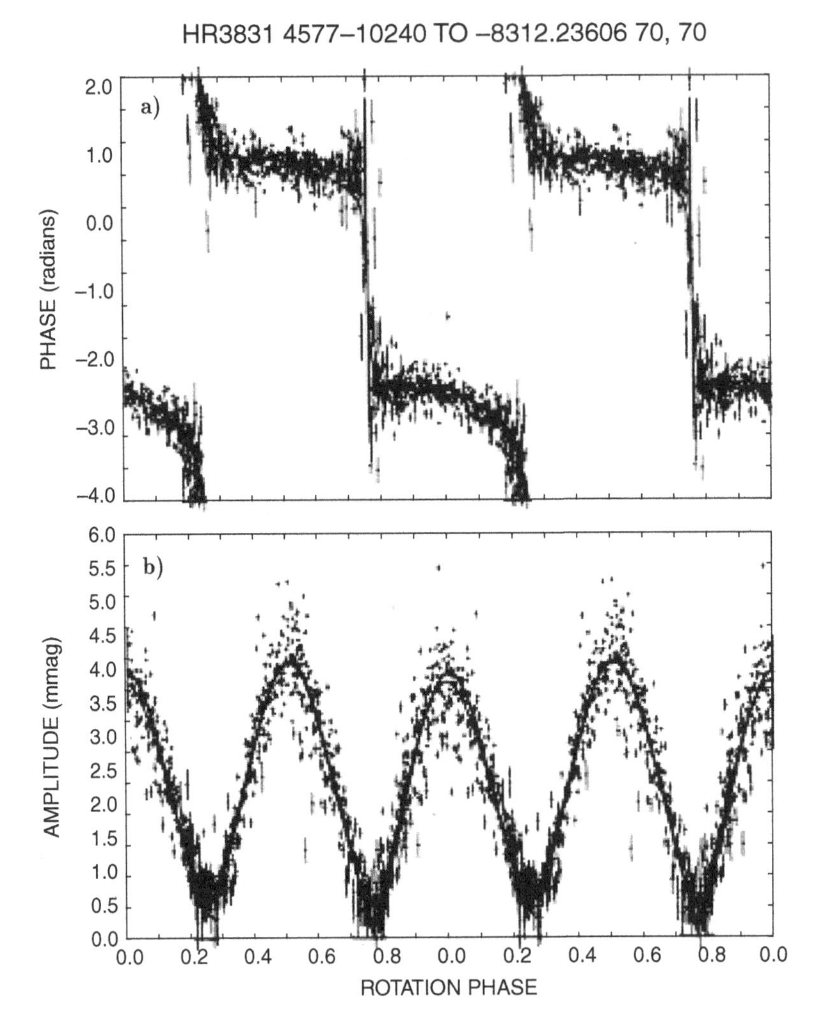

FIG. 6.5. This diagram shows the variation in *(a)* phase and *(b)* pulsation amplitude for HR 3831 with respect to rotational phase. It plots the pulsation amplitude and phase determined from sections of data 1 hr (5 pulsation cycles) long. Two rotation cycles are shown for clarity. This is a clear signature of a magnetically distorted axisymmetric dipole mode. From Kurtz *et al.* (1997).

a normal mode will give rise to a multiplet of $2\ell + 1$ components when observed from variable aspect.

In the case of HR 3831, shown in Fig. 6.4, the septuplet arises from a single pulsation mode where the pulsation axis is inclined to the rotation axis. In this case, the presence of the magnetic field distorts the mode from a normal mode so that it cannot be described as a simple dipole, hence there are higher components giving rise to a septuplet, rather than a simple triplet, even though the mode is thought to be basically an axisymmetric dipole mode. This can be seen graphically in Fig. 6.5 where a least-squares fit of the single pulsation frequency has been done to sections of data 1 hour long (about 5 pulsation cycles) with good coverage of the full rotation period of 2.85 d. It can be seen how the amplitude varies with rotational aspect, and how the pulsation phase flips by π radians at quadrature, when the dominant hemisphere changes – just as expected for a dipole.

Although I originally conceived the oblique pulsator model for the roAp stars with the idea that the pulsation axis coincided with the magnetic axis, deeper insight into the physics has been provided by many theoretical studies. See, for example, Dziembowski and Goode (1985); Shibahashi and Takata (1993); Takata and Shibahashi (1995); Dziembowski and Goode (1996); Gautschy *et al.* (1998); Bigot *et al.* (2000); Cunha and Gough (2000); Balmforth *et al.* (2001); Bigot and Dziembowski (2002); Saio and Gautschy (2004); Saio (2005); Cunha (2005, 2006, 2007); Bigot and Kurtz (2011), and Sousa and Cunha (2011). In particular, note that Bigot and Dziembowski (2002) have shown that the pulsation axis in the presence of a magnetic field and rotation depends on both; it is generally found in the plane defined by the rotation and pulsation axes, and varies over the pulsation cycle. See Bigot and Kurtz (2011) for details and examples, including HR 3831.

The important point to take away from this section is that the pulsation axis of a mode may in some cases not coincide with the rotation axis. This may happen in the presence of a strong magnetic field, as in the roAp stars, or it could be caused by some other distortion of the star, for example, the tidal distortion in a close binary star. It should be kept in mind whenever frequency multiplets are seen that they may be rotationally perturbed m modes, or they may be the representation of oblique pulsation. This must be determined on a case-by-case basis, leading to different astrophysical inferences.

Oblique pulsation has been proposed as an explanation for the Blazhko effect in RR Lyr stars (Shibahashi, 2000); for arguments against this, see Szabó *et al.* (2010) and Kolenberg *et al.* (2011). It has been proposed in conjunction with a possible magnetic field in the β Cep star, β Cep itself (Shibahashi and Aerts, 2000). Strong evidence of oblique pulsation has been found in the presence of a megagauss magnetic field in the DBV white dwarf star GD 358 (Montgomery *et al.*, 2010). Even neutron stars may be oblique pulsators (Rosen *et al.*, 2010).

6.4.2 *Looking into the third dimension*

In roAp stars, the varying aspect of the oblique pulsation modes gives information about the mode geometry, particularly about the magnetic distortion of the modes from normal modes.[5] As we discussed in Section 6.2.2, abundances of many elements and ions are nonuniformly distributed over the stellar surface in the roAp stars, and they are also vertically stratified, both as a consequence of atomic diffusion. As shown schematically in Fig. 6.2, lines of Fe I and Fe II typically form around continuum optical depth $\log \tau_{5000} \sim -0.5$. The abnormally narrow cores of the Hα lines form between optical depths $-4 \leq \log \tau_{5000} \leq -2$ and the first and second ionization states of Nd and Pr form at or above optical depths $\log \tau_{5000} \leq -4$, as they have been radiatively levitated. Thus, by studying lines of different ions, and by studying the line profile variations of individual lines, it is possible to probe the pulsation as a function of atmospheric depth in some detail.

Sousa and Cunha (2011) give a detailed theoretical analysis of the behavior of the pulsation amplitude and phase with atmospheric height, showing that a combination of standing waves, running waves, and evanescent waves leads to the variety seen in the examples shown in Fig. 6.6 and other roAp stars. Moreover, they show that the inability to resolve the surface of the star leads to what they call "false nodes". Some of the roAp stars show what appears to be evidence of resolved radial nodes in their atmospheres. One example is 33 Lib where lines of Nd II and Nd III pulsate in antiphase, suggesting that they form on opposite sides of a radial node in the upper atmosphere (Mkrtichian *et al.*, 2003; Kurtz *et al.*, 2005). A clear graphical example is shown in Fig. 6.7, where pulsation amplitude goes through a zero with a π radian phase flip at a certain depth in the line, suggesting the presence of a radial node. Sousa and Cunha (2011) argue that this is a "false" node that is caused by a combination of running, standing, and evanescent waves that are unresolved for the observer.

[5] Based in part on Kurtz *et al.* (2006c).

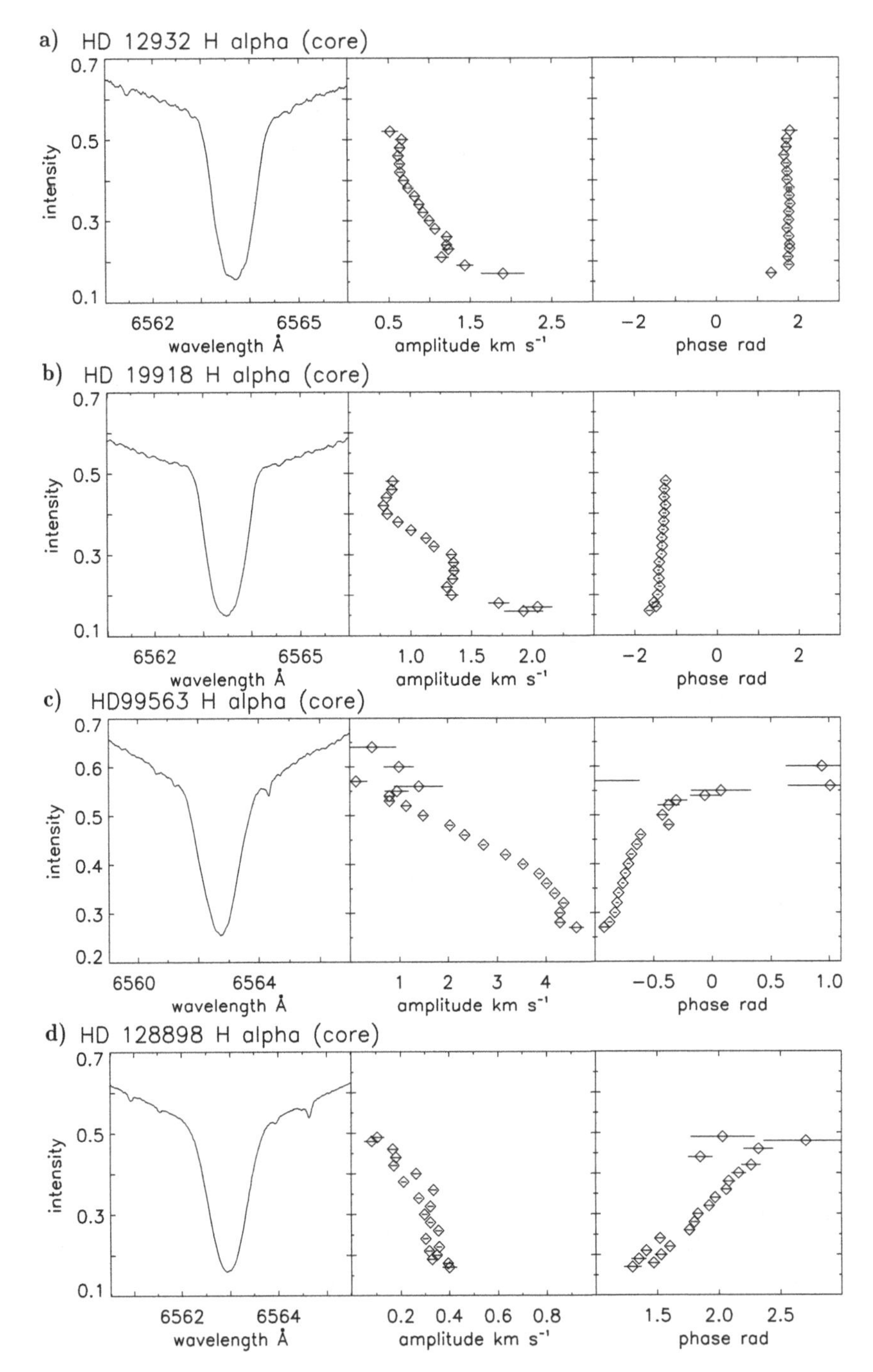

FIG. 6.6. This figure shows the line profiles and pulsation amplitudes and phases determined as a function of line depth, or height in the stellar atmosphere, for four roAp stars. Other members of the class show similar, and even more complex behavior. Sousa and Cunha (2011) argue that all of the observed behavior can be understood in terms of combinations of running, standing, and evanescent magneto-acoustic waves. In (a) through (d), the three boxes show the line profile of the narrow core of the Hα line; the middle box shows the pulsation amplitude as a function of depth in the line (height in the atmosphere); the third box shows the same for pulsation phase. The amplitudes and phases have been determined from the time series of radial velocity variations in the line bisector at each line depth. (e) through (h), show the same for the Nd III 6145 Å line, which forms at higher in the atmosphere than the Hα core (see Fig. 6.2) A vertical line in the phase versus line depth boxes means a standing wave; other shapes are indicative of running or evanescent waves.

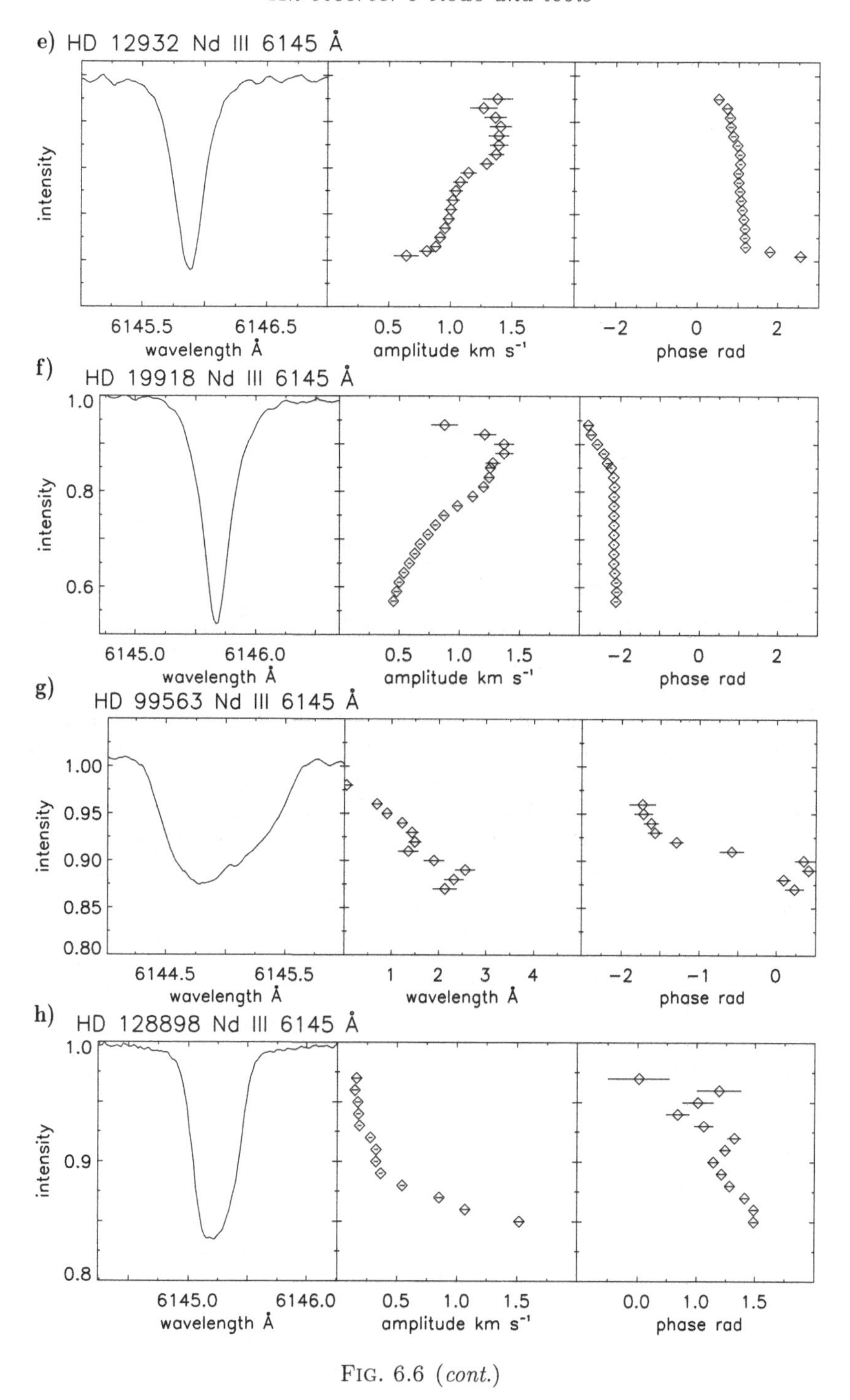

FIG. 6.6 (*cont.*)

Our view of the complex environment of the high atmospheres of the roAp stars is thus becoming clearer. No other stars allow us to resolve such high atmospheric levels, since normally they are transparent to our view, lying at such low optical depth. It is the radiative levitation of certain elements into high-lying stratified layers that gives us the ability to see a realm of rich astrophysical interest.

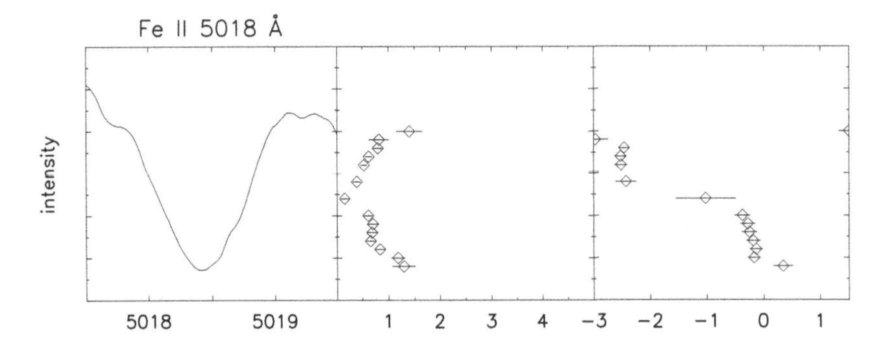

FIG. 6.7. This figure shows the line profile for the Fe II 5108 Å line in the roAp star HD 99563, along with the pulsation amplitude as a function of depth in the line, or height in the atmosphere. The amplitude goes through zero at a certain atmospheric height, while reversing phase by π radians, an apparent signature of a resolved radial node. Sousa and Cunha (2011) argue that this is a "false" node that results from a combination of running, standing, and evanescent waves that are not resolved over the visible stellar hemisphere. From Elkin *et al.* (2005).

6.5 The variable star zoo: the identification of variable types

Asteroseismology allows us to see the interiors of stars by studying the surface manifestations of global pulsation modes. The many different classes of pulsating variables allow different physical inferences about their interiors, depending on the modes that are excited to observable amplitudes. From an observer's point of view, there are different approaches to interesting problems. One is to choose an astrophysical problem to be tackled and then find stars that show the kind of pulsational information that allow progress to be made. For example, you may be interested in studying the internal rotation of stars, leading you to search for nearly equally spaced multiplets characteristic of rotationally split m modes in the amplitude spectra of pulsating stars. Another approach is to study the light curves and amplitude spectra of many stars, searching for novel insight of any kind. Most observers find themselves doing both of these to some measure. Whatever the goal is, to recognize something new, you need to have studied what is already known. With the unprecedented precision of the photometric data coming from the CoRoT and Kepler missions for up to 150,000 stars known types of pulsating stars are seen in bewildering new detail, and new types of variability are seen that need new understanding. It is therefore important to know what the many types of stellar variability look like. This includes geometric variability – star spots, ellipsoidal variability, eclipses – as well, because these can be confused with stellar pulsation, and vice versa. A good example of that was the early controversy over whether 51 Peg b was an exoplanet or a g mode (see, Aerts *et al.*, 2010, for a discussion of this).

A first step in understanding the variety of stellar variability is to study the light curves of the various classes. The monograph of Sterken and Jaschek (2005) is a good starting point. The new space photometry gives precision 100–1,000 times better than the light curves shown in that monograph, so after using that for an overview, studying Kepler light curves and amplitude spectra is recommended. Automated classification of 150,000 stars in the Kepler database is discussed by Debosscher *et al.* (2011) and several examples of Kepler light curves are shown. Some of the classes of pulsating stars were introduced in Section 6.2.1 and in Fig. 6.1. An extensive discussion of the various classes of pulsating stars can be found in Aerts *et al.* (2010); see, particularly, Tables A.1 and A.2 in that book.

Here are some examples of the kinds of remarkable light curves with the μmag precision data of the Kepler Mission, where the variability is not, in general, pulsation, but could be confused as such. Stars are referenced by their Kepler Input Catalog (KIC) numbers. Figure 6.8 shows a light curve of a red giant star, KIC 7869590, from quarter 1 of the

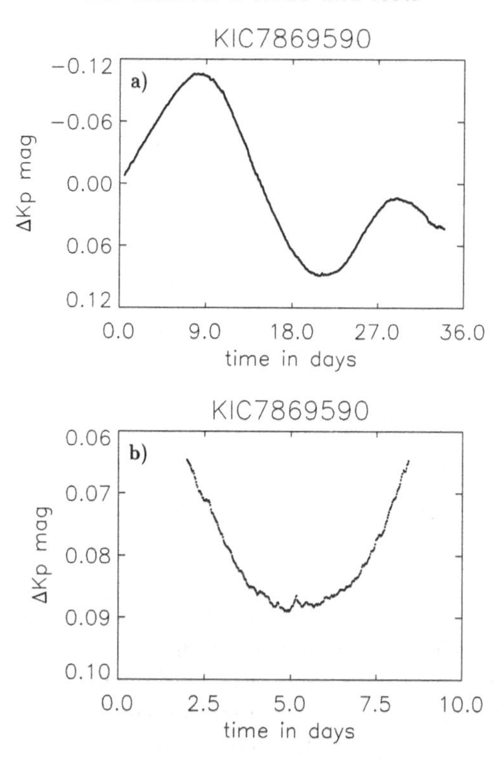

FIG. 6.8. *(a)* This is a 34-d light curve of the red giant star KIC 7869590 taken from the quarter 1 Kepler public data. *(b)* This shows a blow-up of a section of the light curve at higher resolution, so that the supergranulation waves can be seen. The light variations are a combination of spots, supergranulation, and a transit – probably of a main sequence companion – near the end of the light curve. All the variations seen in this light curve are real; the noise level is too low to discern at this resolution. This figure is illustrative of the rich variety of stellar variability that can be seen in *Kepler* data for nonpulsating stars.

Kepler public data. Rotational variation from large spots is obvious in the top panel of the figure, and close inspection of the end of the light curve reveals a transit, which is probably that of a main sequence companion. In the bottom panel, a section of the light curve is shown at higher resolution where waves can be seen that are probably the result of supergranulation. All of the visible variations are real; the noise level is too small to see at this scale. This ability to see directly light variations caused by convection is new to the μmag precision *Kepler* and CoRoT data.

KIC 6437385 was suspected to be a Cepheid variable, but Fig. 6.9 shows that the 14-d variation is more probably rotational variability of a spotted, flaring main sequence star. It is estimated to have $T_{\rm eff} = 5,400$ K, $\log g = 3.7$ from broadband photometry in the Kepler Input Catalog.

Figure 6.10 shows the light curve of a remarkable binary star, HD 187091, also known as KOI-54 (KOI = Kepler Object of Interest). At first, this looks like an upside-down eclipsing binary light curve! Remarkably, it is a binary star in a nearly face-on orbit ($i = 5.5°$ with an eccentricity of 0.83 and an orbital period of 41.8 d; Welsh *et al.*, 2011). With that high eccentricity, at periastron passage the two A stars are only a few stellar radii apart; the 8 mmag spike in the light curve is caused by substellar heating of the components. The two stars are too hot to be in the γ Dor instability strip, so they are not expected to have self-excited g modes of large amplitude, yet the oscillations visible in the light curve are just that: high amplitude g modes. Tidal forcing at periastron passage excites a number of modes that have frequencies that are commensurate with the orbital period. The two highest amplitude modes have 90 and 91 cycles per orbit

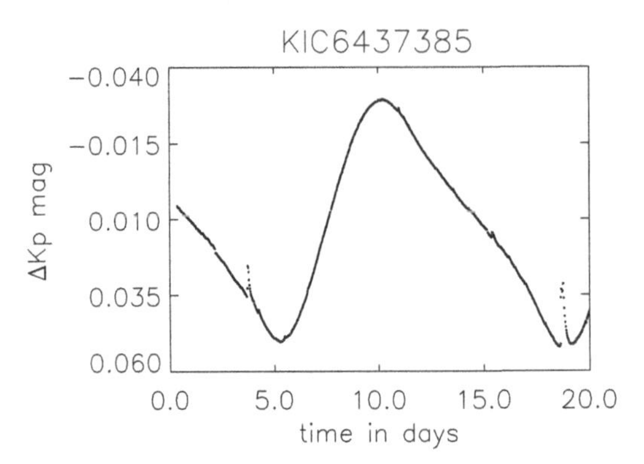

FIG. 6.9. KIC 6437385 was originally thought to be a Cepheid, and, as can be seen in this Kepler light curve from the quarter 2 public data, it does resemble a 14-d, low-amplitude Cepheid. However, the flares suggest otherwise. While the true nature of this star is still to be determined, it seems likely that the 14-d variation is rotational and that this is likely to be a main sequence flare star, rather than a giant Cepheid.

exactly, and they produe the oscillations that are obvious in the light curve. This is an exciting new opportunity to study tidal forcing of oscillations (see Willems and Aerts, 2002; Aerts *et al.*, 2010). There are many other pulsating, eclipsing binary stars in the Kepler database that are now undergoing intensive study.

Figure 6.11 shows a phased light curve for the newly discovered roAp star, KIC 10195926 (Kurtz *et al.*, 2011). The obvious light variations are caused by abundance spots – this is an α^2 CVn Ap star – as discussed in Section 6.2.2. The star also has two oblique pulsations mode that give rise to frequency multiplets. In this case, the frequency separation between the modes appears to be equal to the large separation, yet the geometry of the modes determined from their oblique pulsation multiplets is clearly different, suggesting that they have different pulsation axes. This is expected within the improved oblique pulsator model (Bigot and Dziembowski, 2002; Bigot and Kurtz, 2011) as a consequence of different perturbations from the magnetic field and the rotation of modes that have slightly different pulsation cavities.

While we await a new monograph in the style of Sterken and Jaschek (2005), but with light curve examples using μmag space-based data, a careful study of publicly available

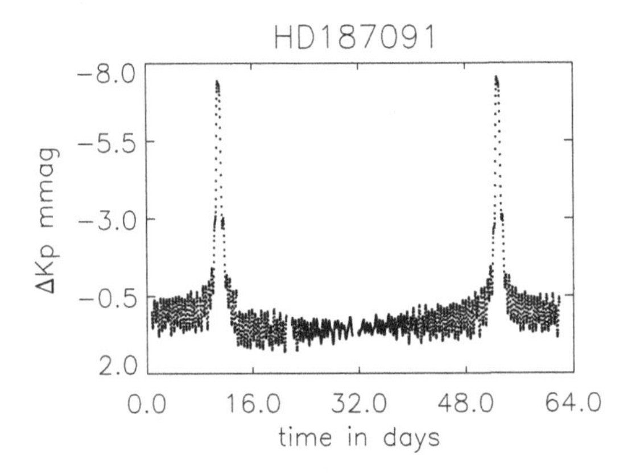

FIG. 6.10. HD 187091, KOI-54, is a remarkable high eccentricity ($e = 0.84$), nearly face-on binary star composed of two early A stars. Substellar heating at periastron passage creates the spikes in the light curve. Tidal forcing excites g modes with periods commensurate with the orbital period. The two highest amplitude pulsation modes have exactly 90 and 91 cycles per orbital period. See Welsh *et al.* (2011).

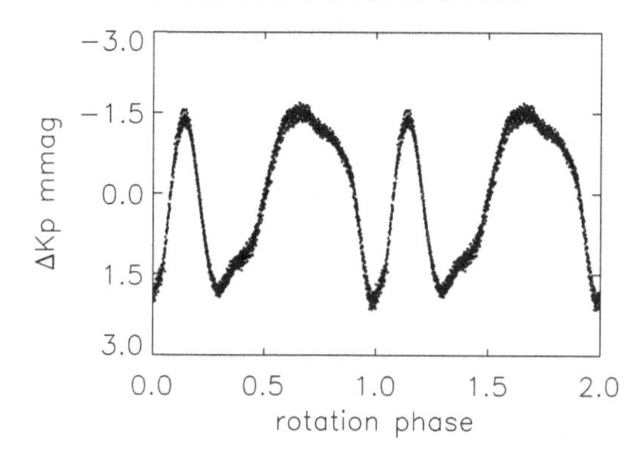

FIG. 6.11. This shows a phased light curve for the roAp star KIC 10195926. The variations seen with a 5.68-d period are the result of spots on the star. At much lower amplitude, two oblique pulsation modes are also present, remarkably with different pulsation axes. See Kurtz *et al.* (2011).

Kepler and CoRoT data is a good exercise to see the rich variety of astrophysical behavior that this new level of photometric precision has given us.

6.6 Miscellaneous useful topics

The first exoplanet to be discovered used the "timing method," in this case Doppler variations in the precise clock of a pulsar, PSR 1829-10 (Bailes *et al.*, 1991). A goal in the study of pulsating stars is to use those that have stable enough pulsations as precise clocks, and search for similar periodic Doppler shifts characteristic of unseen companions; shifts small enough may be indicative of planets. Traditionally, this has been done by the "O−C method"; that is how V391 Peg b was discovered (Silvotti *et al.*, 2007). But it may also be possible to detect planets from frequency multiplets in the amplitude spectrum with the new ultra-precise Kepler data. This has yet to be done, but the search is on.

The first exoplanet PSR 1829-10b has a period of 6 Earth months and a circular orbit. Bailes *et al.* (1991) commented, "It is not clear whether it formed in the aftermath of the supernova that created the neutron star, or was pre-existing and somehow survived through the late stages of stellar evolution ... the finding of such an interesting exoplanet was a driver for discussion of the physics of its formation." Then, a year later, a second paper was published retracting the discovery of the planet (Lyne and Bailes, 1992). An error in the calculation of the barycentric Julian date (BJD) caused by a "small correction to the ellipticity of the Earths orbit in the calculations and a slightly imprecise position of the pulsar had resulted in the discovery of a planet – Earth."

Small errors such as this are easy to make and hard to detect. The lesson here is that you cannot be too careful about your treatment of time, your recording of time, and your reporting of what kind of time you have used in your analyses.

6.6.1 *Time and time traps*

Traditionally, time has been measured by reference to the Earth's rotation rate – the length of the day. Because the length of the day and the year are not commensurate, this leads to intercalation, typically nowadays of a leap day every four years, but in times past of multiple days extracted from the calendar, as in the Gregorian reform, or several months added, as in the Julian reform.

The Earth's rotation rate is not constant. In the long term, the length of the day is increasing by ∼1.7 ms per century as a result of tidal interaction with the Moon. But the rate of change is variable on a shorter time scale; at the time of this writing (2011), the length of the day is *decreasing*. See the US Naval Observatory Web site[6] for a graphical look at this and good discussion of time scales.

Here are some rules to put time scales into some perspective. See McCarthy (2005) for an in-depth, expert discussion of time and time scales.

- **UT**: Universal time (UT) is based on the Earth's rotation. UT1 is based on a mean solar day of 86,400 mean solar seconds. In practice, UT1 is based on Greenwich Mean Sidereal Time (GMST) where $P_s \approx 86,164.0905$ mean solar seconds. This is 0.0084 s shorter than the rotation period because of precession. The mean solar day is about 1 ms longer than 86,400 s. Earth's rotation period is currently 86,164.0999 SI seconds. UT1 is measured with VLBI of quasars; note that GPS is not good enough because of orbital variations of the satellites that make up the GPS system.
- **ET**: Ephemeris Time (ET) is based on the period of the Earth's orbit about the Sun using Newcomb's Tables of the Sun (Newcomb, 1895). The length of the tropical ET year in 1900 was 31,556,925.9747 s. The reciprocal of this defines the ET second. In 1976, Terrestrial Dynamical Time (TDT) and Barycentric Dynamical Time (TDB) were introduced. In 1991 the IAU renamed TDT to Terrestrial Time (TT).
- **TT**: Terrestrial Time (TT) is based on International Atomic Time (TAI) by the definition: TT = TAI + 32.184 s. The offset is the difference between ET and UT1 on 1 January 1958. At the 1991 IAU in Buenos Aires, it was decided that the basis of the time scale includes general relativity to give Geocentric Coordinate Time (TCG) and Barycentric Coordinate Time (TCB). These do not run at the same rate as TT!
- **TAI**: International Atomic Time (TAI) takes the second to be the fundamental unit of time. It is defined by a hyperfine transition in the ground state of Ce^{133}, where the atomic second is 9,192,631,770 periods of the radiation of this transition. Two hundred clocks in fifty countries contribute to the maintenance of TAI.

Does any of this matter to asteroseismologists? Ultimately, yes. On a very short time scale, or for data gathered from only one source over a short time span, then not so much. The difference between TT and UT1 ranges over 60 s since 1900, and the variation between them is not some simple, uniform function. As we now have ultra-precise photometry, in the future we will need ultra-precise time to be sure that what we discover about the stars we study is truly in the physics of the stars. We do not want to rediscover the Earth, yet again, by incorrect use of time scales.

- **UTC**: Coordinated Universal Time (UTC) was introduced in the 1950s. Adjustment to quartz clocks was coordinated between laboratories in the United States and United Kingdom. This coordination is now more international. Both the rate and epoch of UTC are variable to stay in step with astronomical time. TAI–UTC has a range that has increased by over 30 s semi-monotonically (of course, with glitches because of leap seconds) since the 1960s.
- **JD**: Julian Date (JD) was introduced by Joseph Justus Scaliger in 1583 at the time of the Gregorian Calendar reform. The zero point is 4713 BCE (before the Christian era) January 1 at 12:00 at the meridian of Alexandria, Egypt. It takes into account the solar calendar cycle of 28 years, the Metonic cycle of 19 years (235 lunar months = 18.9999 years) and even the Roman indiction (tax) cycle of 15 years. (The baggage we scientists saddle ourselves with!) Thus, the Julian cycle is $15 \times 28 \times 19 = 7{,}980$ years. If you have ever wondered how the zero point for the JD system was chosen, the epoch is the last time they were all in their first year. To give examples of Julian date usage here and in the points below: on the last day of the IAC Winter School on Asteroseismology (for which these are the proceedings), 12:00 UT 2010 November 26 = JD 245,5527.0. Obviously, this date is a large number, it is cumbersome, and in most computers using its full representation will lead to truncation errors at the level of seconds, or greater. Thus, there are many ways people modify Julian date, leading to much confusion.

[6] http://tycho.usno.navy.mil/systime.html.

- **MJD**: Modified Julian Date (MJD) is defined to be MJD = JD2,400,000.5 (watch out for the 0.5!). This is used widely in astronomy and geophysics. As an example, JD 2,455,527.0 = MJD 55,526.5. The advantage is that the number is a little less unwieldy than JD. Why was 0.5 subtracted? Surely, that is such an obvious source of confusion that no one would propose it without very good reason. The answer is, to put the change of date at midnight UT, instead of noon. Why is this considered to be preferable? I haven't a clue.
- **RJD**: Reduced Julian Date (RJD) has a definition very close to that of MJD, but critically is slightly different: RJD = JD2,400,000.0. Thus, for example, JD 2,455,527.0 = RJD 55,527.0. This is MJD without the −0.5! Ouch! This is not used so much, and I have never seen it in the field of asteroseismology. On the other hand, truncated Julian date is widely used, but not everyone means the same thing.
- **TJD**: Truncated Julian date (TJD) means different things to different people. The term is common. To NASA TJD = JD2,440,000.5, so it is a kind of super-truncated MJD. To the U.S. National Institute of Standards and Technology (NIST), TJD = (JD0.5) mod 10,000. In asteroseismology, it is common to define TJD = JD2,440,000.0.
- **HJD**: Because the time of arrival of light from the stars that we study in asteroseismology arrives at our telescopes at times that vary because of the Earth's orbital motion, this injects a periodic signal in our observations that does not originate in the target stars. In effect, the pulsation frequencies of the stars are Doppler shifted periodically by the Earth's orbital motion. One way to correct for this is in a modification of the time scale. The first-order solution is to modify the times of the observations to be the light arrival time of the signal at the center of the Sun, rather than at the telescope. This is done using the Julian date scale and is known as heliocentric Julian date (HJD). Its use is common in variable star research, and it is adequate for most types of variable stars. However, those with short periods must be corrected further, since the Sun itself orbits the barycenter of the solar system.
- **BJD**: Barycentric Julian Date (BJD) corrects the time of observations to the barycenter of the Solar System. This differs from HJD by up to $\sim$2 s, primarily because of the Sun's reflexive motion in its orbit about Jupiter, but even the effect of the asteroids is measurable. BJD is normally used for short period variables observed over long time spans, such as white dwarfs, subdwarf B stars, and, of course, it is used in pulsar research. However, there is yet one more correction, as the JD time scale is based on UTC and that has leap seconds. The stars do not know about the leap seconds.
- **BJED**: Barycentric Julian Ephemeris date (BJED) = BJD with the leap seconds removed from the time scale. Note that the IAU recommends that JD be given in ET, but asteroseismology uses UTC.

Be careful in your use of time scales. It is imperative that you state your definition of time scale clearly to avoid confusing others immediately, and yourself in the future. Be pedantically clear about time.

The potential for mistakes is high, particularly when concatenating data sets obtained with different instruments, from different observatories, and recorded by different people. Those people have not always done what they said they did, and in some cases they do not know what time they used. If you do discover something new – such as an exoplanet – it is better that it not be the Earth.

6.6.2 *A note on computer truncation of time*

A typical time stamp from, for example, the Kepler mission could be, say, BJD 2,451,234.123456. This is precise to a little less than 0.1 s. Leaving alone the accuracy of the mission clock, you need 64-bit precision in your computer to keep track of this time. This number has 11 significant figures, say, of the order of $10^{12} \sim 2^{40}$, so you need 40 bits to keep track of the mantissa. Your computer may keep three decimal exponents, that is, up to 10^{999}; so $999 \sim 2^{10} = 1{,}024$ means you need another 10 bits to record the exponent. Then you need two bits for the signs on the exponent and mantissa. As you can see, this adds up already to 52 bits before you start multiplying your time by some other (possibly large) number, such as a frequency in the computation of an amplitude spectrum. If you are keeping full BJDs and doing 32-bit computing, you will be truncating your times to far, far lower precision than they are recorded. This is an unnecessary additional noise source that can be avoided. Make sure your computational tools have sufficient precision to handle the accuracy of your time data.

6.6.3 *Discovering clocks in the telescope*

When searching for periodic phenomena in astronomy, be careful not to discover the clocks in your telescope. Some examples where this has happened include the pulsar at the core of the SN1987a remnant, and pulsation harmonics in the roAp stars HD 60435 and HR 1217.

SN1987a (February 1987) in the LMC was the first naked-eye supernova since Kepler's supernova of 1604. A team searched hard for a pulsar remnant using the CTIO 4-m telescope. They found what they were looking for in January 1989, a pulsar with $\nu = 1{,}968.629$ Hz, that is, $P \sim 0.5$ ms! Theory argued for spin-up from material falling back onto the remnant, but others could not confirm the observation. It was found a year later that the signal was coming from a television guiding system at the telescope. When bright calibration pulsars were observed to be sure the signal was *not* instrumental, the guiding system was not needed and it was turned off (inadvertently, and unnoticed). This is an easy mistake to make, so is a lesson to all of us about controlling our experiments.

In the cases of the roAp stars mentioned, they have pulsation periods near to 12 min and 6 min where the second and third harmonics of the frequencies lie serendipitously close to the drive frequencies of the telescopes used in their observations; the worm gears of many telescopes have a period of 2 sidereal minutes or 4 sidereal minutes; that is, a drive frequency of 8.36 mHz or 4.18 mHz. If a signal close to those values is found in a Fourier analysis, the telescope should be considered to be the possible source. For further discussion, see Aerts *et al.* (2010).

6.6.4 *Nyquist aliases: a graphical view*

Frequency analysis, Fourier transforms, and the Nyquist frequency are discussed formally by Appourchaux elsewhere in this book. The meaning of the Nyquist frequency is often misunderstood. I want to look at the Nyquist frequency from a visual perspective here and consider two questions about its use. The Nyquist frequency is defined to be half the sampling frequency for equally spaced data. The first question to be answered is why is it half the sampling frequency, and not the sampling frequency? The second question concerns the meaning of the Nyquist frequency. Is it possible in a Fourier frequency analysis to find and identify frequencies above the Nyquist frequency? Or, to put this another way, when there are significant peaks below and above the Nyquist frequency, how do you decide which is the true frequency in the star under analysis?

6.6.4.1 Why is the Nyquist frequency half the sampling frequency?

To answer this question I have generated some artificial data (hereafter, "the data") and produced amplitude spectra for them. The data were sampled once per minute; that is, the observing cadence was 1 min. This is very close to the *Kepler* short cadence sampling time and was chosen for this reason. The data are exactly equally spaced. They have no noise, so the amplitude spectra are also illustrative of the window pattern of the data. With a 1 min cadence, the sampling frequency is $1\,\mathrm{min}^{-1} = 16.\overline{6666}\,\mathrm{mHz} \equiv f_s$, therefore the Nyquist frequency is $8.\overline{3333}\,\mathrm{mHz} \equiv f_N$.

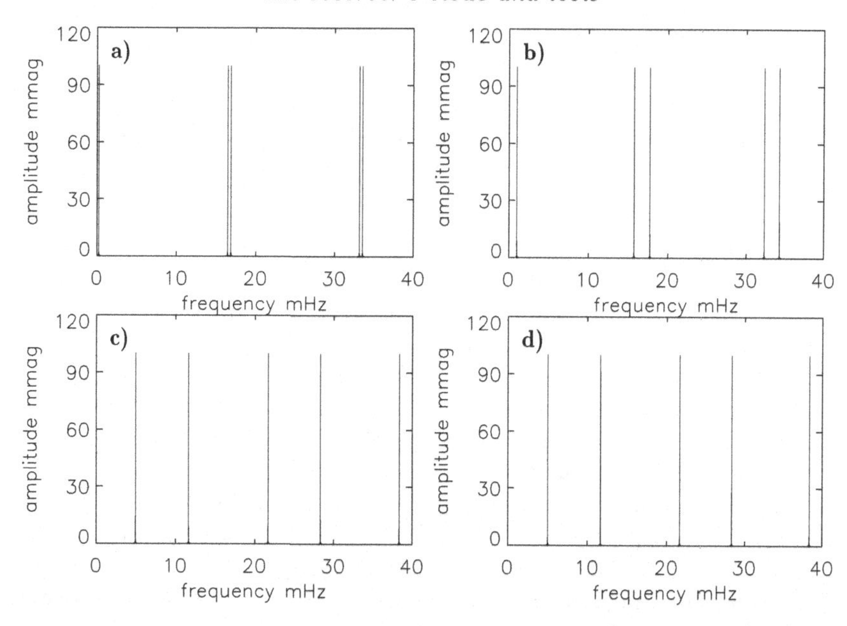

FIG. 6.12. These are amplitude spectra of equally spaced data that have a cadence of 1 min. An artificial frequency has been injected with an amplitude of 100 mmag. The data have no noise and no integration time. *(a)* The left-hand peak at $f_i = 0.2$ mHz is the injected signal. There are four more frequencies visible in the diagram. There is a pair of frequencies at $f_s - f_i$ and $f_s + f_i$ centered on the sampling frequency and split by $2f_i$, and there is a second pair at higher frequency at $2f_s - f_i$ and $2f_s + f_i$. These are all Nyquist aliases. *(b)* This is the same as the upper left panel except $f_i = 1.0$ mHz. Here, the two pairs of Nyquist aliases are separated by $2f_i$, as Nyquist aliases always are, and f_i is closer to the first Nyquist alias at $f_s - f_i$. *(c)* This is the same as the upper left panel except $f_i = 5.0$ mHz. Note how f_i is even closer to the first Nyquist alias at $f_s - f_i$ than in the previous panel. Obviously from this, when $f_i = \frac{1}{2}f_s$, the injected frequency f_i and the first Nyquist alias are equal, that is, $f_i = f_s - f_i$. *(d)* This panel is identical to the lower left panel, but has it not been generated the same way. In this panel $f_i = 11.\overline{6666}$ mHz $= f_s - 5.0$ mHz. Thus the lowest frequency peak is an alias at $f_s - f_i$ and the injected frequency is the next peak at higher frequency. There is no way to distinguish from Fourier analysis alone the difference between the lower left panel and lower right panel. An external constraint is needed.

I have injected into the data a sinusoidal signal with various frequencies and an amplitude of 100 mmag; this is illustrated in the panels of Fig. 6.12. Importantly, the artificial data have no "integration time; that is, they are sampled at 1 min intervals, but the sampling is instantaneous; it has no spread in time, as in real astronomical data.

Look at the upper left panel of the figure. There are four more frequencies visible in addition to f_i. There is a pair of frequencies at $f_s - f_i$ and $f_s + f_i$, that is, 16.4666 mHz and 16.8666 mHz. They are centred on f_s and split by $2f_i$. Then there is a second pair at higher frequency, at $2f_s - f_i$ and $2f_s + f_i$. They are centred at $2f_s$ and again split by f_i. These are all Nyquist aliases. If I had run the amplitude spectrum to higher frequencies, then there would be further pairs of frequencies visible at $nf_s - f_i$ and $nf_s + f_i$ for all n to infinity, since the data are noise-free and have no integration time. All of these pairs are centred at nf_s and split by f_i.

The first point about Nyquist aliases is, therefore, that they occur at frequencies equal to $nf_s - f_i$ and $nf_s + f_i$ for all n. Notice then what happens if the injected frequency is higher. In this case I have chosen $f_i = 1.0$ mHz, as is illustrated in the upper right panel of Fig. 6.12. Note how as the injected frequency f_i moves to higher frequency, the first Nyquist alias moves to a lower frequency. The lower left panel illustrates this further by injected $f_i = 5.0$ mHz. It is obvious that these two will meet and overlap when the injected frequency equal half the sampling frequency, that is, when $f_i = \frac{1}{2}f_s$.

Thus, the answer to the question posed is that the Nyquist frequency is at half the sampling frequency because that is where the first Nyquist aliases coincide with an input frequency of half the sampling frequency. It is thus the frequency where there is confusion over what the Nyquist alias is and where the input frequency begins.

6.6.4.2 *What is the meaning of frequencies above the Nyquist frequency?*

The lower right panel shows the amplitude spectrum for an injected signal equal to $f_i = 11.\overline{6666}\,\mathrm{mHz} = f_s - 5.0\,\mathrm{mHz}$. It is identical to the lower left panel where the injected frequency is $f_i = 5.0$mHz, even though the positions of the injected frequencies and the first Nyquist aliases have changed places. It is therefore often claimed that it is not possible to find a pulsation frequency above the Nyquist frequency. This is not true. A correct statement is that *it is not possible to determine from Fourier analysis alone of equally spaced data which of the peaks in the amplitude spectrum is a real frequency in the star being studied.* This includes the lowest frequency peak.

The only way to determine which of the aliases, including the lowest frequency peak, is the true frequency of the star is to use additional constraints external to the Fourier analysis. In practice, this usually means using T_{eff}, $\log g$ to generate an expectation of the probable frequency range of the star under study based on physical expectations of its pulsation cavity from models.

If a star is pulsating with a frequency higher than the Nyquist frequency, that can be found from Fourier analysis, but there will be a lower frequency peak below the Nyquist frequency as well, and it will be uncertain which is the correct frequency. The standard procedure where this is suspected is to obtain further data at a higher sampling rate to push the Nyquist frequency higher, but even then an external constraint is needed to make a final decision about the reality of any peak.

In real astronomical observations, the situation is somewhat more complicated, but the principle remains the same. External expectations from the physics of the star are needed to decide what probable frequency range is to be expected. With integration times that usually are nearly equal to the sampling time, that is, where there is little dead time between integrations, for higher Nyquist aliases the signal is smeared by the integration time becoming a larger fraction of the pulsation period, hence the Nyquist aliases drop in amplitude. *This does not provide a way to decide which alias is the correct peak.* This will happen for frequencies that lie above the Nyquist frequency, as well as those below.

Finally, when the data are not equally spaced in time, the idea of the Nyquist frequency is much more complex. If enough data are obtained with integration times very short compared to a pulsation period, then even if the true peak is far above the nominal Nyquist frequency, it can be obtained from Fourier analysis when the data cover the pulsation cycle at all phases. A beautiful example of this that I have seen is the recovery of the signal in some roAp stars (periods 5–21 min) in Superwasp data with typical sampling rates of once per day. With data sets stretching over years and variable times of night for the observations of a particular target, the very short periods are well mapped in phase and the true signal stands out in an amplitude spectrum. The lesson from this is that some experiments should be planned to have some staggering of the observation times so that the data are *not* equally spaced.

There is no hard-and-fast rule to decide the reality of a peak or alias. Each case requires consideration of the physics of the star before one makes the decision.

6.6.5 *Pulsation phases*

I have a final short comment about pulsation phases. Of course, these do not appear in a Fourier transform, where either amplitude or power is plotted against frequency. But the information is there in the real and imaginary parts of the transform. Most observers and theoreticians pay scant attention to observed pulsation phase, largely because there is no way to know what sets this for a particular pulsation mode, even where mode lifetimes are long. There are, however, interesting questions to which pulsation phase applies, and some have been discussed in this chapter.

One example we have mentioned is when the pulsator is in a binary with either a stellar or planetary companion. In that case the pulsation phase will be periodically modulated by the Doppler shift of the orbital motion, since phase and frequency are inextricably entangled in a Fourier analysis. Another example is for oblique pulsation. As we have seen, that generates a frequency multiplet. For simple oblique pulsation, all components of the multiplet will have the same phase (or antiphase – it depends on the mode) at the time of pulsation maximum. As this is not expected for a rotationally perturbed multiplet of m modes, it is a way to distinguish those from oblique pulsation.

When pulsation phase matters, it is imperative that it be determined with respect to a zero point in time that is in the center of the data set. If large values of time are used (say, truncated BJD, as mentioned above) and $t = 0.0$ is used for determining phase, then the phase is truly close to meaningless. This is because a small error in frequency leads to a huge change in the determined phase because of the many cycles of pulsation back to time zero. When working with phase, choose the zero point of your time scale to be within your data set (preferably in the middle). You'll note, if you do this, that when using nonlinear least-squares fitting the phase error will be the same as determined from linear least-squares fitting of sinusoids. If you choose $t = 0$ and have large values of t, your phase will have a meaninglessly large error in nonlinear least-squares sinusoid fitting, if the procedure does not fail to converge at all. Note that if you do have a convergence problem, look at the time zero point as a possible source of this.

The message is: pulsation phase contains information. Sometimes it can be useful, so is worth thinking about, even when you are being dazzled by an amplitude spectrum.

REFERENCES

Aerts, C., Christensen-Dalsgaard, J., and Kurtz, D. W. 2010. *Asteroseismology*. Springer. Houten, Netherlands.

Asplund, M., Grevesse, N., Sauval, A. J., and Scott, P. 2009. The chemical composition of the Sun. *ARA&A*, **47**(Sept.), 481–522.

Audard, N., Kupka, F., Morel, P., Provost, J., and Weiss, W. W. 1998. The acoustic cut-off frequency of roAp stars. *A&A*, **335**(July), 954–8.

Bailes, M., Lyne, A. G., and Shemar, S. L. 1991. A planet orbiting the neutron star PSR1829 – 10. *Nature*, **352**(July), 311–13.

Balmforth, N. J., Cunha, M. S., Dolez, N., Gough, D. O., and Vauclair, S. 2001. On the excitation mechanism in roAp stars. *MNRAS*, **323**(May), 362–72.

Baumann, P., Ramírez, I., Meléndez, J., Asplund, M., and Lind, K. 2010. Lithium depletion in solar-like stars: no planet connection. *A&A*, **519**(Sept.), A87–A98.

Bigot, L., and Dziembowski, W. A. 2002. The oblique pulsator model revisited. *A&A*, **391**(Aug.), 235–45.

Bigot, L., and Kurtz, D. W. 2011. Theoretical light curves of dipole oscillations in roAp stars. *A&A***536**(Dec.), A73–A85.

Bigot, L., Provost, J., Berthomieu, G., Dziembowski, W. A., and Goode, P. R. 2000. Non-axisymmetric oscillations of roAp stars. *A&A*, **356**(Apr.), 218–33.

Boesgaard, A. M., and Tripicco, M. J. 1986. Lithium in the Hyades Cluster. *ApJ*, **302**(Mar.), L49–L53.

Chaboyer, B., Demarque, P., Kernan, P. J., and Krauss, L. M. 1996. A lower limit on the age of the Universe. *Science*, **271**(Feb.), 957–61.

Cowley, C. R., Hubrig, S., Ryabchikova, T. A., Mathys, G., Piskunov, N., and Mittermayer, P. 2001. The core-wing anomaly of cool Ap stars: abnormal Balmer Profiles. *A&A*, **367**(Mar.), 939–42.

Cowley, C. R., Hubrig, S., and Kamp, I. 2006. An atlas of K-line spectra for cool magnetic cp stars: the wing-nib anomaly (WNA). *ApJS*, **163**(Apr.), 393–400.

Cunha, M. S. 2005. Asteroseismic theory of rapidly oscillating Ap stars. *Journal of Astrophysics and Astronomy*, **26**(June), 213–21.

Cunha, M. S. 2006. Improved pulsating models of magnetic Ap stars – I. Exploring different magnetic field configurations. *MNRAS*, **365**(Jan.), 153–64.

Cunha, M. S. 2007. Theory of rapidly oscillating Ap stars. *Communications in Asteroseismology*, **150**(June), 48–54.

Cunha, M. S., and Gough, D. 2000. Magnetic perturbations to the acoustic modes of roAp stars. *MNRAS*, **319**(Dec.), 1020–38.

Debosscher, J., Blomme, J., Aerts, C., and De Ridder, J. 2011. Global stellar variability study in the field-of-view of the Kepler satellite. *ArXiv e-prints*, Feb.

Dziembowski, W. A., and Goode, P. R. 1985. Frequency splitting in AP stars. *ApJ*, **296**(Sept.), L27–L30.

Dziembowski, W. A., and Goode, P. R. 1996. Magnetic Effects on Oscillations in roAp Stars. *ApJ*, **458**(Feb.), 338.

Eisenstein, D. J., Liebert, J., Harris, H. C., Kleinman, S. J., Nitta, A., Silvestri, N., Anderson, S. A., Barentine, J. C., Brewington, H. J., Brinkmann, J., Harvanek, M., Krzesiński, J., Neilsen, Jr., E. H., Long, D., Schneider, D. P., and Snedden, S. A. 2006a. A catalog of spectroscopically confirmed white dwarfs from the Sloan Digital Sky Survey data release 4. *ApJS*, **167**(Nov.), 40–58.

Eisenstein, D. J., Liebert, J., Koester, D., Kleinmann, S. J., Nitta, A., Smith, P. S., Barentine, J. C., Brewington, H. J., Brinkmann, J., Harvanek, M., Krzesiński, J., Neilsen, Jr., E. H., Long, D., Schneider, D. P., and Snedden, S. A. 2006b. Hot DB white dwarfs from the Sloan Digital Sky Survey. *AJ*, **132**(Aug.), 676–91.

Elkin, V. G., Kurtz, D. W., and Mathys, G. 2005. The discovery of remarkable 5km/s pulsational radial velocity variations in the roAp star HD99563. *MNRAS*, **364**(Dec.), 864–72.

Fontaine, G., and Wesemael, F. 1987. Recent advances in the theory of white dwarf spectral evolution. Pages 319–326 of: A. G. D. Philip, D. S. Hayes, and J. W. Liebert (ed), *IAU Colloq. 95: Second Conference on Faint Blue Stars.*

Freyhammer, L. M., Kurtz, D. W., Elkin, V. G., Mathys, G., Savanov, I., Zima, W., Shibahashi, H., and Sekiguchi, K. 2009. A 3D study of the photosphere of HD99563. I. Pulsation analysis. *MNRAS*, **396**(June), 325–42.

Gautschy, A., Saio, H., and Harzenmoser, H. 1998. How to drive roAp stars. *MNRAS*, **301**(Nov.), 31–41.

Grigahcène, A., Antoci, V., Balona, L., Catanzaro, G., Daszyńska-Daszkiewicz, J., Guzik, J. A., Handler, G., Houdek, G., Kurtz, D. W., Marconi, M., Monteiro, M. J. P. F. G., Moya, A., Ripepi, V., Suárez, J.-C., Uytterhoeven, K., Borucki, W. J., Brown, T. M., Christensen-Dalsgaard, J., Gilliland, R. L., Jenkins, J. M., Kjeldsen, H., Koch, D., Bernabei, S., Bradley, P., Breger, M., Di Criscienzo, M., Dupret, M.-A., García, R. A., García Hernández, A., Jackiewicz, J., Kaiser, A., Lehmann, H., Martín-Ruiz, S., Mathias, P., Molenda-Żakowicz, J., Nemec, J. M., Nuspl, J., Paparó, M., Roth, M., Szabó, R., Suran, M. D., and Ventura, R. 2010. Hybrid γ Doradus-δ Scuti pulsators: new insights into the physics of the oscillations from Kepler observations. *ApJ*, **713**(Apr.), L192–L197.

Gray, R. O. and Garrison, R. F. 1989a. The early F-type stars. Refined classification, confrontation with Stromgren photometry, and the effects of rotation. *ApJS*, **69**(Feb.), 301–21.

Gray, R. O. and Garrison, R. F. 1989b. The late A-type stars. Refined MK classification, confrontation with Stromgren photometry, and the effects of rotation. *ApJS*, **70**(Jul.), 623–636.

Handler, G. 1999. The domain of γ Doradus variables in the Hertzsprung-Russell diagram. *MNRAS*, **309**(Oct.), L19–L23.

Houk, N. and Cowley, A. P. 1975. *University of Michigan Catalogue of two-dimensional spectral types for the HD stars. Volume I.* University of Michigan, Michigan.

Houk, N. 1978. *University of Michigan Catalogue of two-dimensional spectral types for the HD stars. Volume II.* University of Michigan, Michigan.

Houk, N. 1982. *University of Michigan Catalogue of two-dimensional spectral types for the HD stars. Volume III.* University of Michigan, Michigan.

Houk, N. and Smith-Moore, M. 1988. *University of Michigan Catalogue of two-dimensional spectral types for the HD stars. Volume IV.* University of Michigan, Michigan.

Hubrig, S., Nesvacil, N., Schöller, M., North, P., Mathys, G., Kurtz, D. W., Wolff, B., Szeifert, T., Cunha, M. S., and Elkin, V. G. 2005. Detection of an extraordinarily large magnetic field in the unique ultra-cool Ap star HD 154708. *A&A*, **440**(Sept.), L37–L40.

Hubrig, S., Oskinova, L. M., and Schoeller, M. 2011a. First detection of a magnetic field in the fast rotating runaway Oe star zeta Ophiuchi. *Astronomische Nachrichten*, **332**(Jan.), 147–52.

Hubrig, S., Ilyin, I., Schöller, M., Briquet, M., Morel, T., and De Cat, P. 2011b. First magnetic field models for recently discovered magnetic βCephei and slowly pulsating B stars. *ApJ*, **726**(Jan.), L5–L9.

Kepler, S. O., Kleinman, S. J., Nitta, A., Koester, D., Castanheira, B. G., Giovannini, O., Costa, A. F. M., and Althaus, L. 2007. White dwarf mass distribution in the SDSS. *MNRAS*, **375**(Mar.), 1315–24.

Kleinman, S. J., Harris, H. C., Eisenstein, D. J., Liebert, J., Nitta, A., Krzesiński, J., Munn, J. A., Dahn, C. C., Hawley, S. L., Pier, J. R., Schmidt, G., Silvestri, N. M., Smith, J. A., Szkody, P., Strauss, M. A., Knapp, G. R., Collinge, M. J., Mukadam, A. S., Koester, D., Uomoto, A., Schlegel, D. J., Anderson, S. F., Brinkmann, J., Lamb, D. Q., Schneider, D. P., and York, D. G. 2004. A catalog of spectroscopically identified white dwarf stars in the first data release of the Sloan Digital Sky Survey. *ApJ*, **607**(May), 426–44.

Kochukhov, O. 2006. Pulsational line profile variation of the roAp star HR 3831. *A&A*, **446**(Feb.), 1051–70.

Kochukhov, O., Bagnulo, S., and Barklem, P. S. 2002. Interpretation of the core-wing anomaly of Balmer line profiles of cool Ap stars. *ApJ*, **578**(Oct.), L75–L78.

Kochukhov, O., Drake, N. A., Piskunov, N., and de la Reza, R. 2004. Multi-element abundance Doppler imaging of the rapidly oscillating Ap star HR 3831. *A&A*, **424**(Sept.), 935–50.

Kolenberg, K., Bryson, S., Szabó, R., Kurtz, D. W., Smolec, R., Nemec, J. M., Guggenberger, E., Moskalik, P., Benkő, J. M., Chadid, M., Jeon, Y.-B., Kiss, L. L., Kopacki, G., Nuspl, J., Still, M., Christensen-Dalsgaard, J., Kjeldsen, H., Borucki, W. J., Caldwell, D. A., Jenkins, J. M., and Koch, D. 2011. Kepler photometry of the prototypical Blazhko star RR Lyr: an old friend seen in a new light. *MNRAS*, **411**(Feb.), 878–90.

Kurtz, D. W. 1976. Metallicism and pulsation: an analysis of the Delta Delphini stars. *ApJS*, **32**(Oct.), 651–80.

Kurtz, D. W. 1982. Rapidly oscillating Ap stars. *MNRAS*, **200**(Sept.), 807–859.

Kurtz, D. W., and Martinez, P. 2000. Observing roAp Stars with WET: a primer. *Baltic Astronomy*, **9**, 253–53.

Kurtz, D. W., van Wyk, F., Roberts, G., Marang, F., Handler, G., Medupe, R., and Kilkenny, D. 1997. Frequency variability in the rapidly oscillating Ap star HR 3831: three more years of monitoring. *MNRAS*, **287**(May), 69–78.

Kurtz, D. W., Elkin, V. G., and Mathys, G. 2005. Probing the magnetoacoustic boundary layer in the peculiar magnetic star 33 Lib (HD 137949). *MNRAS*, **358**(Mar.), L6–L10.

Kurtz, D. W., Elkin, V. G., Cunha, M. S., Mathys, G., Hubrig, S., Wolff, B., and Savanov, I. 2006a. The discovery of 8.0-min radial velocity variations in the strongly magnetic cool Ap star HD154708, a new roAp star. *MNRAS*, **372**(Oct.), 286–92.

Kurtz, D. W., Elkin, V. G., and Mathys, G. 2006b. The discovery of a new type of upper atmospheric variability in the rapidly oscillating Ap stars with VLT high-resolution spectroscopy. *MNRAS*, **370**(Aug.), 1274–94.

Kurtz, D. W., Elkin, V. G., and Mathys, G. 2006c. Observations of magneto-acoustic modes in strongly magnetic pulsating roAp stars. In: *Proceedings of SOHO 18/GONG 2006/HELAS I, Beyond the spherical Sun.* ESA Special Publication, vol. 624.

Kurtz, D. W., Elkin, V. G., Mathys, G., and van Wyk, F. 2007. On the nature of the upper atmospheric variability in the rapidly oscillating Ap star HD134214. *MNRAS*, **381**(Nov.), 1301–12.

Kurtz, D. W., Shibahashi, H., Dhillon, V. S., Marsh, T. R., and Littlefair, S. P. 2008. A search for a new class of pulsating DA white dwarf stars in the DB gap. *MNRAS*, **389**(Oct.), 1771–9.

Kurtz, D. W., Cunha, M. S., Saio, H., Bigot, L., Balona, L. A., Elkin, V. G., Shibahashi, H., Brandao, I. M., Uytterhoeven, K., Frandsen, S., Frimann, S., Hatzes, A., Lueftinger, T., Gruberbauer, M., Kjeldsen, H., Christensen-Dalsgaard, J., and Kawaler, S. D. 2011. The first evidence for multiple pulsation axes: a new roAp star in the Kepler field, KIC 10195926. *ArXiv e-prints*, Feb.

Ledoux, P. 1951. The nonradial oscillations of gaseous stars and the problem of Beta Canis Majoris. *ApJ*, **114**(Nov.), 373–84.

Liebert, J. 1986. The origin and evolution of helium-rich white dwarfs. Pages 367–381 of: K. Hunger, D. Schoenberner, and N. Kameswara Rao (ed), *IAU Colloq. 87: Hydrogen Deficient Stars and Related Objects*. Astrophysics and Space Science Library, vol. 128.

Lyne, A. G., and Bailes, M. 1992. No planet orbiting PS R1829-10. *Nature*, **355**(Jan.), 213.

Mashonkina, L., Ryabchikova, T., and Ryabtsev, A. 2005. NLTE ionization equilibrium of Nd II and Nd III in cool A and Ap stars. *A&A*, **441**(Oct.), 309–18.

Mathys, G., and Lanz, T. 1990. The magnetic field of the AM star Omicron Pegasi *A&A*, **230**(Apr.), L21–L24.

Mathys, G. and Kharchenko, N. and Hubrig, S. 1996. A kinematical study of rapidly oscillating AP stars. *A&A*, **311**(July), 901–10.

McCarthy, D. D. 2005. Precision time and the rotation of the Earth. Pages 180–197 of: D. W. Kurtz (ed), *IAU Colloq. 196: Transits of Venus: New Views of the Solar System and Galaxy*.

McCook, G. P., and Sion, E. M. 1999. A catalog of spectroscopically identified white dwarfs. *ApJS*, **121**(Mar.), 1–130.

Mkrtichian, D. E., Hatzes, A. P., and Kanaan, A. 2003. Radial velocity variations in pulsating Ap stars. II. 33 Librae. *MNRAS*, **345**(Nov.), 781–94.

Montgomery, M. H., Provencal, J. L., Kanaan, A., Mukadam, A. S., Thompson, S. E., Dalessio, J., Shipman, H. L., Winget, D. E., Kepler, S. O., and Koester, D. 2010. Evidence for temperature change and oblique pulsation from light curve fits of the pulsating white dwarf GD 358. *ApJ*, **716**(June), 84–96.

Moya, A., Amado, P. J., Barrado, D., Hernández, A. G., Aberasturi, M., Montesinos, B., and Aceituno, F. 2010. The planetary system host HR8799: on its λ Bootis nature. *MNRAS*, **406**(July), 566–75.

Newcomb, S. 1895. Tables of the motion of the earth on its axis and around the sun. *Astronomical papers prepared for the use of the American ephemeris and nautical almanac*. Washington, Bureau of Equipment, Navy Dept., **6**, 2.

Preston, G. W. 1974. The chemically peculiar stars of the upper main sequence. *ARA&A*, **12**, 257–77.

Rosen, R., Clemens, C., Demorest, P., and Wyman, K. 2010. Non-radial Oscillations in Radio Pulsars. Page 453.16 of: *American Astronomical Society Meeting Abstracts number 215*. Bulletin of the American Astronomical Society, vol. 42.

Saio, H. 2005. A non-adiabatic analysis for axisymmetric pulsations of magnetic stars. *MNRAS*, **360**(July), 1022–32.

Saio, H., and Gautschy, A. 2004. Axisymmetric p-mode pulsations of stars with dipole magnetic fields. *MNRAS*, **350**(May), 485–505.

Shibahashi, H. 2000. The oblique pulsator model for the Blazhko effect in RR Lyrae stars. Theory of amplitude modulation I. Pages 299–306 of: L. Szabados and D. Kurtz (ed), *IAU Colloq. 176: The Impact of Large-Scale Surveys on Pulsating Star Research*. Astronomical Society of the Pacific Conference Series, vol. 203.

Shibahashi, H. 2005. The DB gap and pulsations of white dwarfs. Pages 143–148 of: G. Alecian, O. Richard, and S. Vauclair (ed), *EAS Publications Series*. EAS Publications Series, vol. 17.

Shibahashi, H. 2007. The DB gap and pulsations of white dwarfs. Pages 35–42 of: R. J. Stancliffe, G. Houdek, R. G. Martin, and C. A. Tout (ed), *Unsolved Problems in Stellar Physics: A Conference in Honor of Douglas Gough*. American Institute of Physics Conference Series, vol. 948.

Shibahashi, H., and Aerts, C. 2000. Asteroseismology and oblique pulsator model of β Cephei. *ApJ*, **531**(Mar.), L143–L146.

Shibahashi, H., and Takata, M. 1993. Theory for the distorted dipole modes of the rapidly oscillating Ap stars: a refinement of the oblique pulsator model. *PASJ*, **45**(Aug.), 617–41.

Silvotti, R., Schuh, S., Janulis, R., Solheim, J.-E., Bernabei, S., Østensen, R., Oswalt, T. D., Bruni, I., Gualandi, R., Bonanno, A., Vauclair, G., Reed, M., Chen, C.-W., Leibowitz, E., Paparo, M., Baran, A., Charpinet, S., Dolez, N., Kawaler, S., Kurtz, D., Moskalik, P., Riddle, R., and Zola, S. 2007. A giant planet orbiting the "extreme horizontal branch" star V391 Pegasi. *Nature*, **449**(Sept.), 189–91.

Sousa, J. C., and Cunha, M. 2011. On the understanding of pulsations in the atmosphere of roAp stars: phase diversity and false nodes. *ArXiv e-prints*, Mar.

Sterken, C., and Jaschek, C. 2005. Light Curves of Variable Stars. C., Sterken and C., Jaschek (ed), *Light Curves of Variable Stars*. Cambridge University Press. Cambridge, United Kingdom.

Szabó, R., Kolláth, Z., Molnár, L., Kolenberg, K., Kurtz, D. W., Bryson, S. T., Benkő, J. M., Christensen-Dalsgaard, J., Kjeldsen, H., Borucki, W. J., Koch, D., Twicken, J. D., Chadid, M., di Criscienzo, M., Jeon, Y.-B., Moskalik, P., Nemec, J. M., and Nuspl, J. 2010. Does Kepler unveil the mystery of the Blazhko effect? First detection of period doubling in Kepler Blazhko RR Lyrae stars. *MNRAS*, **409**(Dec.), 1244–52.

Takata, M., and Shibahashi, H. 1995. Effects of the quadrupole component of magnetic fields on the rapid oscillations of Ap stars. *PASJ*, **47**(Apr.), 219–31.

Welsh, W. F., Orosz, J. A., Aerts, C., Brown, T., Brugamyer, E., Cochran, W., Gilliland, R. L., Guzik, J. A., Kurtz, D. W., Latham, D., Marcy, G. W., Quinn, S. N., Zima, W., Allen, C., Batalha, N., Bryson, S., Buchhave, L., Caldwell, D. A., Gautier, T. N., Howell, S., Kinemuchi, K., Ibrahim, K. A., Isaacson, H., Jenkins, J., Prsa, A., Still, M., Street, R., Wohler, B., Koch, D. G., and Borucki, W. J. 2011. KOI-54: The Kepler discovery of tidally-excited pulsations and brightenings in a highly eccentric binary. *ApJS*, **197**(Feb.), 4–18.

Willems, B., and Aerts, C. 2002. Tidally induced radial-velocity variations in close binaries. *A&A*, **384**(Mar.), 441–51.

York, D. G., Adelman, J., Anderson, Jr., J. E., Anderson, S. F., Annis, J., Bahcall, N. A., Bakken, J. A., Barkhouser, R., Bastian, S., Berman, E., Boroski, W. N., Bracker, S., Briegel, C., Briggs, J. W., Brinkmann, J., Brunner, R., Burles, S., Carey, L., Carr, M. A., Castander, F. J., Chen, B., Colestock, P. L., Connolly, A. J., Crocker, J. H., Csabai, I., Czarapata, P. C., Davis, J. E., Doi, M., Dombeck, T., Eisenstein, D., Ellman, N., Elms, B. R., Evans, M. L., Fan, X., Federwitz, G. R., Fiscelli, L., Friedman, S., Frieman, J. A., Fukugita, M., Gillespie, B., Gunn, J. E., Gurbani, V. K., de Haas, E., Haldeman, M., Harris, F. H., Hayes, J., Heckman, T. M., Hennessy, G. S., Hindsley, R. B., Holm, S., Holmgren, D. J., Huang, C.-h., Hull, C., Husby, D., Ichikawa, S.-I., Ichikawa, T., Ivezić, Ž., Kent, S., Kim, R. S. J., Kinney, E., Klaene, M., Kleinman, A. N., Kleinman, S., Knapp, G. R., Korienek, J., Kron, R. G., Kunszt, P. Z., Lamb, D. Q., Lee, B., Leger, R. F., Limmongkol, S., Lindenmeyer, C., Long, D. C., Loomis, C., Loveday, J., Lucinio, R., Lupton, R. H., MacKinnon, B., Mannery, E. J., Mantsch, P. M., Margon, B., McGehee, P., McKay, T. A., Meiksin, A., Merelli, A., Monet, D. G., Munn, J. A., Narayanan, V. K., Nash, T., Neilsen, E., Neswold, R., Newberg, H. J., Nichol, R. C., Nicinski, T., Nonino, M., Okada, N., Okamura, S., Ostriker, J. P., Owen, R., Pauls, A. G., Peoples, J., Peterson, R. L., Petravick, D., Pier, J. R., Pope, A., Pordes, R., Prosapio, A., Rechenmacher, R., Quinn, T. R., Richards, G. T., Richmond, M. W., Rivetta, C. H., Rockosi, C. M., Ruthmansdorfer, K., Sandford, D., Schlegel, D. J., Schneider, D. P., Sekiguchi, M., Sergey, G., Shimasaku, K., Siegmund, W. A., Smee, S., Smith, J. A., Snedden, S., Stone, R., Stoughton, C., Strauss, M. A., Stubbs, C., SubbaRao, M., Szalay, A. S., Szapudi, I., Szokoly, G. P., Thakar, A. R., Tremonti, C., Tucker, D. L., Uomoto, A., Vanden Berk, D., Vogeley, M. S., Waddell, P., Wang, S.-i., Watanabe, M., Weinberg, D. H., Yanny, B., and Yasuda, N. 2000. The Sloan Digital Sky Survey: technical summary. *AJ*, **120**(Sept.), 1579–87.

Zverko, J., Ziznovsky, J., Adelman, S. J., and Weiss, W. W. 2004 (Dec.). IAU S224. Pages 829–834 of: J. Zverko, J. Ziznovsky, S. J. Adelman, and W. W. Weiss (ed), *The A-Star Puzzle*. IAU Symposium, vol. 224.

7. Asteroseismology of red giants

JØRGEN CHRISTENSEN-DALSGAARD

7.1 Introduction

Asteroseismology, and hence the study of stellar properties, is being revolutionized by the extremely accurate and extensive data from the CoRoT (Baglin *et al.*, 2009) and *Kepler* (Borucki *et al.*, 2009) space missions. Analysis of time series of unprecedented extent, continuity, and sensitivity has allowed the study of a broad range of stellar variability, including oscillations of a variety of pulsating stars (see, e.g., Gilliland *et al.*, 2010; Christensen-Dalsgaard and Thompson, 2011, for reviews), and leading to comparative asteroseismology (or *synasteroseismology*[1]) for main-sequence stars showing solar-like oscillation (Chaplin *et al.*, 2011). Also, early analyses of *Kepler* data have demonstrated the power of asteroseismology in characterizing the central stars in planetary systems (Christensen-Dalsgaard *et al.*, 2010; Batalha *et al.*, 2011), and such investigations will undoubtedly play a major role in the continuing *Kepler* exploration of extrasolar planetary systems.

However, perhaps the most striking results of space asteroseismology have come from the investigation of red giants. Given the extensive outer convection zones of red giants, solar-like oscillations were predicted quite early (Christensen-Dalsgaard and Frandsen, 1983). Ground-based observations have been carried out in a few cases (e.g., Frandsen *et al.*, 2002; De Ridder *et al.*, 2006) but, owing to the very long periods of these huge stars, such observations are extremely demanding in terms of observing and observer's time. Space observations, on the other hand, allow nearly continuous observations over very extended periods, as demonstrated by early observations by the WIRE (Retter *et al.*, 2003) and MOST (Barban *et al.*, 2007) satellites. With CoRoT and *Kepler*, we have obtained extensive data for literally thousands of red giants, providing precise characterization of their overall characteristics, as well as, in some cases, detailed information on the properties of their deep interiors. This promises completely to change our understanding of these late phases of stellar evolution.

Here, I provide an overview of the oscillation properties of red giants and discuss some of the initial results of the analysis of data from CoRoT and *Kepler*. An introduction to the properties of solar-like oscillations, emphasizing the observational aspects, was provided by Bedding (Chapter 3). Further introductions to solar-like stellar oscillations were provided by Christensen-Dalsgaard (2004) and, in particular, by the monograph by Aerts *et al.* (2010). To illustrate the relevant properties, computations of red giant models and oscillations are presented. These probably do not fully cover the complexities of these late stages of stellar evolution and hence are not intended as models of specific observations, but it is hoped that they at least capture the main features of these stars. The modeling was carried out using the Aarhus Stellar Evolution Code (ASTEC) (Christensen-Dalsgaard, 2008b), with OPAL equation of state (Rogers *et al.*, 1996) and opacities (Iglesias and Rogers, 1996), and NACRE (Angulo *et al.*, 1999) nuclear reaction rates. Convection was treated with the Böhm-Vitense (1958) mixing-length formalism, with a mixing length proportional to the pressure scale height and approximately calibrated to a solar model. Diffusion and settling were neglected and no mass loss was included. Adiabatic oscillations were calculated using the Aarhus adiabatic pulsation code ADIPLS (Christensen-Dalsgaard, 2008a). To resolve the very rapid variations in the eigenfunctions in the stellar core, the models were transferred from the original mesh used in the model calculation to a much finer mesh, designed to capture adequately the

[1] A term coined by D. O. Gough.

properties of the eigenfunctions. However, it is probably fair to say that the mesh distribution, and more generally the numerical procedures to deal with this type of oscillation, requires further optimization and testing.

7.2 Simple properties of stellar oscillations

This section provides a brief background on the theory of stellar oscillations, as far as the oscillation frequencies and their relation to stellar properties are concerned. Some emphasis is given to the oscillation properties of red giants.

To understand the properties of stellar oscillation frequencies, it is a good approximation to assume that the oscillations are adiabatic, and hence ignore the processes that damp or excite the modes. This approximation breaks down in the near-surface layers, where there are strong departures from adiabaticity. Also, the modeling of this region is highly uncertain, owing to the detailed effects of convection on the mean structure of the star and the dynamics of the oscillations. These near-surface processes have a significant effect on the frequencies, as is evident in the analysis of solar observations, and they must be taken into account in detailed asteroseismic analyses of solar-like oscillation frequencies (e.g., Kjeldsen *et al.*, 2008). However, for the present discussion, we can ignore them.

With this approximation, the problem of computing stellar oscillation frequencies for a given stellar model is relatively straightforward.[2] This feature, together with the ability to determine the observed frequencies with very high accuracy, makes stellar oscillations such excellent diagnostics of the stellar interior. For the purpose of computing adiabatic frequencies, and assuming that the equilibrium model satisfies hydrostatic equilibrium (see Kawaler, Chapter 2), the structure of the star is characterized by the density $\rho(r)$ as a function of distance r to the center, as well as the adiabatic exponent $\Gamma_1(r)$, where

$$\Gamma_1 = \left(\frac{\partial \ln p}{\partial \ln \rho}\right)_{\rm ad} , \tag{7.1}$$

where p is pressure and the derivative corresponds to an adiabatic change. Given $\rho(r)$, the mass distribution in the star can immediately be determined, and $p(r)$ can then be computed from the equation of hydrostatic equilibrium, with a suitable boundary condition. This essentially defines all that is needed to complete the equations of adiabatic oscillation. For the predominantly acoustic solar-like oscillations, the most important quantity is the adiabatic sound speed c, given by

$$c^2 = \frac{\Gamma_1 p}{\rho} . \tag{7.2}$$

In many relevant cases, the equation of state can be approximated by the ideal gas law:

$$p = \frac{k_{\rm B} \rho T}{\mu m_{\rm u}} , \tag{7.3}$$

where T is temperature, μ the mean molecular weight, $k_{\rm B}$ Boltzmann's constant, and $m_{\rm u}$ the atomic mass unit. We then obtain

$$c^2 \simeq \frac{\Gamma_1 k_{\rm B} T}{\mu m_{\rm u}} . \tag{7.4}$$

We consider small-amplitude oscillations in a slowly rotating star. Then, the dependence of the eigenfunctions on co-latitude θ and longitude ϕ can be separated as spherical harmonics $Y_l^m(\theta, \phi)$; the degree l measures the total number of nodal lines on the stellar

[2] Although, as discussed in the following, some care is needed when considering red giants.

surface and provides a measure of the local horizontal wavenumber k_h:

$$k_\mathrm{h}^2 = \frac{l(l+1)}{r^2}, \tag{7.5}$$

while the azimuthal order m, with $|m| \leq l$, measures the number of nodal lines crossing the equator. For each (l, m), the star has a set of eigenfrequencies ω_{nlm} labeled by the radial order n. For adiabatic oscillations ω_{nlm} is real; the oscillation depends on time t and ϕ as $\cos(m\phi - \omega t)$, that is, for $m \neq 0$, as a running wave in longitude. For a nonrotating star, the frequencies are independent of m. Slow rotation with angular velocity Ω introduces a frequency splitting

$$\omega_{nlm} = \omega_{nl0} + m\beta_{nlm}\langle\Omega\rangle_{nlm}, \tag{7.6}$$

where $\langle\Omega\rangle_{nlm}$ is an average over the star with a weight determined by the oscillation eigenfunction. If $\omega = \Omega(r)$ depends on r alone, $\beta_{nlm} = \beta_{nl}$ and the average are independent of m and the rotational splitting is linear in m.

After separation of Y_l^m, the equations of adiabatic oscillations reduce to a set of four linear differential equations in r, with suitable boundary conditions at the center and surface, which can be solved numerically, with the frequencies ω_{nlm} as eigenvalues. However, a great deal of insight into the properties of the solution and the relation of the frequencies to the structure of the star can be obtained from asymptotic analysis of the equations. Deubner and Gough (1984) derived an asymptotic equation, applicable for modes of high radial order, in terms of the quantity $X = c^2\rho^{1/2}\mathrm{div}\,\boldsymbol{\delta r}$, where $\boldsymbol{\delta r}$ is the displacement vector:

$$\frac{\mathrm{d}^2X}{\mathrm{d}r^2} = -K(r)X, \tag{7.7}$$

where

$$K = \frac{1}{c^2}\left[S_l^2\left(\frac{N^2}{\omega^2} - 1\right) + \omega^2 - \omega_\mathrm{c}^2\right]. \tag{7.8}$$

Thus, K is controlled by three characteristic frequencies of the star: the *Lamb frequency* S_l, with

$$S_l^2 = \frac{l(l+1)c^2}{r^2}, \tag{7.9}$$

the *buoyancy frequency* (or *Brunt-Väisälä frequency*) N,

$$N^2 = g\left(\frac{1}{\Gamma_1}\frac{\mathrm{d}\ln p}{\mathrm{d}r} - \frac{\mathrm{d}\ln\rho}{\mathrm{d}r}\right), \tag{7.10}$$

where g is the local gravitational acceleration, and the *acoustic cutoff frequency* ω_c,

$$\omega_\mathrm{c}^2 = \frac{c^2}{4H^2}\left(1 - 2\frac{\mathrm{d}H}{\mathrm{d}r}\right), \tag{7.11}$$

where $H = -(\mathrm{d}\ln\rho/\mathrm{d}r)^{-1}$ is the density scale height.

Equation 7.7 determines the local properties of the eigenfunction. In regions where $K(r) > 0$, the solution locally oscillates as a function of r, while the behavior is exponentially increasing or decreasing where $K(r) < 0$. The intermediate points, where $K(r) = 0$, are called *turning points*. Often, there is an interval $[r_1, r_2]$ where $K(r) > 0$, with K being negative just outside and the solution decreasing exponentially in the direction away from that interval. In that case the mode is said to be trapped between r_1 and r_2; from JWKB analysis (e.g., Gough, 2007), one finds that the frequency approximately satisfies the dispersion relation

$$\int_{r_1}^{r_2} K^{1/2}\mathrm{d}r = (n - 1/2)\pi. \tag{7.12}$$

The behavior of the Lamb frequency and the acoustic cutoff frequency is relatively simple. The Lamb frequency generally decreases with increasing r;[3] for the low degrees that are relevant to stellar observations it is small near the surface. Here, typically, ω_c^2 dominates in Equation 7.8. In the atmosphere, where the temperature is approximately constant, H can be approximated by the (constant) pressure scale height and we obtain

$$\omega_c \simeq \omega_{c,atm} = \frac{1}{2} g \sqrt{\Gamma_1 \rho / p} \propto M R^{-2} T_{eff}^{-1/2}, \tag{7.13}$$

evaluated in the atmosphere; in the last proportionality we assumed the ideal gas law (see Equation 7.3) and that the atmospheric temperature is proportional to the effective temperature T_{eff}. Below the surface, ω_c decreases rapidly with increasing depth. When $\omega < \omega_{c,atm}$ the solution decreases with height in the atmosphere; this corresponds to reflection of the mode at the surface so that it is trapped in the stellar interior. Waves with $\omega > \omega_{c,atm}$ can propagate out through the atmosphere and hence lose energy, leading to strong damping. Thus, $\omega_{c,atm}$ defines the upper limit in frequency to trapped modes, at least in the adiabatic approximation.

The behavior of the buoyancy frequency is more complex. This is seen most clearly by assuming the ideal gas approximation, Equation 7.3. Then, we obtain, from Equation 7.10, that

$$N^2 = g^2 \frac{\rho}{p} (\nabla_{ad} - \nabla + \nabla_\mu), \tag{7.14}$$

where, following the usual convention (see Kawaler, Chapter 2),

$$\nabla = \frac{d \ln T}{d \ln p}, \qquad \nabla_{ad} = \left(\frac{\partial \ln T}{\partial \ln p} \right)_{ad}, \qquad \nabla_\mu = \frac{d \ln \mu}{d \ln p}. \tag{7.15}$$

Because typically μ increases toward the center and hence with increasing p, as a result of nuclear burning, the term in ∇_μ typically gives a positive contribution to N^2 in the deep interior of the star. It should also be noted that the condition for convective instability[4] is that $N^2 < 0$. In convection zones the composition is uniform and $\nabla_\mu = 0$; also, except near the surface, ∇ is nearly adiabatic and N^2 is only slightly negative.

Solar-like oscillations are believed to be excited stochastically by the vigorous motion in the near-surface layers of convective envelopes (see Section 7.3). Thus stars showing such oscillations have outer convection zones and hence the buoyancy frequency is positive only in the deeper interior. Here, we illustrate the characteristic frequencies and their dependence on the evolutionary state by considering selected models in a 1 $M_\odot$ evolution sequence, illustrated in an HR diagram in Fig. 7.1. A typical example of a rather unevolved star, at roughly solar age, is shown in Fig. 7.2. Here, there is a fairly clear separation between S_l and N. As a result there are essentially two different scenarios for having a positive K, also illustrated in the figure: by having $\omega > S_l$ (and hence typically $\omega \gg N$) or by having $\omega < N$ (and hence typically $\omega \ll S_l$). The former case corresponds to acoustic modes (or p modes), where the restoring force is predominantly pressure, whereas the latter case corresponds to internal gravity waves (or g modes), where the restoring force is buoyancy.

In discussing the properties of these modes in more detail, we neglect ω_c in Equation 7.8, except insofar as it causes the required reflection at the surface of the p modes. For the p modes, assuming that $\omega \gg N$, we then obtain from Equation 7.8 that

$$K \simeq \frac{1}{c^2} (\omega^2 - S_l^2), \tag{7.16}$$

[3] Possibly with the exception of discontinuities in composition, caused by nuclear burning in the stellar core, which leads to discontinuities in density and hence sound speed.

[4] Strictly speaking, the Ledoux criterion.

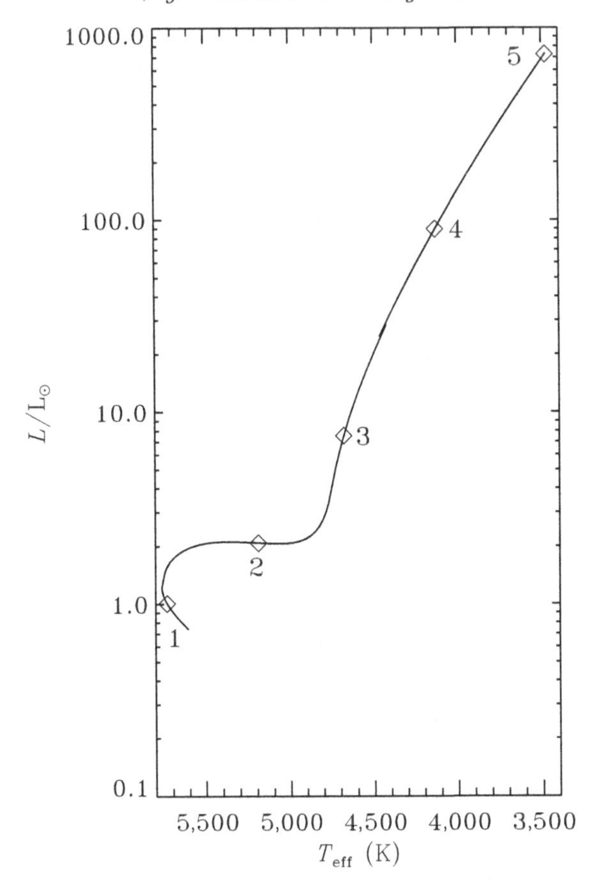

FIG. 7.1. Evolution track for a 1 $M_\odot$ model, approximately calibrated to the Sun at the present solar age.

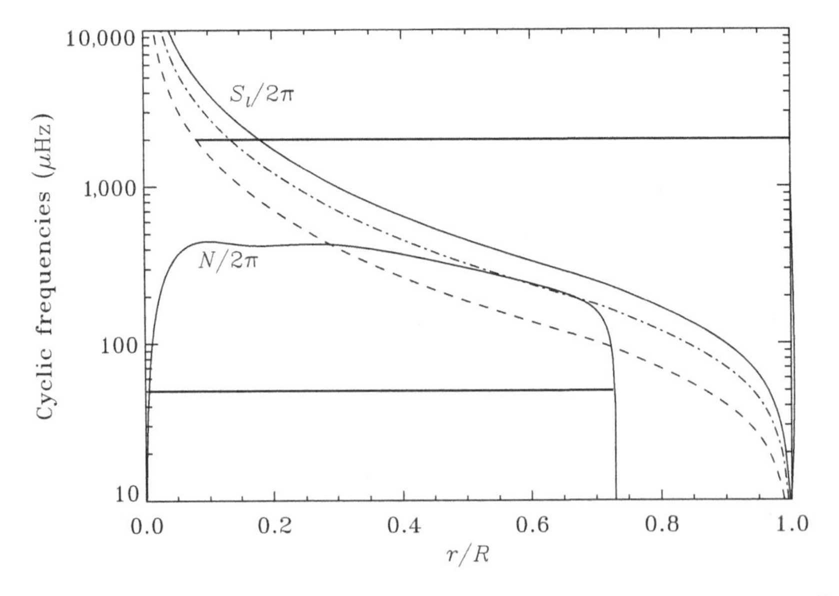

FIG. 7.2. Characteristic cyclic frequencies for Model 1 in the 1 $M_\odot$ evolution sequence illustrated in Fig. 7.1. The solid curve shows $N/2\pi$, where N is the buoyancy frequency (see Equation 7.10), whereas the dashed, dot-dashed, and triple-dot-dashed curves show the Lamb frequency $S_l/2\pi$ (see Equation 7.9) for $l = 1$, 2, and 3, respectively. The horizontal solid lines show typical propagation regions for a p mode (with $l = 1$, at high frequency) and a g mode (at low frequency).

with a lower turning point $r_{\rm t}$ defined by $S_l = \omega$ or

$$\frac{c(r_{\rm t})}{r_{\rm t}} = \frac{\omega}{L}, \tag{7.17}$$

where $L^2 = l(l+1)$. Also, the corresponding approximation to Equation 7.12 can be rearranged to yield

$$\int_{r_{\rm t}}^{R} \left(1 - \frac{L^2 c^2}{r^2 \omega^2}\right)^{1/2} \frac{{\rm d}r}{c} = \frac{(n+\alpha)\pi}{\omega}. \tag{7.18}$$

Here, we replaced the $-1/2$ in Equation 7.12 by α, which generally is a function of frequency, to account for the phase change in the reflection at the surface. Equation 7.18 has played a major role in the analysis of solar oscillation frequencies; a relation of this form was first obtained by Duvall (1982) from analysis of observed frequencies. It explicitly demonstrates that in this limit the frequencies are determined by the sound speed and by α.

In unevolved stars, $r_{\rm t}$ is close to the center where c varies relatively slowly. Equation 7.18 can then be expanded (Gough, 1986, 1993) to yield, in terms of the cyclic frequencies $\nu_{nl} = \omega_{nl}/2\pi$,

$$\nu_{nl} = \Delta\nu \left(n + \frac{l}{2} + \alpha + \frac{1}{4}\right) - d_{nl}, \tag{7.19}$$

where

$$\Delta\nu = \left(2 \int_0^R \frac{{\rm d}r}{c}\right)^{-1}, \tag{7.20}$$

and we replaced L by $l + 1/2$ (see also Vandakurov, 1967; Tassoul, 1980, 1990). Neglecting the small correction term d_{nl}, this shows that the frequencies are uniformly spaced in radial order n, with a spacing given by the *large separation* $\Delta\nu$ and such that there is degeneracy between ν_{nl} and $\nu_{n-1\,l+2}$. This degeneracy is lifted by d_{nl} which asymptotically, for main sequence models, can be approximated by

$$d_{nl} \simeq -\frac{\Delta\nu}{4\pi^2 \nu_{nl}} L^2 \int_0^R \frac{{\rm d}c}{{\rm d}r} \frac{{\rm d}r}{r}. \tag{7.21}$$

To this approximation d_{nl}, and hence the small separations $\delta\nu_{l\,l+2}(n) = \nu_{nl} - \nu_{n-1\,l+2}$, are very sensitive to the sound-speed gradient in the stellar core and hence, according to Equation 7.4, to the composition profile that has resulted from the nuclear burning. Thus, for main sequence stars the small frequency separations provide a measure of the evolutionary stage and hence the age of the star. Interestingly, although the derivation sketched above and the expression for d_{nl} do not hold for highly evolved stars such as red giants, the frequency pattern reflected in Equation 7.19 is still found (see also Section 7.6).

In the opposite extreme, where $\omega \ll S_l$ and $\omega < N$, we approximate K by

$$K \simeq \frac{L^2}{r^2}\left(\frac{N^2}{\omega^2} - 1\right). \tag{7.22}$$

With a relatively simple behavior of N, such as is illustrated in Fig. 7.2, there are typically just two turning points $[r_1, r_2]$ where $\omega = N$, and Equation 7.12 yields

$$L \int_{r_1}^{r_2} \left(\frac{N^2}{\omega^2} - 1\right)^{1/2} \frac{{\rm d}r}{r} = (n - 1/2)\pi. \tag{7.23}$$

If $\omega \ll N$ almost everywhere on $[r_1, r_2]$, we can approximate Equation 7.23 further by neglecting 1 compared with N^2/ω^2, to obtain finally an approximate expression for the period $\Pi = 2\pi/\omega$:

$$\Pi = \frac{\Pi_0}{L}(n + \alpha_{\rm g}), \tag{7.24}$$

where

$$\Pi_0 = 2\pi^2 \left(\int_{r_1}^{r_2} N \frac{{\rm d}r}{r} \right)^{-1}, \tag{7.25}$$

assuming that $N^2 \geq 0$ on $[r_1, r_2]$, with $N = 0$ at r_1 and r_2 (e.g., Tassoul, 1980). Also, $\alpha_{\rm g}$, which may depend on l, reflects the actual phase shift at the turning points. Equation 7.24 shows that in this case we get periods that are uniformly spaced in the radial order n, with a spacing that is inversely proportional to L.

The dependence of the characteristic frequencies and hence the oscillation properties on stellar parameters can to a large extent be characterized by simple scaling relations. An important example is the scaling of the acoustic cutoff frequency in Equation 7.13. Under many circumstances, the structure of a star, as a function of the fractional radius $x = r/R$, approximately satisfies so-called *homology scaling relations* (e.g., Kippenhahn and Weigert, 1990).[5] According to these pressure scales as GM^2/R^4, G being the gravitational constant, and density scales as M/R^3. It follows that the squared sound speed c^2 scales as GM/R and the squared characteristic frequencies scale as GM/R^3. This motivates introducing the dimensionless scaled frequencies by

$$\hat{S}_l^2 = \frac{R^3}{GM} S_l^2, \qquad \hat{N}^2 = \frac{R^3}{GM} N^2, \qquad \hat{\omega}_{\rm c}^2 = \frac{R^3}{GM} \omega_{\rm c}^2, \tag{7.26}$$

as well as $\hat{c}^2 = (R/GM)c^2$. Then the asymptotic Equations, 7.7 and 7.8, can be expressed as

$$\frac{{\rm d}^2 X}{{\rm d}x^2} = -\hat{K}(r) X, \tag{7.27}$$

where

$$\hat{K} = \frac{1}{\hat{c}^2} \left[\hat{S}_l^2 \left(\frac{\hat{N}^2}{\sigma^2} - 1 \right) + \sigma^2 - \hat{\omega}_{\rm c}^2 \right], \tag{7.28}$$

in terms of the dimensionless frequency

$$\sigma^2 = \frac{R^3}{GM} \omega^2 . \tag{7.29}$$

With similar scalings, the full equations of adiabatic oscillations can be expressed in a form that is homologously invariant. It follows that homologous models have the same dimensionless frequencies σ_{nl}, and that the actual frequencies satisfy the scaling

$$\omega_{nl} \propto (M/R^3)^{1/2} \propto \langle \rho \rangle^{1/2}, \tag{7.30}$$

where $\langle \rho \rangle$ is the mean density of the star. The same scaling obviously applies, for example, to the large frequency separation $\Delta\nu$.

Actual stellar models are never strictly homologous, and in particular the changing structure of the core as nuclear burning proceeds gives strong departures from these scalings. On the other hand, they are approximately satisfied in the outer layers that predominantly determine the frequencies of acoustic modes, and hence for such modes the scaling in Equation 7.30 is a reasonable approximation. To illustrate this, we consider

[5] A simple example, where the scalings are exact, is polytropic models.

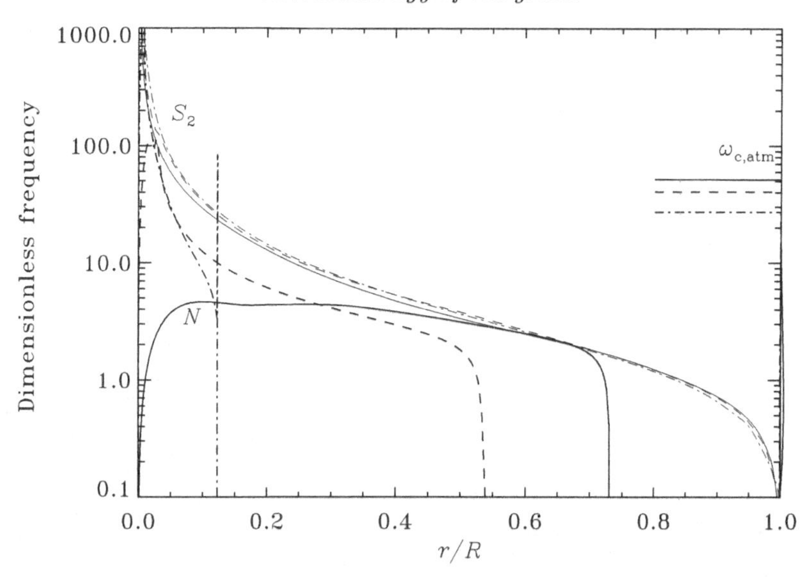

FIG. 7.3. Dimensionless characteristic frequencies (in units of $(GM/R^3)^{1/2}$) for models along the 1 $M_\odot$ evolutionary track illustrated in Fig. 7.1. Solid, dashed, and dot-dashed curves are for Models 1, 2, and 3, respectively. The thick curves show the buoyancy frequency N and the thinner curves show the Lamb frequency S_l, for $l = 2$. The horizontal lines at the right edge of the plot similarly show the dimensionless value of the acoustic cutoff frequency (see Equation 7.13) in the atmosphere of the models.

the evolution of the frequencies along a 1 $M_\odot$ evolution sequence, illustrated in Fig. 7.1. The dimensionless frequencies for Models 1, 2, and 3 are shown in Fig. 7.3. Evidently, there is relatively little change in $\hat{S}_2$. On the other hand, the changes in $\hat{N}$ are dramatic, showing very strong departures from the homologous scaling. The increase in the depth of the outer convection zone, where $|N| \simeq 0$, is evident. Also, N increases strongly in the core. This is mainly the result of the increase in the gravitational acceleration (see Equation 7.14) as the star develops a very compact helium core. Thus, Model 3 has a helium core with a mass of 0.18 $M_\odot$, contained within the inner 0.7 percent of the stellar radius. In this model, the convective envelope is so deep that it extends into a region where the composition has been changed by hydrogen fusion. This causes a discontinuity in the composition and hence the density, giving rise to a spike in the buoyancy frequency, as shown in the figure.

In Fig. 7.3 are also indicated the dimensionless values $\hat{\omega}_{\rm c,atm}$ of the acoustic cutoff frequency in the atmospheres of the three models. Since this is determined by the atmospheric properties of the star, it does not follow the homologous frequency scaling. In fact, using Equation 7.13 for $\nu_{\rm c,atm} = \omega_{\rm c,atm}/2\pi$ and the scaling in Equation 7.30 for $\Delta\nu$ it follows that

$$\frac{\nu_{\rm c,atm}}{\Delta\nu} \propto M^{1/2} R^{-1/2} T_{\rm eff}^{-1/2}. \tag{7.31}$$

Since $\nu_{\rm c,atm}$ is an upper limit to the frequencies of modes trapped in the stellar interior, this ratio, according to Equation 7.19, provides a measure of the number of acoustic modes that are trapped. This obviously decreases as the star evolves up the red giant branch.

Figure 7.3 also shows that in Models 2 and 3, the maximum of $\hat{N}$ in the core is substantially higher than $\hat{\omega}_{\rm c,atm}$. This means that *all* trapped non-radial modes satisfy $\omega < N, S_l$ in the core and hence behave like g modes there, although they may behave as acoustic modes in the outer parts of the star where $\omega > N, S_l$. The overall character of the

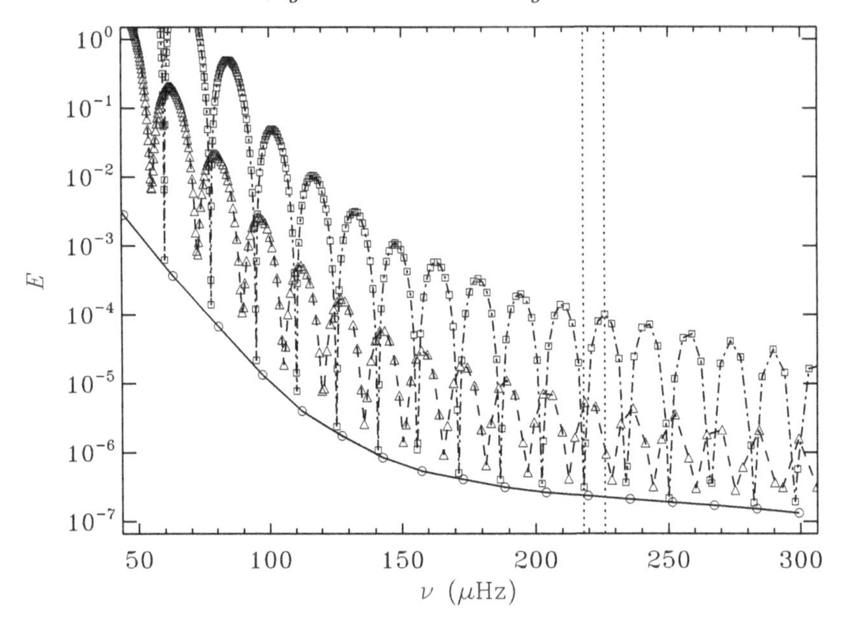

FIG. 7.4. Dimensionless inertia (see Eq 7.32) for Model 3 in the 1 M$_\odot$ evolution sequence illustrated in Fig. 7.1. Radial modes are shown with circles connected by solid lines, $l = 1$ modes with triangles connected by dashed lines, and $l = 2$ modes with squares connected by dot-dashed lines. The vertical dotted lines mark the $l = 2$ modes for which the eigenfunctions are illustrated in Fig. 7.5.

mode then depends on whether it has the largest amplitude in the g-mode or the p-mode region, and hence, effectively, on the behavior in the intermediate region. If the eigenfunction decreases exponentially with depth in this region, the mode has its largest amplitude in the outer region of the star and it may be said to be *p-dominated.* In the opposite case, with an eigenfunction increasing with depth, the amplitude is largest in the core and the mode is *g-dominated.* These two conditions essentially require the frequency to be determined such that the mode resonates with the outer acoustic, and the inner buoyancy-driven, cavities, respectively. For the p-dominated modes, this leads to Equation 7.18 for the frequencies, which may be cast, at least approximately, in the form of Equation 7.19. For the g-dominated modes we similarly find that the frequencies satisfy Equations 7.23 or 7.24. For evolved red giants like Model 3, with a very high buoyancy frequency in the core, the period spacing Π_0/L (see Eq. 7.25) becomes very small and hence the density of the g-dominated modes is much higher than the density of p-dominated modes, whose frequency spacing is given by $\Delta\nu$. It should also be noted that the high density of g-dominated modes corresponds to an extremely high number of nodes in the eigenfunctions in the core, requiring substantial care in the numerical computation of the oscillations.

The properties of the modes can be illustrated by considering the normalized mode inertia

$$E = \frac{\int_V \rho |\boldsymbol{\delta r}|^2 \mathrm{d}V}{M |\boldsymbol{\delta r}|_\mathrm{s}^2}, \tag{7.32}$$

where the integral is over the volume V of the star and $|\boldsymbol{\delta r}|_\mathrm{s}^2$ is the surface value of the squared norm of the displacement vector $\boldsymbol{\delta r}$. This is shown in Fig. 7.4 for modes of degree $l = 0 - 2$ in Model 3. The radial modes ($l = 0$) are purely acoustic, with an inertia that decreases with increasing frequency. For $l = 1$ and 2, on the other hand, there is a dense set of g-dominated modes, with acoustic resonances where the inertia approaches the value for the radial modes at the corresponding frequency. The pattern

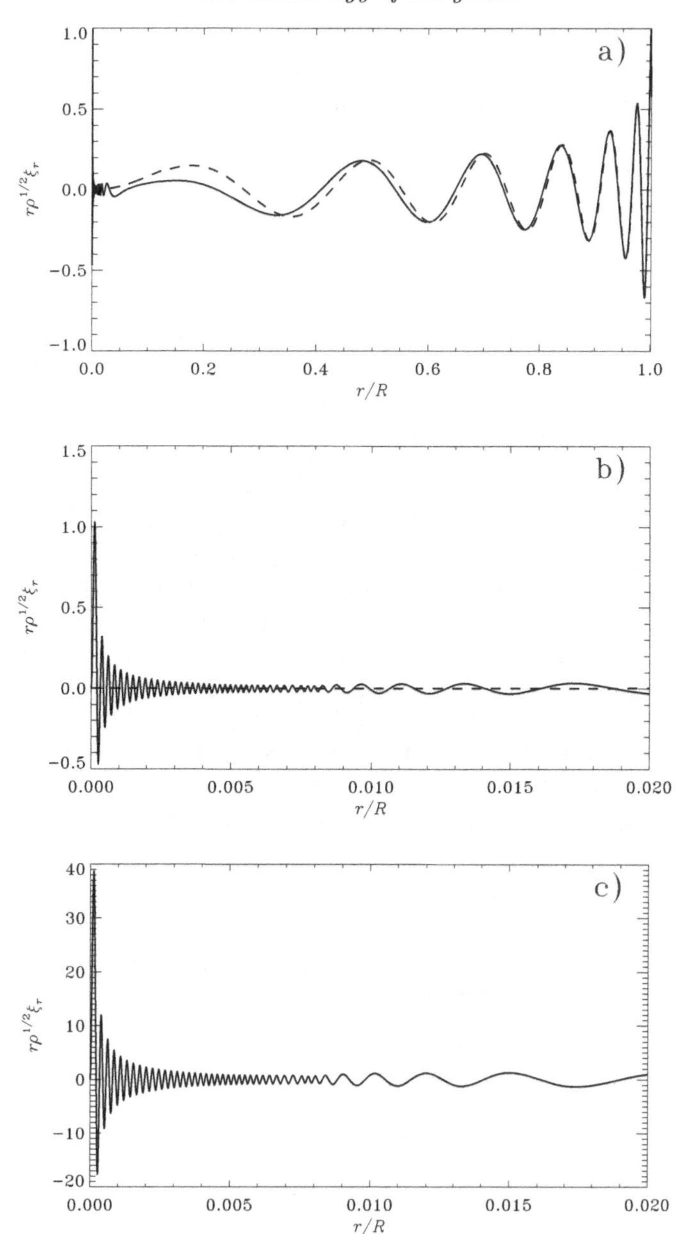

FIG. 7.5. Scaled eigenfunctions, normalized at the surface, for Model 3 in the 1 $M_\odot$ evolution sequence shown in Fig. 7.1. The quantity shown is $r\rho^{1/2}\xi_r$, where ξ_r is the amplitude of the vertical displacement. In panels (a) and (b) the dashed line shows a radial mode with cyclic frequency $\nu = 219.6\,\mu$Hz and mode inertia (see Equation 7.32) $E = 2.35 \times 10^{-7}$, and the solid line shows a p-dominated $l = 2$ mode with $\nu = 218.1\,\mu$Hz and $E = 3.12 \times 10^{-7}$. In panel (c) is illustrated a neighboring g-dominated $l = 2$ mode with $\nu = 226.1\,\mu$Hz and $E = 1.00 \times 10^{-4}$.

of these resonances is very similar to Equation 7.19, with the $l = 1$ resonances roughly halfway between the $l = 0$ frequencies and a small separation between an $l = 2$ resonance and the neighboring $l = 0$ mode.

Figure 7.5 shows a few selected eigenfunctions, in the form of scaled radial component of the displacement, normalized to unity at the surface. The comparison between the neighboring $l = 0$ and 2 modes in panel (a) shows that they are very similar in the bulk of the star where the $l = 2$ mode is predominantly of acoustic character. In the core the eigenfunction for $l = 2$ varies rapidly but at generally small amplitude for this p-dominated mode, as seen in panel (b). The behavior in the core for the mode shown in panel (c) is very similar, but at a much higher amplitude relative to the surface normalization, and corresponding to the local maximum in the inertia. We discuss the consequences for the observable visibility of the modes in Section 7.5.

7.3 Damping and excitation of solar-like oscillations

When nonadiabatic effects are taken into account, the oscillation frequencies are no longer real. Assuming again a time dependence as $\exp(-i\omega t)$ and writing $\omega = \omega_r + i\omega_i$ in terms of real and imaginary parts, the real part ω_r determines the frequency of the mode whereas the imaginary part ω_i determines the growth or decay of the amplitude, $\omega_i < 0$ corresponding to a damped mode. The determination of the damping or excitation of the modes requires a solution of the full nonadiabatic oscillation equations, introducing also the energy equation and equations for the energy generation and transport. The last is given by the perturbations to the flux $\mathbf{F} = \mathbf{F}_{rad} + \mathbf{F}_{con}$ which, as indicated, has both a radiative ($\mathbf{F}_{rad}$) and convective ($\mathbf{F}_{con}$) contribution. The radiative part can be dealt with in a relatively straightforward manner, although the treatment of the convective part remains highly uncertain. Furthermore, a significant contribution to the damping may come from the dynamical effects of convection through the perturbation to the Reynolds stresses, often approximated by a turbulent pressure p_t, which again is uncertain (e.g., Houdek, 2010a).

A convenient way to discuss the damping or excitation of the modes is to express the growth rate, approximately in terms of the *work integral* $W = W_g + W_t$, as

$$\omega_i \simeq \frac{W_g}{J} + \frac{W_t}{J}, \tag{7.33}$$

with

$$\begin{aligned} W_g &= \mathrm{Re}\left[\int_V \frac{\delta\rho^*}{\rho}(\Gamma_3 - 1)\delta(\rho\epsilon - \mathrm{div}\,\mathbf{F})\mathrm{d}V\right], \\ W_t &= -\omega_r \mathrm{Im}\left[\int_V \frac{\delta\rho^*}{\rho}\delta p_t \mathrm{d}V\right], \\ J &= 2\omega_r^2 \int_V \rho|\boldsymbol{\delta r}|^2 \mathrm{d}V = 2\omega_r^2 M |\boldsymbol{\delta r}_s|^2 E. \end{aligned} \tag{7.34}$$

Here, δ denotes the Lagrangian perturbation, that is, the perturbation following the motion, $\Gamma_3 - 1 = (\partial \ln T/\partial \ln \rho)_{ad}$, and ϵ is the rate of energy generation per unit mass; the star indicates the complex conjugate.

Equations 7.33 and 7.34 have a simple physical meaning: In the expression for W_g $\delta(\rho\epsilon - \mathrm{div}\,\mathbf{F})$ gives the perturbation to the heating rate; thus W_g gets a positive contribution, and hence a contribution to the excitation of the mode, where heating and compression, described by $\delta\rho/\rho$, are in phase. Heating at compression is the basis for the operation of any heat engine, and hence modes that are unstable because of this can be said to be driven by the *heat-engine* mechanism.[6] Depending on the phase relation between compression and δp_t, the term in W_t may contribute to the damping or driving of the mode.

As already indicated, the treatment of the convection–pulsation interaction is a major uncertainty in nonadiabatic calculations for stars with significant outer convection zones,

[6] Because the perturbation to the heating is often determined by the perturbation to the opacity κ, this is also known as the κ mechanism.

where the convective timescales and the pulsation periods are typically similar. Two generalizations of mixing-length theory have seen fairly widespread use. In one, based on an original model by Unno (1967), steady convective eddies are considered, with a balance between buoyancy and turbulent viscous drag. In the second, developed by Gough (1977), convective eddies are continually created and destroyed. The effects of the perturbations associated with the pulsations on these physical models can then be described, leading to expressions for the perturbations to the convective flux and Reynolds stresses. Unno's description has been further developed (see Grigahcène *et al.*, 2005, and references therein), and Gough's description was developed and applied by, for example, Balmforth (1992) and Houdek *et al.* (1999). With judicious (but not unreasonable) choices of parameters, both formulations can predict the transition to stability at the red edge of the Cepheid instability strip, as well as yield results for solar models that are consistent with the observationally inferred damping rates.[7] Interestingly, the dominant mechanism resulting in damping of the modes differs between the two formulations: in the calculations based on Gough's formulation, damping is dominated by the perturbation to the turbulent pressure, whereas the results based on Unno's formulation show damping through the perturbation to the convective flux. A more detailed comparison of these two sets of results is clearly called for.

The cause of the solar oscillations is now almost universally thought to be stochastic excitation by near-surface convection of the intrinsically stable modes. Stochastic excitation of waves in the solar atmosphere was first considered by Stein (1968), followed by the application to the excitation of normal modes by Goldreich and Keeley (1977). The basic properties of damped oscillations excited by turbulence were considered by Batchelor (1956) (see also Christensen-Dalsgaard *et al.*, 1989). The result is that the power spectrum of a single mode, of natural frequency ω_0, and damping rate ω_i, is

$$P(\omega) \simeq \frac{1}{4\omega_0^2} \frac{P_\mathrm{f}(\omega)}{(\omega - \omega_0)^2 + \omega_\mathrm{i}^2}, \tag{7.35}$$

where $P_\mathrm{f}(\omega)$ is the power spectrum of turbulent forcing. Assuming that P_f varies slowly with frequency the result is a stochastic function modulated by a Lorentzian envelope, with a full width at half maximum of $2|\omega_\mathrm{i}|$. An example, from BiSON observations of a solar radial mode, is shown in Fig. 7.6. From the fitted Lorentzian, it is obviously possible to determine the damping rate $|\omega_\mathrm{i}|$. Further analyses of the statistical properties of the stochastically excited oscillators were carried out by Kumar *et al.* (1988) and Chang and Gough (1998). It was shown by Chaplin *et al.* (1997) that the observed distribution of solar oscillation amplitudes was consistent with the predictions of these analyses.

An early, rough prediction of the excitation of solar-like modes in other stars was carried out by Christensen-Dalsgaard and Frandsen (1983). It was shown by Kjeldsen and Bedding (1995) that the resulting velocity amplitudes approximately scaled as L/M, where L is the surface luminosity of the star. Later, more careful investigations (e.g., Houdek *et al.*, 1999; Samadi *et al.*, 2007) confirm that the amplitudes increase with L/M, although generally to a power slightly less than one.

The excitation of a specific mode clearly depends on the detailed property of the mode, the amplitude being determined by the balance between the stochastic energy input from convection and the damping. The energy input increases as a high power of the convective velocity (Stein, 1968) and hence is dominated by the near-surface layers where convection is most vigorous. Here, the properties of the oscillations depend predominantly on the frequency, with little dependence on the degree (at least for the low-degree modes that

[7] A different formulation by Xiong *et al.* (1997), based on equations for the second- and third-order correlations, similarly predicts the red edge but apparently finds that some of the observed solar modes are unstable (Xiong and Deng, 2010).

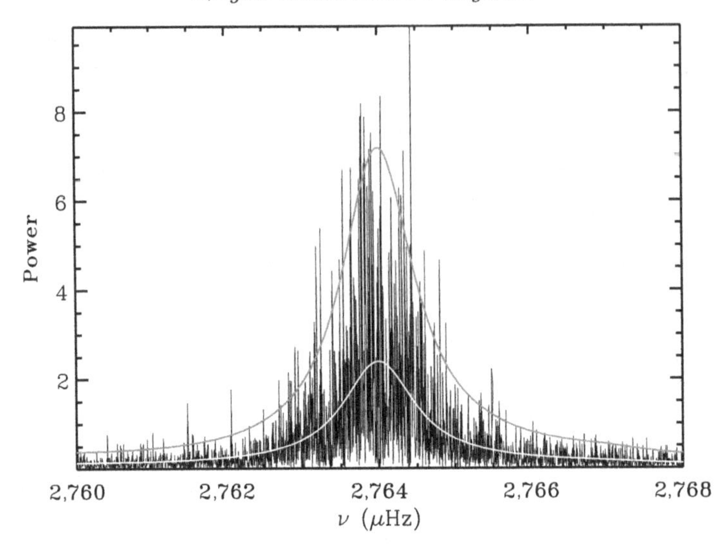

FIG. 7.6. Power spectrum of a solar radial mode, from eight years of BiSON observations. The white curve shows a Lorentzian fit (see Equation 7.35) to the observed spectrum, while the smooth gray curve shows the same fit, but multiplied by a factor of three. Data courtesy of W. J. Chaplin; see Chaplin *et al.* (2002).

are relevant in distant stars). The outcome is that the mean square amplitude can be written as (e.g., Chaplin *et al.*, 2005)

$$\langle A^2 \rangle \simeq \frac{1}{E|\omega_{\rm i}|} \frac{\mathcal{P}_{\rm f}(\omega)}{E}, \qquad (7.36)$$

where A might be the surface velocity or the relative intensity fluctuation; here $\mathcal{P}_{\rm f}(\omega)$, which depends on frequency but in general not on degree, describes the energy input from convection (see, e.g., Houdek, 2010b, for a review). Also, the damping rate is given by Equation 7.33 and 7.34. In the common case where the integrals in $W_{\rm g}$ and $W_{\rm t}$ are dominated by the near-surface layers, it follows that $|\omega_{\rm i}|E$ is independent of E at constant frequency; consequently, in this case, $\langle A^2 \rangle \propto E^{-1}$ at constant frequency. This is important in the discussion of the excitation of solar-like oscillations in red giants where we see large variations in E over a narrow frequency range (see Fig. 7.4): the g-dominated modes, with large inertias, are clearly expected to have a much smaller root-mean-square amplitude than the p-dominated modes with low inertia. However, the question of the mode *visibility*, and hence the likelihood to detect them, is more complicated; we return to this in Section 7.5.

The balance between energy input and damping leads to a characteristic bell-shaped amplitude distribution of the mean amplitude, although with fluctuations around the mean reflecting the stochastic nature of the excitation. The shape of the amplitude distribution was discussed by Goldreich *et al.* (1994). They found that the increase in amplitude with frequency at low frequency was predominantly caused by the change in the shape of the eigenfunction as the upper reflection point, where $\omega = \omega_{\rm c}$ moves closer to the surface with increasing ω. At high frequency, this upper reflection point is very close to the photosphere, and the amplitude decreases with increasing frequency due to an increasing mismatch between the oscillation periods and the time scales of the convective eddies that dominate the excitation. As the frequency approaches $\omega_{\rm c,atm}$ beginning energy loss through the stellar atmosphere may also contribute to the damping and hence further decrease the amplitudes. Indeed, the variation with frequency of the damping rate clearly also plays a role for the variation in amplitude.

It has been found observationally that the cyclic frequency ν_{max} of maximum amplitude appears to scale with the acoustic cutoff frequency $\nu_{\mathrm{c,atm}}$ (e.g., Brown *et al.*, 1991; Kjeldsen and Bedding, 1995; Bedding and Kjeldsen, 2003; Stello *et al.*, 2008) (see also Bedding, Chapter 3), and hence follows the scaling in Equation 7.13. This scaling has been a very useful tool in characterizing the observed properties of stars showing solar-like oscillations (e.g., Kallinger *et al.*, 2010), but the relation between ν_{max} and $\nu_{\mathrm{c,atm}}$ still defies a definite theoretical understanding. Belkacem *et al.* (2011) noted that the location of ν_{max} is closely related to the plateau in damping rates at intermediate frequency, apparently caused by a resonance between the pulsation period and the local thermal timescale, a feature already noticed by Gough (1980) and Christensen-Dalsgaard *et al.* (1989). However, this does not fully explain the observed tight relation.

7.4 Do red giants have non-radial oscillations?

As pointed out by Kjeldsen and Bedding (1995), the amplitude predictions by Christensen-Dalsgaard and Frandsen (1983) scaled as L/M. Thus one could expect to see solar-like oscillations in red giants with quite substantial amplitudes. Early evidence was found for such oscillations in Arcturus (Smith *et al.*, 1987; Innis *et al.*, 1988) while Edmonds and Gilliland (1996) identified oscillations in K giants in the globular cluster 47 Tuc. Christensen-Dalsgaard *et al.* (2001) noted that the statistics of very long-period semi-regular variables observed by the American Association of Variable Star Observers appeared to obey amplitude statistics consistent with stochastic excitation, strongly suggesting that the variability of these stars was also solar-like, although at month-long periods and with amplitudes allowing simple visual observations. Photometric observations of the oscillations in Arcturus were carried out by Retter *et al.* (2003) with the WIRE satellite. Analyses of large samples of red giants observed by OGLE[8] (e.g., Kiss and Bedding, 2003, 2004; Soszyński *et al.*, 2007) showed oscillations in several modes, and Dziembowski and Soszyński (2010) confirmed that the most likely cause was solar-like oscillations. In a further analysis of ground-based survey observations of red giants in the Galaxy and the Large Magellanic Cloud, Tabur *et al.* (2010) demonstrated a continuous transition between pulsations in G and K giants and those seen in very luminous M giants.

The first detailed observations of the oscillations of a red giant were carried out by Frandsen *et al.* (2002) for the G7 giant ξ Hya. The resulting power spectrum is shown in Fig. 7.7. This clearly shows peaks at uniformly spaced frequency, in accordance with Equation 7.19, and a power envelope very similar to what is observed in the Sun, strongly supporting the solar-like nature of the oscillations. From analysis of the spectrum, the separation between the peaks was determined to be $\simeq 7\,\mu\mathrm{Hz}$.

Modelling of ξ Hya to match the well-determined luminosity and effective temperature is shown in Fig. 7.8. Stars ascend the red giant branch while burning hydrogen in a shell around an inert and almost isothermal helium core. As the mass of the core increases, so does the luminosity and the temperature of the core. At the tip of the red giant branch the temperature reaches a sufficiently high value, around $10^8\,\mathrm{K}$, where helium fusion becomes significant. The resulting change in structure, with an expansion of the core, leads to a reduction in the surface radius and luminosity, with the star entering a phase of stable helium burning. Even during this phase, however, a substantial part of the total energy production takes place in the hydrogen-burning shell. As the phase is relatively extended, with a duration of around 20 percent of that of the central hydrogen-burning main sequence, the chances of observing a star in this phase are fairly high; in stellar clusters such stars populate the so-called "red clump."

[8] Optical Gravitational Lensing Experiment.

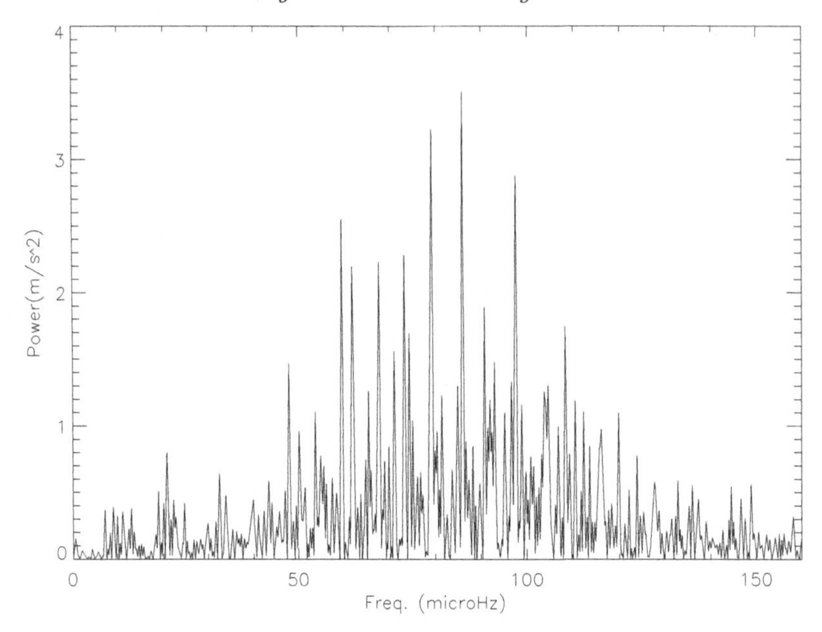

FIG. 7.7. Power spectrum of the G7 giant ξ Hydrae, from one month of Doppler velocity measurements. Adapted from Frandsen *et al.* (2002).

As noted by Teixeira *et al.* (2003), ξ Hya could be identified as being in the ascending phase on the red giant branch. However, because this phase is extremely brief, a more likely identification would be with the core helium burning phase. In either case, a match to the observed frequencies was possible only if the observed frequency spacing was identified with $\Delta\nu$, such that only modes of one degree, presumably radial modes, were observed. In contrast, from a strict application of Equation 7.19 one would have expected the spacing to be $\Delta\nu/2$, and hence $\Delta\nu \simeq 14\,\mu$Hz. However, given the scaling in Equation 7.30, this would be entirely inconsistent with the location of the star in the HR diagram.

To account for this identification, Christensen-Dalsgaard (2004) considered the possibility that all non-radial modes in red giants would likely be heavily damped and hence have small amplitudes. He noted that $W_{\rm g}$ (see Equation 7.34) depends on $\delta(\mathrm{div}\,\mathbf{F})$, that is, on the derivative of $\delta\mathbf{F}$. In turn, roughly $\delta\mathbf{F} \propto \nabla(\delta T)$ and $\delta T/T \propto \delta\rho/\rho \propto \mathrm{div}\,\boldsymbol{\delta r}$. It follows that $W_{\rm g}$ depends on the third derivative of $\boldsymbol{\delta r}$. Thus even if the amplitude of $\boldsymbol{\delta r}$ in the core is quite small, as is the case for p-dominated modes (see Fig. 7.5), $\delta(\mathrm{div}\,\mathbf{F})$ could be quite large for a mode varying very rapidly as a function of r, leading to a substantial contribution to the damping from the core for all non-radial modes. Although the analysis was entirely qualitative, Christensen-Dalsgaard (2004) found some support for this conjecture in the analyses by Dziembowski (1977) and Dziembowski *et al.* (2001).

This pessimistic expectation for red giant seismology has been completely overturned by observations and an improved theoretical understanding. Hekker *et al.* (2006) found evidence for non-radial pulsation in the line-profile variation of a few red giants showing solar-like oscillations. Even more dramatically, De Ridder *et al.* (2009) clearly demonstrated the presence of non-radial solar-like oscillations in a large number of red giants from early CoRoT observations. Figure 7.9 shows resulting power spectra. The power enhancement from the solar-like oscillations is evident; as stars on the red giant branch have roughly the same effective temperature, the frequency at maximum power decreases with increasing radius and luminosity in accordance with Equation 7.13. The observed frequencies satisfy a relation similar to Equation 7.19, with definite evidence for non-radial modes. This is most clearly illustrated in an *échelle diagram* (see Bedding, Chapter 3),

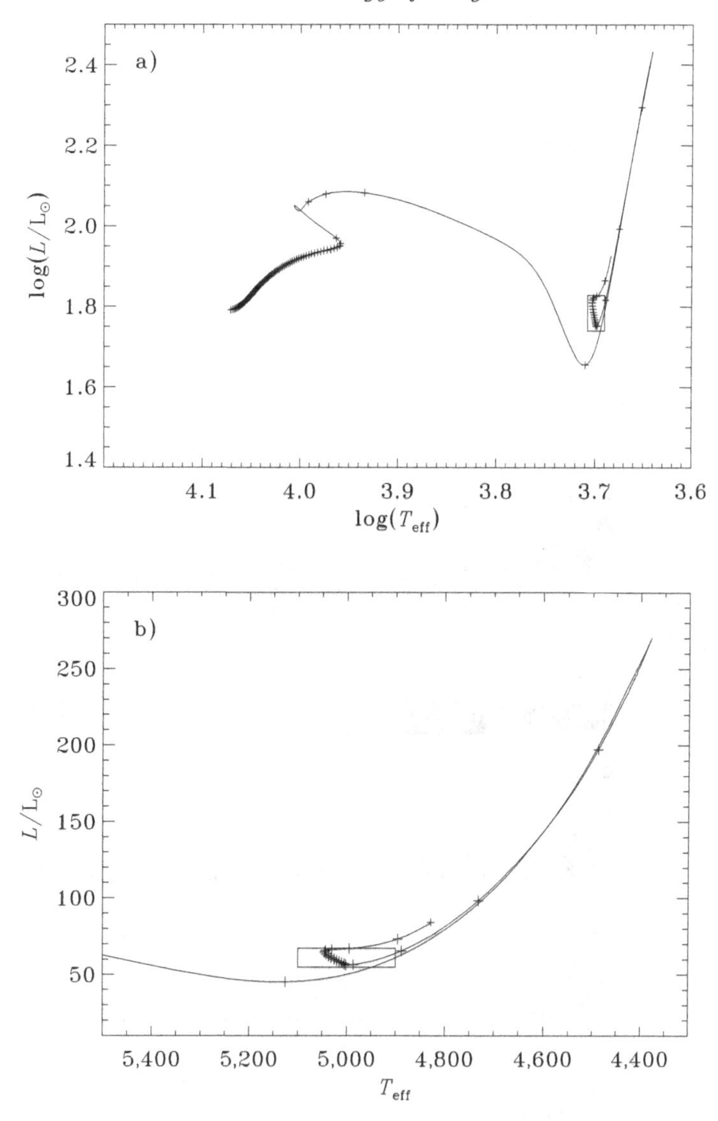

FIG. 7.8. Evolution track for a model of ξ Hydrae, with mass $\mathrm{M} = 2.8\ \mathrm{M}_\odot$ and heavy-element abundance $Z = 0.019$. The pluses are placed at 5-Myr intervals along the track. The box marks the $1-\sigma$ error box for the observed effective temperature and luminosity.

where the frequencies are shown against the frequency reduced modulo $\Delta\nu$, effectively dividing the spectrum into segments of length $\Delta\nu$ and stacking the segments. This has been done for one of the stars in Fig. 7.10. In accordance with Equation 7.19, the three columns of points correspond to modes of degree $l = 0$ and 2 to the right, separated by the small separation, and modes with $l = 1$ to the left. Analyses of red giants observed by CoRoT and *Kepler* (e.g., Huber *et al.*, 2010; Mosser *et al.*, 2011b) have shown that this frequency pattern is universal among red giants.

The very long time series obtained by CoRoT and in particular *Kepler* have allowed fits to the power spectra in terms of the Lorentzian profiles (see Equation 7.35) and hence a determination of the mode damping rates or, equivalently, the lifetimes $\tau = 1/|\omega_\mathrm{i}|$. For red giants, this yields lifetimes near the maximum amplitude of around 15 days (Huber *et al.*, 2010; Baudin *et al.*, 2011). Interestingly, this is very similar to the value obtained for ξ Hya by Houdek and Gough (2002), using a convection treatment based on Gough (1977).

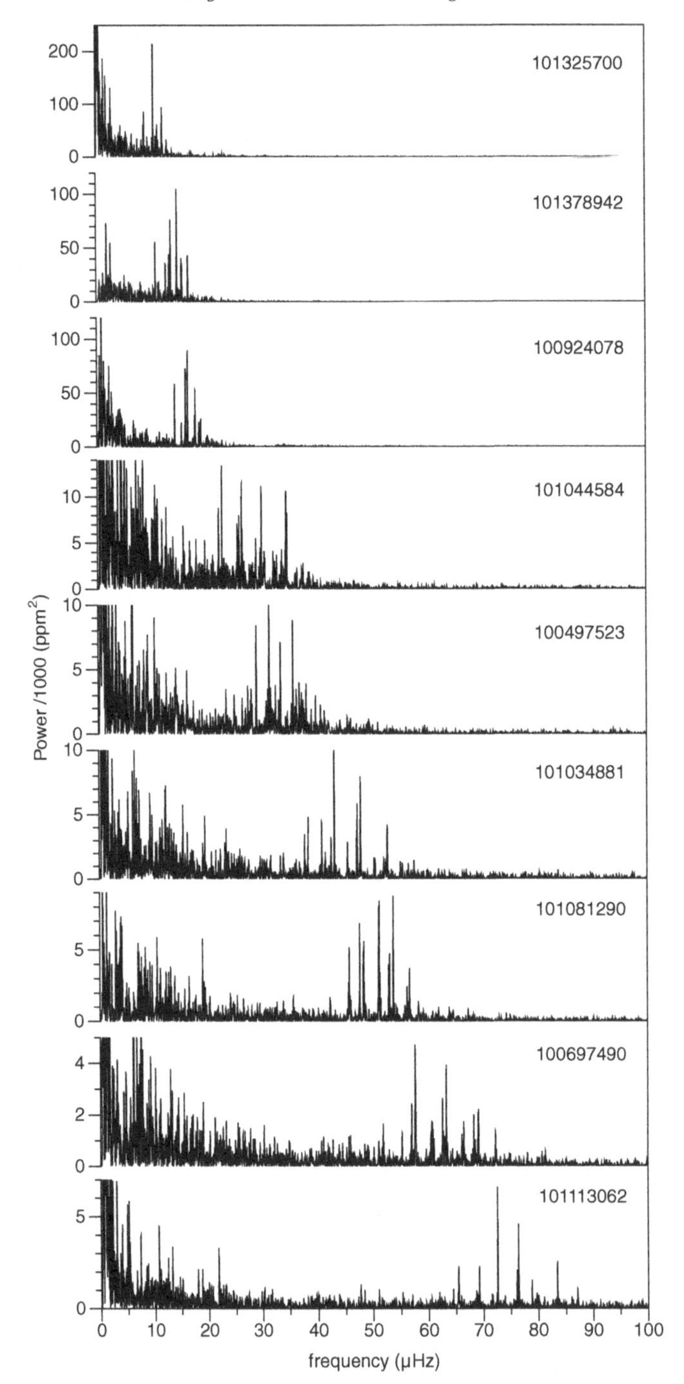

FIG. 7.9. Power spectra of solar-like oscillations in red giants, from five months of observations with the CoRoT satellite. The panels are labeled by the CoRoT identification number. Radius and luminosity increase toward the top. Figure courtesy of J. De Ridder; from De Ridder *et al.* (2009).

7.5 Amplitudes of solar-like oscillations in red giants

To interpret the observations, a more detailed analysis of the mode excitation and visibility is required. The analysis leading to Equation 7.36 determines the mean square amplitude of the mode, corresponding essentially to the area under the corresponding peak in the power spectrum. However, it was noted by Chaplin *et al.* (2005) that ability

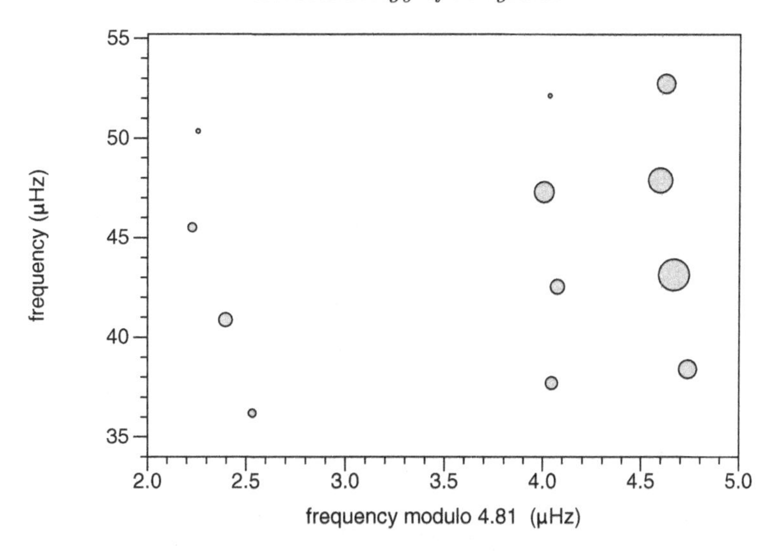

FIG. 7.10. Échelle diagram of the star 101034881 in Fig. 7.9. This corresponds to stacking slices of the power spectrum of a length given the large separation $\Delta\nu = 4.81\,\mu$Hz: the abscissa is the frequency modulo $\Delta\nu$ whereas the ordinate is the frequency. Figure courtesy of J. De Ridder; from De Ridder *et al.* (2009).

to detect a mode in the power spectrum depends instead on the *peak height* H, related to A^2 by[9]

$$A^2 = |\omega_{\rm i}|H, \tag{7.37}$$

because $|\omega_{\rm i}|$ determines the width of the peak. If the damping is dominated by the near-surface layers, it was argued above that both A^2 and $|\omega_{\rm i}|$ are proportional to E^{-1} at a given frequency. It then follows from Equation 7.37 that H is independent of E, that is, that all modes in a given frequency interval are excited to roughly the same height. For the dense spectrum of modes in a red giant, such as those illustrated in Fig. 7.4, the result would be a very dense observed power spectrum, with no indication of the acoustic resonances approximately satisfying Equation 7.19. This is in obvious contradiction to, for example, the observations by De Ridder *et al.* (2009).

This argument, however, neglects the fact that the oscillations are observed for a finite time $\mathcal{T}$ which, for the g-dominated modes, is typically shorter than the lifetime τ of the mode. The resulting peaks in the power-density spectrum have a width that, for $\mathcal{T} \ll \tau$, is proportional to $\mathcal{T}^{-1}$ and hence a correspondingly smaller peak height, very likely making the modes invisible. As a convenient interpolation between the two extremes, Fletcher *et al.* (2006) proposed the relation

$$H \propto \frac{A^2}{|\omega_{\rm i}| + 2/\mathcal{T}}. \tag{7.38}$$

A detailed and very illuminating analysis of the excitation of modes in red giants was carried out by Dupret *et al.* (2009). They considered several red giant models, with locations in the HR diagram as illustrated in Fig. 7.11, carrying out nonadiabatic calculations of non-radial modes using the convection formalism of Grigahcène *et al.* (2005). From a model of the stochastic energy input (e.g., Belkacem *et al.*, 2006) they then computed the expected amplitudes of the modes and the resulting power-density diagrams. They assumed an observing period $\mathcal{T}$ of 150 days, corresponding to CoRoT long runs. Rather than using Equation 7.38, they effectively adopted $H \propto \tau A^2$ for $\tau > \mathcal{T}/2$ and $H \propto \mathcal{T}A^2/2$

[9] For simplicity, we drop $\langle \ldots \rangle$ in the following.

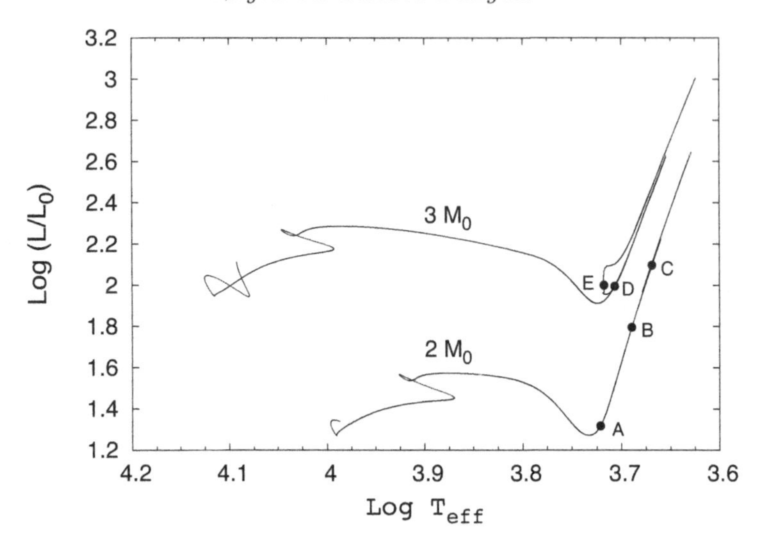

FIG. 7.11. Evolution tracks for models of mass 2 and 3 $M_\odot$, for which Dupret *et al.* (2009) considered the stochastic excitation of solar-like oscillations. Figure courtesy of M.-A. Dupret.

otherwise, with a continuous transition between the two cases. Here, we briefly consider their results for Models A and B.

Figure 7.12 shows the mode lifetimes for modes of degree $l = 0 - 2$ in Model A. In this case the damping is dominated by the near-surface layers and hence, as is obvious in the figure, the variation of τ reflects the behavior of E, which is qualitatively similar to the case illustrated in Fig. 7.4. For $l = 2$, the g-dominated modes have lifetimes substantially higher than the assumed $\mathcal{T}/2$ indicated in the figure by a horizontal dashed line. On the other hand, the $l = 1$ modes at higher frequency all have lifetimes at or below the observing time, and hence would be expected to reach similar peak heights in a given frequency interval. This is confirmed by Fig. 7.13, which shows the predicted peak heights.

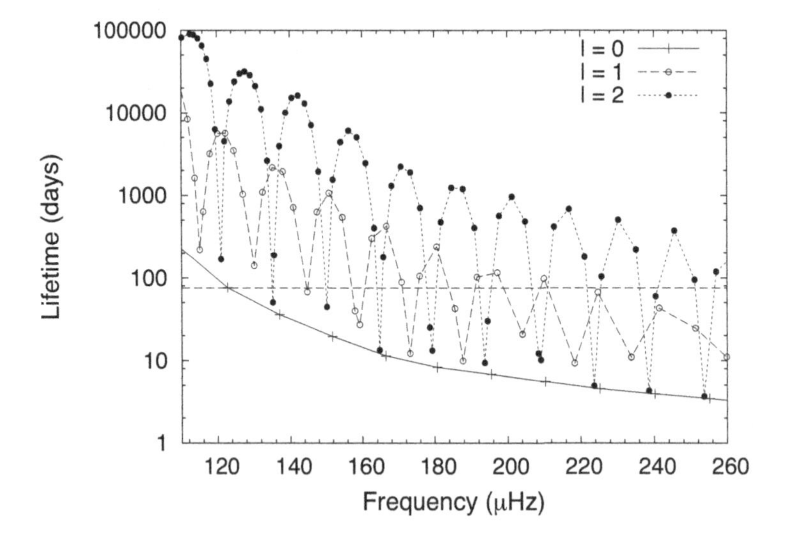

FIG. 7.12. Computed lifetimes of Model A in Fig. 7.11, at the base of the 2 $M_\odot$ red giant branch. Modes of degree $l = 0$, 1, and 2 are shown, respectively, by pluses connected by solid lines, open circles connected by dashed lines, and closed circles connected by dotted lines. The horizontal dashed line marks a damping time of 75 d, corresponding to the border between unresolved and resolved modes for a CoRoT long run of 150 days. Figure courtesy of M.-A. Dupret; from Dupret *et al.* (2009).

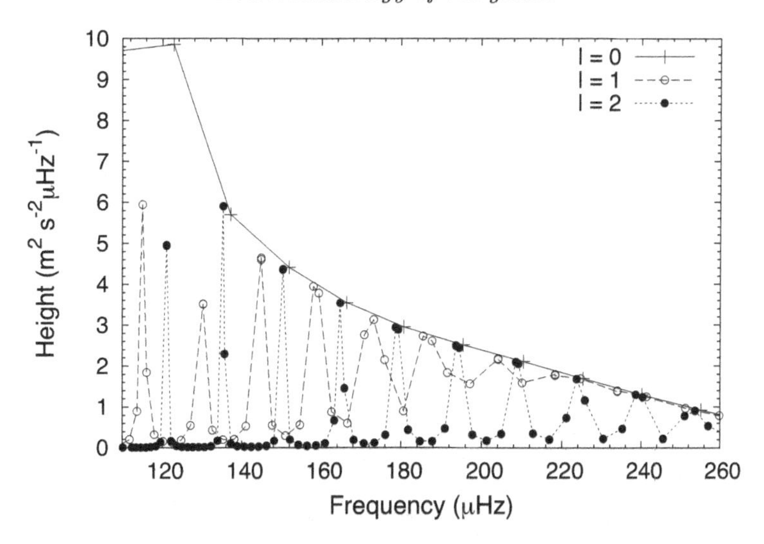

FIG. 7.13. Estimated peak heights in the power spectrum of solar-like oscillations in Model A in Fig. 7.11, at the base of the red giant branch; for symbol and line styles, see caption to Fig. 7.12. Figure courtesy of M.-A. Dupret; from Dupret *et al.* (2009).

The overall envelope seems to be somewhat skewed toward a lower frequency, relative to what is expected from the observed scaling with the acoustic cutoff frequency; given the uncertainties in modeling both the damping and the energy input, this is hardly surprising. However, it is striking that although the $l = 2$ modes show strong acoustic resonances with small peak heights for the g-dominated modes, the $l = 1$ modes at high frequency all reach peak heights comparable with those for the radial modes. It is obvious that such a spectrum can be quite complex (see also Di Mauro *et al.*, 2011).

The situation is rather different for the more evolved Model B. As illustrated in Fig. 7.14, there is now strong damping in the core for the g-dominated modes with $l = 2$; for $l = 1$ the same is the case at relatively low frequency, while the surface layers dominate at high frequency. The damping in the core is a result of the very rapidly oscillating eigenfunctions in the core, as discussed earlier. Owing to the increased central condensation, the buoyancy frequency is substantially higher than in Model A, leading to a higher vertical wave number and hence stronger damping in the core. For p-dominated modes, on the other hand, the damping is still dominated by the near-surface layers. Together with a strong trapping of the g-dominated modes in the core, this leads to a much stronger contrast in lifetime between the p- and g-dominated modes, with the latter having lifetimes very substantially higher than the observing time (see Fig. 7.15). The effect of this on the predicted power density spectrum is evident in Fig. 7.16: there are now obvious acoustic resonances for both $l = 1$ and 2. The $l = 2$ resonances are confined to very narrow frequency intervals, caused also by the more extended evanescent region between the peak in the buoyancy frequency and the region of acoustic propagation, and typically involve only 1–2 modes at non-negligible peak heights. For $l = 1$, on the other hand, there is typically a cluster of peaks with heights that may make the modes observable. As discussed by Bedding (Chapter 3), this was found to be the case by Beck *et al.* (2011) and Bedding *et al.* (2011). We discuss the significance of these results in the following section.

7.6 Asteroseismic diagnostics in red giants

Detection of solar-like oscillations in red giants, such as is illustrated in Fig. 7.9, almost immediately provides a measure of the large frequency separation $\Delta\nu$ and the frequency $\nu_{\max}$ of maximum power. These basic quantities already have a substantial diagnostic

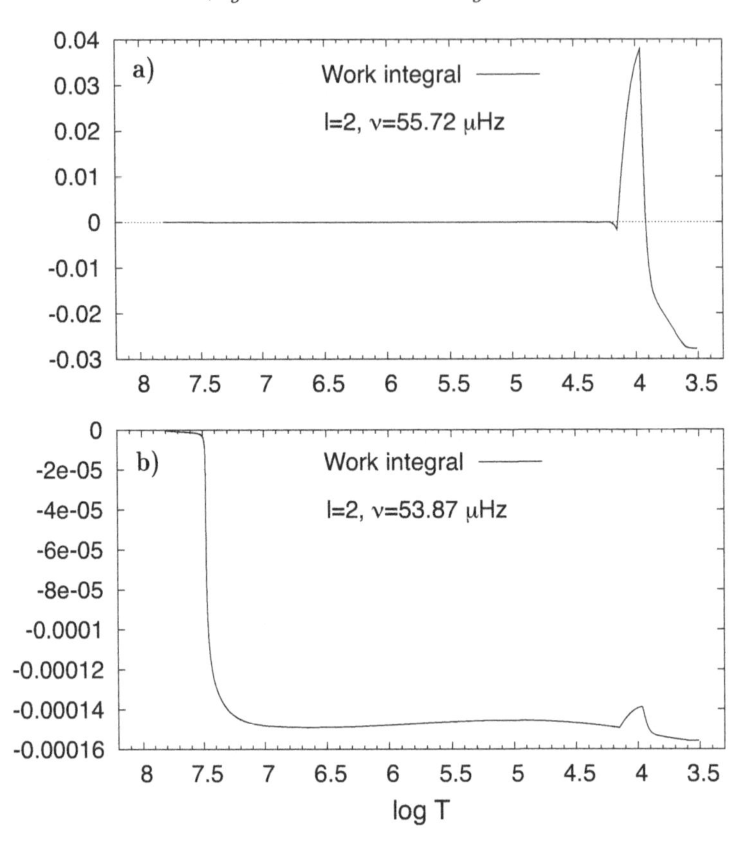

FIG. 7.14. Work integrals for the intermediate red giant branch 2 $M_\odot$ Model B in Fig. 7.11; this essentially corresponds to the partial integrals of $W_g + W_t$, defined in Equation 7.34, and here plotted against temperature T in the model. (a) is for a p-dominated mode while (b) is for a g-dominated mode largely trapped in the core. Figure courtesy of M.-A. Dupret; from Dupret *et al.* (2009).

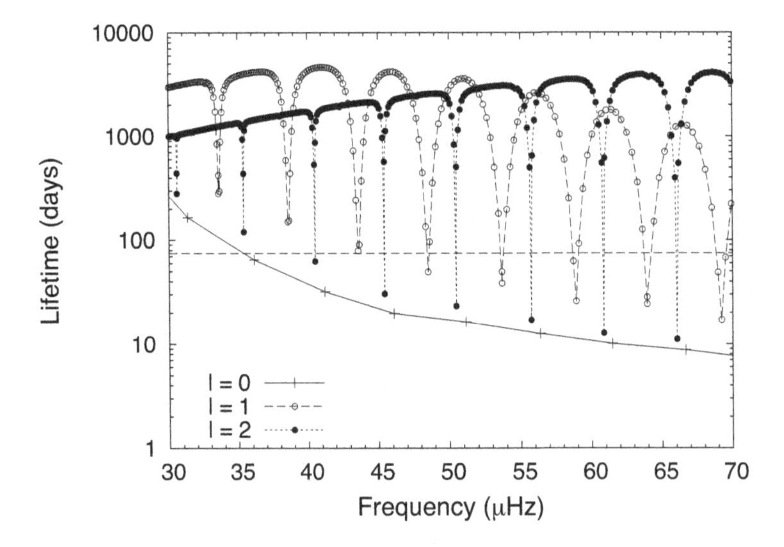

FIG. 7.15. Computed lifetimes for the intermediate red giant branch 2 $M_\odot$ Model B in Fig. 7.11; see caption to Fig. 7.12. Figure courtesy of M.-A. Dupret; from Dupret *et al.* (2009).

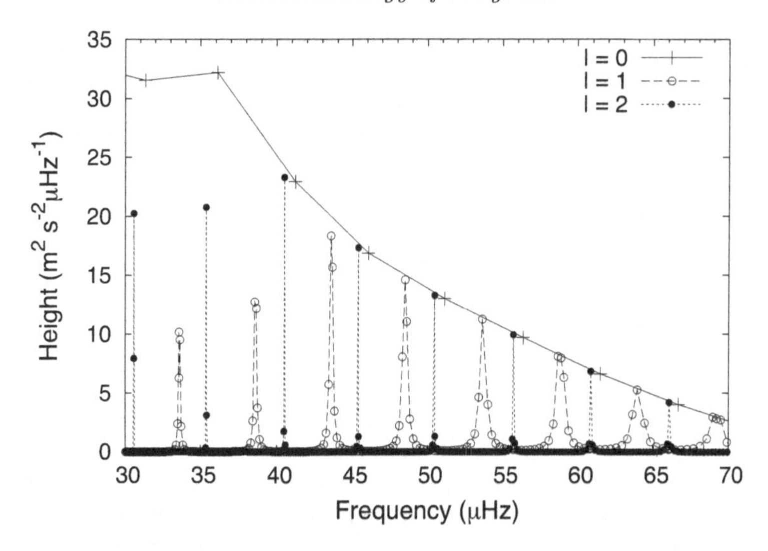

FIG. 7.16. Estimated peak heights in the power spectrum of solar-like oscillations in the intermediate red giant branch 2 $M_\odot$ Model B in Fig. 7.11; for symbol and line styles, see caption to Fig. 7.12. Figure courtesy of M.-A. Dupret; from Dupret *et al.* (2009).

potential: $\Delta\nu$ approximately satisfies the scaling (Equation 7.30) with the mean density and it can be assumed that $\nu_{\rm max}$ scales as the atmospherical acoustic cutoff frequency (Equation 7.13). If $T_{\rm eff}$ is known from photometric or spectroscopic observations, one can then determine the radius and mass of the star as

$$\frac{R}{R_\odot} = \frac{\nu_{\rm max}}{\nu_{\rm max,\odot}} \left(\frac{\Delta\nu}{\Delta\nu_\odot}\right)^{-2} \left(\frac{T_{\rm eff}}{T_{\rm eff,\odot}}\right)^{1/2} \tag{7.39}$$

and

$$\frac{M}{M_\odot} = \left(\frac{R}{R_\odot}\right)^3 \left(\frac{\Delta\nu}{\Delta\nu_\odot}\right)^2 , \tag{7.40}$$

where $\odot$ denotes solar values (e.g., Kallinger *et al.*, 2010). Such analyses provide unique possibilities for population studies of field red giants (e.g., Miglio *et al.*, 2009; Mosser *et al.*, 2010; Hekker *et al.*, 2011). Stellar modeling provides further constraints on the relation between $T_{\rm eff}$, M, and R and hence a more precise determination of stellar properties (Gai *et al.*, 2011), as used in an analysis of red giants in two of the open clusters in the *Kepler* field by Basu *et al.* (2011).

As discussed in connection with Equation 7.21, for main sequence stars the small frequency separations $\delta\nu_{l\,l+2}(n) = \nu_{nl} - \nu_{n-1\,l+2}$ provide a measure of the age of the star. This no longer holds for red giants. Here, in most cases, $\delta\nu_{l\,l+2}$ scales essentially as the large frequency separation and hence mainly reflects the variation of the mean density of the star (e.g., Bedding *et al.*, 2010; Huber *et al.*, 2010; Mosser *et al.*, 2011b). This is a natural consequence of the fact that the propagation region of the non-radial acoustic modes for such stars lies in the convective envelope, which probably changes in a largely homologous fashion as the properties of the star change. A detailed analysis of the diagnostic potential of acoustic modes in red giants was carried out by Montalbán *et al.* (2010). They also considered other types of small frequency separations, such as $\delta\nu_{01}(n) = (\nu_{n0} + \nu_{n+1\,0})/2 - \nu_{n1}$. Interestingly, they found that for models with somewhat shallower convective envelopes, particularly amongst core helium-burning stars, where the lower turning point of the $l = 1$ modes was in the radiative region, the average $\langle\delta\nu_{01}\rangle/\Delta\nu$ showed a marked dependence on the distance from the base of the convective envelope to the turning point.

The asymptotic analysis leading to Equation 7.19 for the acoustic-mode frequencies assumes that the equilibrium structure varies slowly compared with the eigenfunctions. There are regions in the star where this does not hold, with variations in the sound speed on a scale that is comparable with or smaller than the local wavelength. This causes periodic variations in the frequencies with properties that depend on the depth and nature of the sound-speed feature. In red giants, the most important example of such an *acoustic glitch* is related to the second ionization of helium, which causes a dip in Γ_1 and hence (see Equation 7.2) a localized variation in the sound speed. This variation provides a diagnostics of the helium abundance in the star (e.g., Gough, 1990; Vorontsov *et al.*, 1991; Basu *et al.*, 2004; Monteiro and Thompson, 2005; Houdek and Gough, 2007). A clear signature of this effect was found by Miglio *et al.* (2010) in a red giant observed by CoRoT. They noted, following an earlier discussion by Mazumdar (2005) of the diagnostic potential of such glitches, that the inferred depth of the sound-speed feature together with the large frequency separation provides a purely asteroseismic determination of the mass and radius of the star. This feature can probably be detected in a substantial fraction of the red giants observed with CoRoT and *Kepler*, promising a determination of the so far poorly known helium abundance in these stars and hence important constraints on the chemical evolution of the Galaxy.

From the analysis by Dupret *et al.* (2009) discussed in Section 7.5, we noted that for $l = 1$ several modes might be excited to observable amplitudes in the vicinity of an acoustic resonance. This is consistent with the rather larger scatter for these dipolar modes observed by Bedding *et al.* (2010) in an collapsed échelle diagram based on early *Kepler* data and, as noted by Montalbán *et al.* (2010), has important diagnostic potential. As discussed by Bedding (Chapter 3), this potential was dramatically realized in the analyses by Beck *et al.* (2011) and Bedding *et al.* (2011). The groups of dipolar peaks near the acoustic resonances were found to have a regular structure that allowed determination of well-defined period spacings, reminiscent of the behavior of g modes (see Equation 7.24) and matching a corresponding behavior in the computed frequencies of stellar models. Such behavior has also been found in analysis of CoRoT data (Mosser *et al.*, 2011a).

To illustrate the properties of the computed period spacings, Fig. 7.17 shows the difference in period between adjacent dipolar modes in Model 3 in the 1 $M_\odot$ evolution sequence illustrated in Fig. 7.1. It is clear that, particularly in the dense region of g-dominated modes at relatively low frequency, there is an almost uniform spacing, interspersed with decreases that are associated with the acoustic resonances. To illustrate this relation, the figure also shows $\log Q_{nl}$ where Q_{nl} is the mode inertia normalized by the inertia of a radial mode at the same frequency. Specifically,

$$Q_{nl} = \frac{E_{nl}}{\bar{E}_0(\omega_{nl})}, \tag{7.41}$$

where E_{nl} is the inertia of the mode and $\bar{E}_0(\omega_{nl})$ is the inertia of radial modes, interpolated to the frequency ω_{nl} of the mode. It is obvious that the variation in the period spacing is closely related to the variation in the inertia.

The uniform period spacing is clearly only realized for the most g-dominated modes whose high inertia makes detection unlikely.[10] However, it was demonstrated by Bedding *et al.* (2011) that with the *Kepler* data, the pure g-mode period spacing can be determined from the observed frequencies in data of sufficient quality; this was illustrated in a "g-mode échelle diagram," dividing the spectrum into segments of fixed length in period (see also Bedding, Chapter 3). In other cases, the period spacings can be inferred from analysis of the power spectra and extrapolated to the value for pure g modes (see also Mosser *et al.*, 2011a). Bedding *et al.* (2011) showed that the inferred period spacing is substantially higher for clump stars in the core helium burning phase than for stars on

[10] Except perhaps in extremely long time series, exceeding the lifetimes of even these modes.

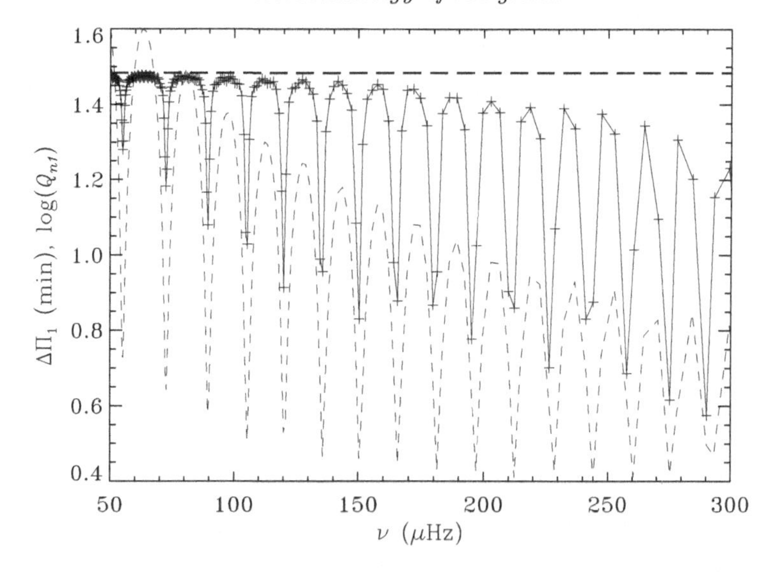

FIG. 7.17. The solid curve shows the period spacing between adjacent $l = 1$ modes in Model 3 in the 1 $M_\odot$ sequence in Fig. 7.1. The horizontal dashed line shows the asymptotic value for pure g modes, from Equation 7.42. For comparison, the short-dashed curve shows the logarithm of the normalized inertia Q_{nl} (see Equation 7.41) on an arbitrary scale.

the ascending red giant branch, providing a clear separation into these two groups of stars that are superficially very similar.

From the asymptotic results in Equations 7.24 and 7.25, the period spacing for pure dipolar g modes is approximately given by

$$\Delta\Pi_1 = \sqrt{2}\pi^2 \left(\int_{r_1}^{r_2} N \frac{\mathrm{d}r}{r} \right)^{-1} . \tag{7.42}$$

This is illustrated by the dashed horizontal line in Fig. 7.17; it is clearly in good agreement with the value obtained from the most strongly trapped modes. Given the $\Delta\Pi_1$, as discussed above, can be obtained from the observations, we therefore obtain diagnostics of the buoyancy frequency in the core of the star.

This can be used to illustrate the evolution of the star in an "HR" diagram, plotting L against $\Delta\Pi_1$, as done in Fig. 7.18. It is evident that $\Delta\Pi_1$ varies little as the star moves up the red giant branch. This behavior can be understood in terms of the variation in the buoyancy frequency, illustrated for three representative models in Fig. 7.19. On this logarithmic r scale the main effect is a shift toward smaller radii as the core contracts, with little change in the shape of N. Noting that the integral in Equation 7.42 is in terms of $\log r$, it follows that there is little change in $\Delta\Pi_1$ with evolution. It should also be noticed that N is depressed in the core of the star. This is probably caused by the increasing electron degeneracy of the gas, leading to $\mathrm{d}\ln p/\mathrm{d}\ln\rho$ approaching 5/3.[11]

As discussed by Bedding (Chapter 3), there is a striking difference in $\Delta\Pi_1$ between stars in the hydrogen shell-burning phase, as the 1 $M_\odot$ model illustrated above, and models that burn helium in the core. Unfortunately, ASTEC does not allow computation of a model through a helium flash, and hence the evolution of the 1 $M_\odot$ model cannot be followed beyond helium ignition. For more massive stars, the helium ignition takes place in a more quiet manner that can be followed by ASTEC. Fig. 7.20 shows the evolution track of a 2.5 $M_\odot$ model up to a point near exhaustion of helium at the center.

[11] In white dwarfs, this leads to a very small buoyancy frequency in the interior of the star, confining the g modes to the outer layers (Fontaine and Brassard, 2008).

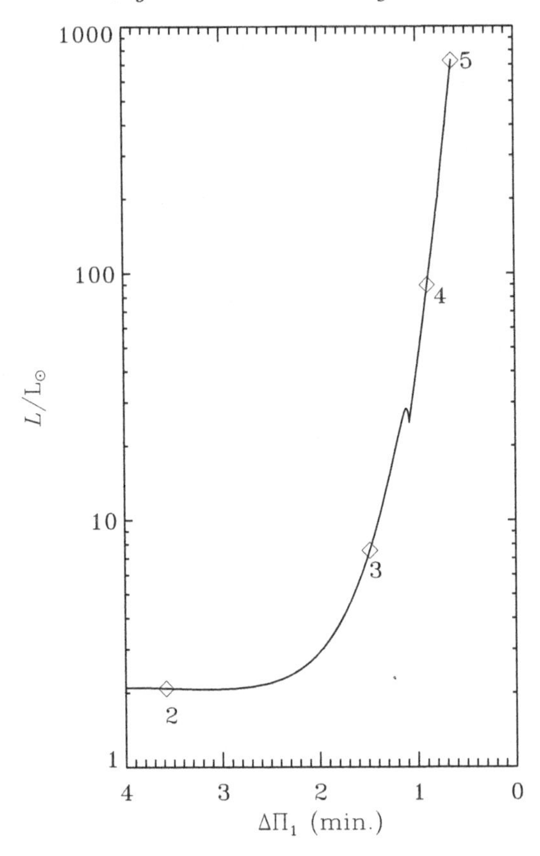

FIG. 7.18. "Hertzsprung-Russell diagram," in terms of the asymptotic dipolar g-mode spacing $\Delta\Pi_1$ (see Equation 7.42) and luminosity, for the 1 $M_\odot$ evolution track illustrated in Fig. 7.1.

In the $(\Delta\Pi_1, L)$ diagram, this leads to a rather more complex evolution, illustrated in Fig. 7.21. The variation on the ascending red giant branch is very similar to the 1 $M_\odot$ evolution illustrated in Fig. 7.18. With the helium ignition, the star moves through a somewhat convoluted path, shown dotted in the figure, toward larger $\Delta\Pi_1$, before the

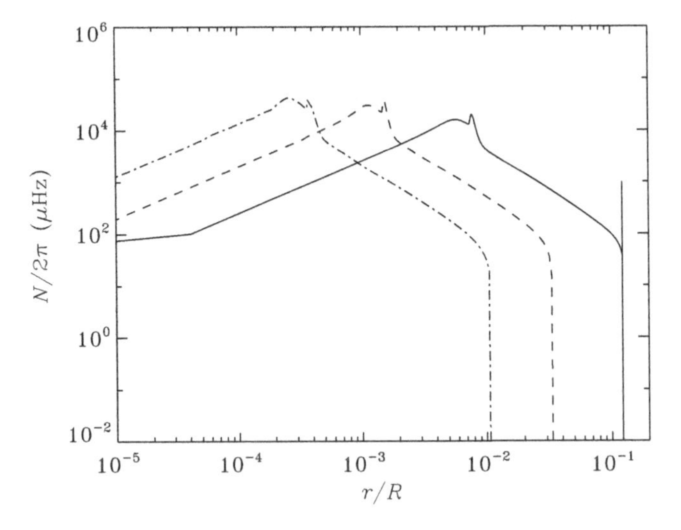

FIG. 7.19. Cyclic buoyancy frequencies $N/2\pi$ in Model 3 (solid), Model 4 (dashed), and Model 5 (dot-dashed) in the 1 $M_\odot$ evolution sequence illustrated in Figs. 7.1 and 7.18.

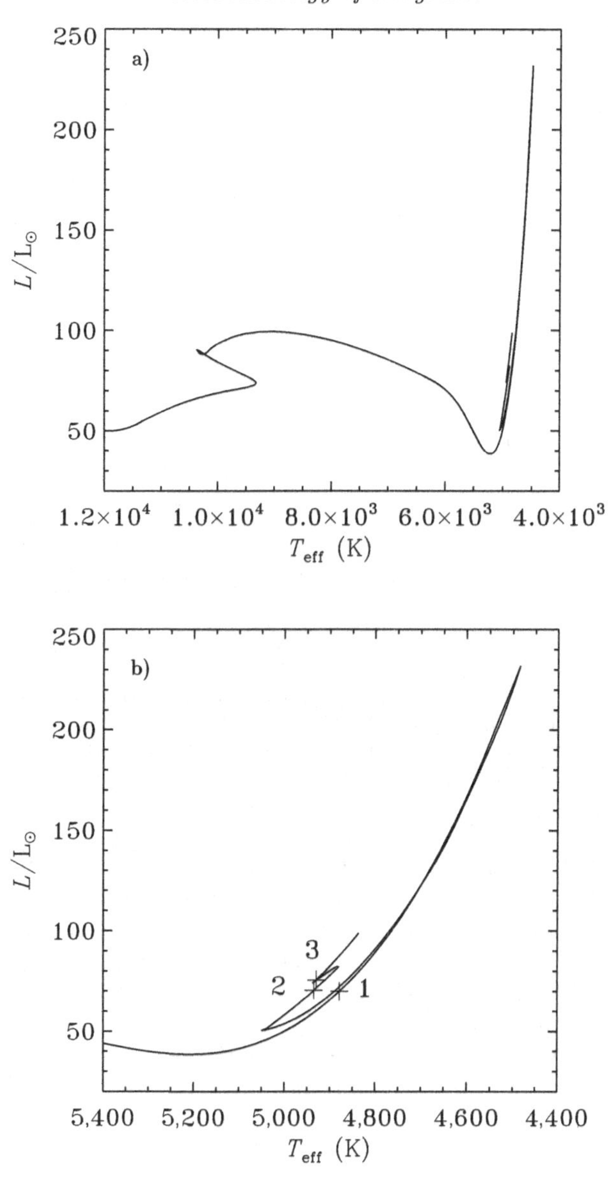

FIG. 7.20. Evolution track for a 2.5 $M_\odot$ model. The track goes through helium ignition at the tip of the red giant branch and ends near central helium exhaustion.

star settles down to quiet helium burning.[12] During the central helium burning, $\Delta\Pi_1$ undergoes a further jump to larger values, which is also reflected in the evolution track in Fig. 7.20.

Again, the variation in $\Delta\Pi_1$ can be understood by considering the evolution of the buoyancy frequency, illustrated in Fig. 7.22. Comparing Models 1 and 2, at roughly the same luminosity, the former model is obviously similar to the red giant 1 $M_\odot$ model in Fig. 7.19. In Model 2, on the other hand, helium burning causes a convective core where the g modes are excluded; because the behavior of N in the rest of the core is quite similar to that of Model 1, the effect is to decrease the integral in Equation 7.42 and

[12] The additional variations along this path appear to be related to modest oscillations in the central properties, similar to, but of smaller amplitude than, the oscillations seen in a full-fledged helium flash (Serenelli and Weiss, 2005).

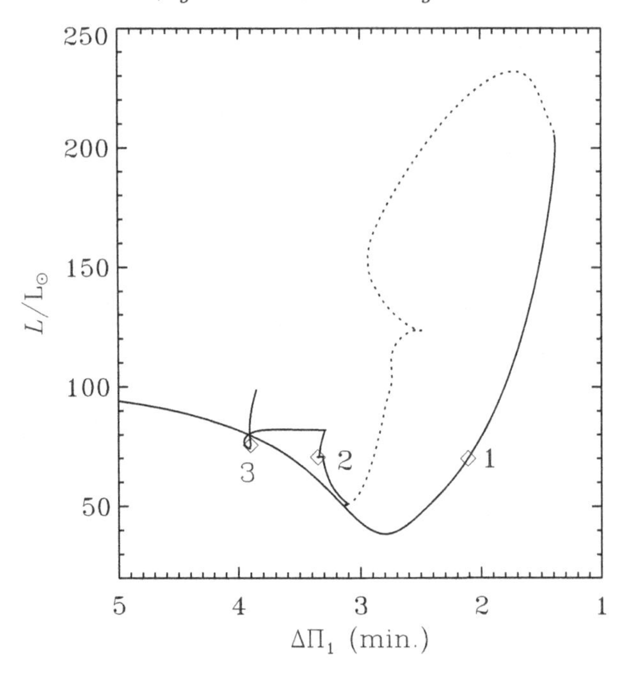

FIG. 7.21. "Hertzsprung-Russell diagram," in terms of the asymptotic dipolar g-mode spacing $\Delta\Pi_1$ (see Equation 7.42) and luminosity, for the 2.5 $M_\odot$ evolution track illustrated in Fig. 7.20. The dotted part of the curve extends from helium ignition, at a model age of 0.421 Gyr, to the establishment of stable helium burning, at 0.439 Gyr. (The end of the track is at 0.565 Gyr.)

hence to increase $\Delta\Pi_1$. This is the dominant cause of the remarkable difference in the period spacing between stars on the red giant branch and in the red clump observed by Bedding *et al.* (2011) and discussed further by Bedding in this volume. The subsequent jump in $\Delta\Pi_1$ and L is caused by a jump in the size of the convective core, as illustrated in Fig. 7.22 by Model 3, probably associated with changes in the composition profile in the core (e.g., Castellani *et al.*, 1971).

This analysis provides a simple explanation for the behavior found by Bedding *et al.* (2011), and furthermore indicates how this may allow the period spacing to be used as a

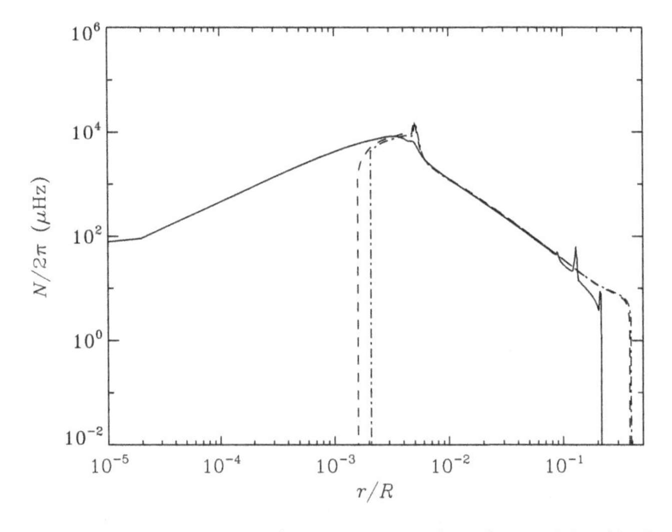

FIG. 7.22. Cyclic buoyancy frequencies $N/2\pi$ in Model 1 (solid), Model 2 (dashed), and Model 3 (dot-dashed) in the 2.5 $M_\odot$ evolution sequence illustrated in Figs. 7.20 and 7.21.

sensitive diagnostics of the buoyancy frequency in the stellar core. Further investigations will be required to obtain a deeper understanding of the diagnostic power of such data, including the ability to probe effects of convective overshoot and rotationally induced mixing in the core.

Acknowledgments

I am very grateful to Pere Pallé and his colleagues at the IAC for the organization of an excellent Winter School. I look forward to the next opportunity to participate in such a school. J. De Ridder and M.-A. Dupret are thanked for providing several key figures.

REFERENCES

Aerts, C., Christensen-Dalsgaard, J. and Kurtz, D. W. 2010. *Asteroseismology.* Springer, Heidelberg.

Angulo, C., Arnould, M., Rayet, M., Descouvemont, P., Baye, D., Leclercq-Willain, C., Coc, A., Barhoumi, S., Aguer, P., Rolfs, C., Kunz, R., Hammer, J. W., Mayer, A., Paradellis, T., Kossionides, S., Chronidou, C., Spyrou, K., Degl'Innocenti, S., Fiorentini, G., Ricci, B., Zavatarelli, S., Providencia, C., Wolters, H., Soares, J., Grama, C., Rahighi, J., Shotter, A., and Lamehi Rachti, M. 1999. A compilation of charged-particle induced thermonuclear reaction rates. *Nucl. Phys. A*, **656**, 3–183.

Baglin, A., Auvergne, M., Barge, P., Deleuil, M., Michel, E., and the CoRoT Exoplanet Science Team. 2009. CoRoT: description of the mission and early results. Pages 71–81 of: F. Pont, D. Sasselov, and M. Holman (ed), *Proc. IAU Symp. 253, Transiting Planets.* IAU and Cambridge University Press, p. 71–81.

Balmforth, N. J. 1992. Solar pulsational stability. I. Pulsation-mode thermodynamics. *MNRAS*, **255**, 603–31.

Barban, C., Matthews, J. M., De Ridder, J., Baudin, F., Kuschnig, R., Mazumdar, A., Samadi, R., Guenther, D. B., Moffat, A. F. J., Rucinski, S. M., Sasselov, D., Walker, G. A. H. and Weiss, W. E. 2007. Detection of solar-like oscillations in the red giant ϵ Ophiuchi by MOST spacebased photometry. *A&A*, **468**, 1033–8.

Basu, S., Mazumdar, A., Antia, H. M., and Demarque, P. 2004. Asteroseismic determination of helium abundance in stellar envelopes. *MNRAS*, **350**, 277–86.

Basu, S., Grundahl, F., Stello, D., Kallinger, T., Hekker, S., Mosser, B., García, R. A., Mathur, S., Brogaard, K., Bruntt, H., Chaplin, W. J., Gai, N., Elsworth, Y., Esch, L., Ballot, J., Bedding, T. R., Gruberbauer, M., Huber, D., Miglio, A., Yildiz, M., Kjeldsen, H., Christensen-Dalsgaard, J., Gilliland, R. L., Fanelli, M. M., Ibrahim, K. A., and Smith, J. C. 2011. Sounding open clusters: asteroseismic constraints from *Kepler* on the properties of NGC 6791 and NGC 6819. *ApJ*, **729**, L10.

Batalha, N. M., Borucki, W. J., Bryson, S. T., Buchhave, L. A., Caldwell, D. A., Christensen-Dalsgaard, J., Ciardi, D., Dunham, E. W., Fressin, F., Gautier, T. N., III, Gilliland, R. L., Haas, M. R., Howell, S. B., Jenkins, J. M., Kjeldsen, H., Koch, D. G., Latham, D. W., Lissauer, J. J., Marcy, G. W., Rowe, J. F., Sasselov, D. D., Seager, S., Steffen, J. H., Torres, G., Basri, G. S., Brown, T. M., Charbonneau, D., Christiansen, J., Clarke, B., Cochran, W. D., Dupree, A., Fabrycky, D. C., Fischer, D., Ford, E. B., Fortney, J., Girouard, F. R., Holman, M. J., Johnson, J., Isaacson, H., Klaus, T. C., Machalek, P., Moorehead, A. V., Morehead, R. C., Ragozzine, D., Tenenbaum, P., Twicken, J., Quinn, S., VanCleve, J., Walkowicz, L. M., Welsh, W. F., Devore, E., and Gould, A. 2011. *Kepler*'s first rocky planet: Kepler-10b. *ApJ*, **729**, 27.

Batchelor, G. K. 1956. *The theory of homogeneous turbulence.* Cambridge University Press. Cambridge.

Baudin, F., Barban, C., Belkacem, K., Hekker, S., Morel, T., Samadi, R., Benomar, O., Goupil, M. -J., Carrier, F., Ballot, J., Deheuvels, S., De Ridder, J., Hatzes, A. P., Kallinger, T., and Weiss, W. W. 2011. Amplitudes and lifetimes of solar-like oscillations observed by CoRoT: red giant versus main-sequence stars. *A&A*, **529**, A84.

Beck, P. G., Bedding, T. R., Mosser, B., Stello, D., Garcia, R. A., Kallinger, T., Hekker, S., Elsworth, Y., Frandsen, S., Carrier, F., De Ridder, J., Aerts, C., White, T. R., Huber, D., Dupret, M.-A., Montalbán, J., Miglio, A., Noels, A., Chaplin, W. J., Kjeldsen, H.,

Christensen-Dalsgaard, J., Gilliland, R. L., Brown, T. M., Kawaler, S. D., Mathur, S., and Jenkins, J. M. 2011. Kepler detected gravity-mode period spacings in a red giant. *Science*, **332**, 205.

Bedding, T. R. and Kjeldsen, H. 2003. Solar-like oscillations. *PASA*, **20**, 203–212.

Bedding, T. R., Huber, D., Stello, D., Elsworth, Y. P., Hekker, S., Kallinger, T., Mathur, S., Mosser, B., Preston, H. L., Ballot, J., Barban, C., Broomhall, A. M., Buzasi, D. L., Chaplin, W. J., García, R. A., Gruberbauer, M., Hale, S. J., De Ridder, J., Frandsen, S., Borucki, W. J., Brown, T., Christensen-Dalsgaard, J., Gilliland, R. L., Jenkins, J. M., Kjeldsen, H., Koch, D., Belkacem, K., Bildsten, L., Bruntt, H., Campante, T. L., Deheuvels, S., Derekas, A., Dupret, M.-A., Goupil, M.-J., Hatzes, A., Houdek, G., Ireland, M. J., Jiang, C., Karoff, C., Kiss, L. L., Lebreton, Y., Miglio, A., Montalbán, J., Noels, A., Roxburgh, I. W., Sangaralingam, V., Stevens, I. R., Suran, M. D., Tarrant, N. J., and Weiss, A. 2010. Solar-like oscillations in low-luminosity red giants: first results from *Kepler*. *ApJ*, **713**, L176–L181.

Bedding, T. R., Mosser, B., Huber, D., Montalbán, J., Beck, P., Christensen-Dalsgaard, J., Elsworth, Y. P., García, R. A., Miglio, A., Stello, D., White, T. R., De Ridder, J., Hekker, S., Aerts, C., Barban, C., Belkacem, K., Broomhall, A.-M., Brown, T. M., Buzasi, D. L., Carrier, F., Chaplin, W. J., Di Mauro, M. P., Dupret, M.-A., Frandsen, S., Gilliland, R. L., Goupil, M.-J., Jenkins, J. M., Kallinger, T., Kawaler, S., Kjeldsen, H., Mathur, S., Noels, Silva Aguirre, V., and Ventura, P. 2011. Gravity modes as a way to distinguish between hydrogen- and helium-burning red giant stars. *Nature*, **471**, 608–11.

Belkacem, K., Samadi, R., Goupil, M. J., Kupka, F., and Baudin, F. 2006. A closure model with plumes. II. Application to the stochastic excitation of solar p modes. *A&A*, **460**, 183–90.

Belkacem, K., Goupil, M. J., Dupret, M. A., Samadi, R., Baudin, F., Noels, A., and Mosser, B. 2011. The underlying physical meaning of the $\nu_{\rm max} - \nu_{\rm c}$ relation. *A&A*, **530**, A142.

Böhm-Vitense, E. 1958. Über die Wasserstoffkonvektionszone in Sternen verschiedener Effektivtemperaturen und Leuchtkräfte *Z. Astrophys.*, **46**, 108–43.

Borucki, W., Koch, D., Batalha, N., Caldwell, D., Christensen-Dalsgaard, J., Cochran, W. D., Dunham, E., Gautier, T. N., Geary, J., Gilliland, R., Jenkins, J., Kjeldsen, H., Lissauer, J. J., and Rowe, J. 2009. *KEPLER*: Search for Earth-size planets in the habitable zone. Pages 289–299 of: F. Pont, D. Sasselov, and M. Holman (ed), *Proc. IAU Symp. 253, Transiting Planets*. IAU and Cambridge University Press.

Brown, T. M., Gilliland, R. L., Noyes, R. W., and Ramsey, L. W. 1991. Detection of possible p-mode oscillations of Procyon. *ApJ*, **368**, 599–609.

Castellani, V., Giannone, P., and Renzini, A. 1971. Overshooting of convective cores in helium-burning horizontal-branch stars. I. *Ap&SS*, **10**, 340–9.

Chang, H.-Y. and Gough, D. O. 1998. On the power distribution of solar p modes. *Solar Phys.*, **181**, 251–63.

Chaplin, W. J., Elsworth, Y., Howe, R., Isaak, G. R., McLeod, C. P., Miller, B. A., and New, R. 1997. The observation and simulation of stochastically excited solar p modes. *MNRAS*, **287**, 51–6.

Chaplin, W. J., Elsworth, Y., Isaak, G. R., Marchenkov, K. I., Miller, B. A., New, R., Pinter, B., and Appourchaux, T. 2002. Peak finding at low signal-to-noise: low-ℓ solar acoustic eigenmodes at $n \leq 9$ from the analysis of BiSON data. *MNRAS*, **336**, 979–91.

Chaplin, W. J., Houdek, G., Elsworth, Y., Gough, D. O., Isaak, G. R., and New, R. 2005. On model predictions of the power spectral density of radial solar p modes. *MNRAS*, **360**, 859–68.

Chaplin, W. J., Kjeldsen, H., Christensen-Dalsgaard, J., Basu, S., Miglio, A., Appourchaux, T., Bedding, T. R., Elsworth, Y., García, R. A., Gilliland, R. L., Girardi, L., Houdek, G., Karoff, C., Kawaler, S. D., Metcalfe, T. S., Molenda-Żakowicz, J., Monteiro, M. J. P. F. G., Thompson, M. J., Verner, G. A., Ballot, J., Bonanno, A., Brandão, I. M., Broomhall, A.-M., Bruntt, H., Campante, T. L., Corsaro, E., Creevey, O. L., Doğan, G., Esch, L., Gai, N., Gaulme, P., Hale, S. J., Handberg, R., Hekker, S., Huber, D., Jiménez, A., Mathur, S., Mazumdar, A., Mosser, B., New, R., Pinsonneault, M. H., Pricopi, D., Quirion, P.-O., Régulo, C., Salabert, D., Serenelli, A. M., Silva Aguirre, V., Sousa, S. G., Stello, D., Stevens, I. R., Suran, M. D., Uytterhoeven, K., White, T. R., Borucki, W. J., Brown, T. M., Jenkins, J. M., Kinemuchi, K., Van Cleve, J. and, Klaus, T. C. 2011. Ensemble asteroseismology of solar-type stars with the NASA Kepler mission. *Science*, **332**, 213–16.

Christensen-Dalsgaard, J. 2004. Physics of solar-like oscillations. *Solar Phys.*, **220**, 137–68.

Christensen-Dalsgaard, J. 2008a. ADIPLS – the Aarhus adiabatic pulsation package. *Ap&SS*, **316**, 113–20.

Christensen-Dalsgaard, J. 2008b. ASTEC – the Aarhus STellar Evolution Code. *Ap&SS*, **316**, 13–24.

Christensen-Dalsgaard, J. and Frandsen, S. 1983. Stellar 5 min oscillations. *Solar Phys.*, **82**, 469–86.

Christensen-Dalsgaard, J. and Thompson, M. J. 2011. Stellar hydrodynamics caught in the act: asteroseismology with CoRoT and Kepler. In: N. Brummell, A. S. Brun, M. S. Miesch, and Y. Ponty (ed), *Proc. IAU Symposium 271: Astrophysical dynamics: from stars to planets.* IAU and Cambridge University Press. In press `[arXiv:1104.5191]`.

Christensen-Dalsgaard, J., Gough, D. O., and Libbrecht, K. G. 1989. Seismology of solar oscillation line widths. *ApJ*, **341**, L103–L106.

Christensen-Dalsgaard, J., Kjeldsen, H., and Mattei, J. A. 2001. Solar-like oscillations of semiregular variables. *ApJ*, **562**, L141–L144.

Christensen-Dalsgaard, J., Kjeldsen, H., Brown, T. M., Gilliland, R. L., Arentoft, T., Frandsen, S., Quirion, P.-O., Borucki, W. J., Koch, D., and Jenkins, J. M. 2010. Asteroseismic investigation of known planet hosts in the *Kepler* field. *ApJ*, **713**, L164–L168.

De Ridder, J., Barban, C., Carrier, F., Mazumdar, A., Eggenberger, P., Aerts, C., Deruyter, S., and Vanautgaerden, J. 2006. Discovery of solar-like oscillations in the red giant ϵ Ophiuchi. *A&A*, **448**, 689–95.

De Ridder, J., Barban, C., Baudin, F., Carrier, F., Hatzes, A. P., Hekker, S., Kallinger, T., Weiss, W. W., Baglin, A., Auvergne, M., Samadi, R., Barge, P., and Deleuil, M. 2009. Non-radial oscillation modes with long lifetimes in giant stars. *Nature*, **459**, 398–400.

Deubner, F.-L. and Gough, D. O. 1984. Helioseismology: oscillations as a diagnostic of the solar interior. *ARAA*, **22**, 593–619.

Di Mauro, M. P., Cardini, D., Catanzaro, G., Ventura, R., Barban, C., Bedding, T. R., Christensen-Dalsgaard, J., De Ridder, J., Hekker, S., Huber, D., Kallinger, T., Miglio, A., Montalban, J., Mosser, B., Stello, D., Uytterhoeven, K., Kinemuchi, K., Kjeldsen, H., Mullally, F., and Still, M. 2011. Solar-like oscillations from the depths of the red giant star KIC 4351319 observed with *Kepler*. *MNRAS*, in press `[arXiv:1105.1076]`.

Dupret, M.-A., Belkacem, K., Samadi, R., Montalban, J., Moreira, O., Miglio, A., Godart, M., Ventura, P., Ludwig, H.-G., Grigahcène, A., Goupil, M.-J., Noels, A. and Caffau, E. 2009. Theoretical amplitudes and lifetimes of non-radial solar-like oscillations in red giants. *A&A*, **506**, 57–67.

Duvall, T. L. 1982. A dispersion law for solar oscillations. *Nature*, **300**, 242–3.

Dziembowski, W. 1977. Oscillations of giants and supergiants. *AcA*, **27**, 95–126.

Dziembowski, W. A. and Soszyński, I. 2010. Acoustic oscillations in stars near the tip of the red giant branch. *A&A*, **524**, A88.

Dziembowski, W. A., Gough, D. O., Houdek, G., and Sienkiewicz, R. 2001. Oscillations of α UMa and other red giants. *MNRAS*, **328**, 601–10.

Edmonds, P. D. and Gilliland, R. L. 1996, *ApJ*, 464, L157.

Fletcher, S. T., Chaplin, W. J., Elsworth, Y., Schou, J., and Buzasi, D. 2006. Frequency, splitting, linewidth and amplitude estimates of low-ℓ p modes of α Cen A: analysis of Wide-Field Infrared Explorer photometry. *MNRAS*, **371**, 935–44.

Fontaine, G. and Brassard, P. 2008. The pulsating white dwarf stars. *PASP*, **120**, 1043–96.

Frandsen, S., Carrier, F., Aerts, C., Stello, D., Maas, T., Burnet, M., Bruntt, H., Teixeira, T. C., de Medeiros, J. R., Bouchy, F., Kjeldsen, H., Pijpers, F., and Christensen-Dalsgaard, J. 2002. Detection of solar-like oscillations in the G7 giant star ξ Hya. *A&A*, **394**, L5–L8.

Gai, N., Basu, S., Chaplin, W. J. and Elsworth, Y. 2011. An in-depth study of grid-based asteroseismic analysis. *ApJ*, **730**, 63.

Gilliland, R. L., Brown, T. M., Christensen-Dalsgaard, J., Kjeldsen, H., Aerts, C., Appourchaux, T., Basu, S., Bedding, T. R., Chaplin, W. J., Cunha, M. S., De Cat, P., De Ridder, J., Guzik, J. A., Handler, G., Kawaler, S., Kiss, L., Kolenberg, K., Kurtz, D. W., Metcalfe, T. S., Monteiro, M. J. P. F. G., Szabó, R., Arentoft, T., Balona, L., Debosscher, J., Elsworth, Y. P., Quirion, P.-O., Stello, D., Suárez, J. C., Borucki, W. J., Jenkins, J. M., Koch, D.,

Kondo, Y., Latham, D. W., Rowe, J. F., and Steffen, J. H. 2010. *Kepler* asteroseismology program: Introduction and first results. *PASP*, **122**, 131–43.

Goldreich, P. and Keeley, D. A. 1977. Solar seismology. II. The stochastic excitation of the solar p-modes by turbulent convection. *ApJ*, **212**, 243–51.

Goldreich, P., Murray, N., and Kumar, P. 1994. Excitation of solar p-modes. *ApJ*, **424**, 466–479.

Gough, D. O. 1977. Mixing-length theory for pulsating stars. *ApJ*, **214**, 196–213.

Gough, D. O. 1980. Some theoretical remarks on solar oscillations. Pages 273–299 of: H. A. Hill and W. Dziembowski (ed), *Lecture Notes in Physics*, vol. **125**. Springer, Berlin.

Gough, D. O. 1986. EBK quantization of stellar waves. Pages 117–143 of: Y. Osaki (ed), *Hydrodynamic and magnetohydrodynamic problems in the Sun and stars.* University of Tokyo.

Gough, D. O. 1990. Comments on helioseismic inference. Pages 283–318 of: Y. Osaki and H. Shibahashi (ed), *Progress of seismology of the sun and stars. Lecture Notes in Physics*, vol. **367**. Springer, Berlin.

Gough, D. O. 1993. Course 7. Linear adiabatic stellar pulsation. Pages of 399–560 of: J.-P. Zahn and J. Zinn-Justin (ed), *Astrophysical fluid dynamics, Les Houches Session XLVII.* Elsevier, Amsterdam.

Gough, D. O. 2007. An elementary introduction to the JWKB theory. *AN*, **328**, 273–85.

Grigahcène, A., Dupret, M.-A., Gabriel, M., Garrido, R., and Scuflaire, R. 2005. Convection-pulsation coupling. I. A mixing-length perturbative theory. *A&A*, **434**, 1055–62.

Hekker, S., Caerts, C., De Ridder, J., and Carrier, F. 2006. Pulsations detected in the line profile variations of red giants: modeling of line moments, line bisector and line shape. *A&A*, **458**, 931–40.

Hekker, S., Gilliland, R. L., Elsworth, Y., Chaplin, W. J., De Ridder, J., Stello, D., Kallinger, T., Ibrahim, K. A., Klaus, T. C., and Li, J. 2011. Characterisation of red giant stars in the public *Kepler* data. *MNRAS*, **414**, 2594–601.

Houdek, G. 2010a. Convection and oscillations. Pages 998–1003 of: *Proc. 4th HELAS International Conference: Seismological Challenges for Stellar Structure. AN*, 331.

Houdek, G. 2010b. Stellar turbulence and mode physics. Pages 237–244 of: M. Marconi, D. Cardini and M. P. Di Mauro (ed), *Proc. HELAS Workshop on "Synergies between solar and stellar modeling,"* Rome 22–26 June 2009, *Ap&SS*, **328**.

Houdek, G. and Gough, D. O. 2002. Modelling pulsation amplitudes of ξ Hydrae. *MNRAS*, **336**, L65–L69.

Houdek, G. and Gough, D. O. 2007. An asteroseismic signature of helium ionization. *MNRAS*, **375**, 861–80.

Houdek, G., Balmforth, N. J., Christensen-Dalsgaard, J., and Gough, D. O. 1999. Amplitudes of stochastically excited oscillations in main-sequence stars. *A&A*, **351**, 582–96.

Huber, D., Bedding, T. R., Stello, D., Mosser, B., Mathur, S., Kallinger, T., Hekker, S., Elsworth, Y. P., Buzasi, D. L., De Ridder, J., Gilliland, R. L., Kjeldsen, H., Chaplin, W. J., García, R. A., Hale, S. J., Preston, H. L., White, T. R., Borucki, W. J., Christensen-Dalsgaard, J., Clarke, B. D., Jenkins, J. M., and Koch, D. 2010. Asteroseismology of red giants from the first four months of *Kepler* data: global oscillation parameters for 800 stars. *ApJ*, **723**, 1607–17.

Iglesias, C. A. and Rogers, F. J. 1996. Updated OPAL opacities. *ApJ*, **464**, 943–53.

Innis, J. L., Isaak, G. R., Brazier, R. I., Belmonte, J. A., Palle, P. L., Roca Cortes, T., and Jones, A. R. 1988. High precision velocity observations of Arcturus using the 7699 Å line of potassium. Pages 569–573 of: V. Domingo and E. J. Rolfe (ed), *Seismology of the Sun and Sun-like Stars.* ESA SP-286. ESA Publications Division. Noordwijk, The Netherlands.

Kallinger, T., Weiss, W. W., Barban, C., Baudin, F., Cameron, C., Carrier, F., De Ridder, J., Goupil, M.-J., Gruberbauer, M., Hatzes, A., Hekker, S., Samadi, R., and Deleuil, M. 2010. Oscillating red giants in the CoRoT exofield: asteroseismic mass and radius determination. *A&A*, **509**, A77.

Kippenhahn, R. and Weigert, A. 1990. *Stellar structure and evolution.* Springer-Verlag. Berlin.

Kiss, L. L. and Bedding, T. R. 2003. Red variables in the OGLE-II data base. I. Pulsations and period-luminosity relations below the tip of the red giant branch of the Large Magellanic Cloud. *MNRAS*, **343**, L79–L83.

Kiss, L. L. and Bedding, T. R. 2004. Red variables in the OGLE-II data base. II. Comparison of the Large and Small Magellanic Clouds. *MNRAS*, **347**, L83–L87.

Kjeldsen, H. and Bedding, T. R. 1995. Amplitudes of stellar oscillations: the implications for asteroseismology. *A&A*, **293**, 87–106.

Kjeldsen, H., Bedding, T. R., and Christensen-Dalsgaard, J. 2008. Correcting stellar oscillation frequencies for near-surface effects. *ApJ*, **683**, L175–L178.

Kumar, P., Franklin, J., and Goldreich, P. 1988. Distribution function for the time-averaged energies of stochastically excited solar *p*-modes. *ApJ*, **328**, 879–87.

Mazumdar, A. 2005. Asteroseismic diagrams for solar-type stars. *A&A*, **441**, 1079–86.

Miglio, A., Montalbán, J., Baudin, F., Eggenberger, P., Noels, A., Hekker, S., De Ridder, J., Weiss, W., and Baglin, A. 2009. Probing populations of red giants in the galactic disk with CoRoT. *A&A*, **503**, L21–L24.

Miglio, A., Montalbán, J., Carrier, F., De Ridder, J., Mosser, B., Eggenberger, P., Scuflaire, R., Ventura, P., D'Antona, F., Noels, A., and Baglin, A. 2010. Evidence for a sharp structure variation inside a red giant star. *A&A*, **520**, L6.

Montalbán, J., Miglio, A., Noels, A., Scuflaire, R., and Ventura, P. 2010. Seismic diagnostics of red giants: first comparison with stellar models. *ApJ*, **721**, L182–L188.

Monteiro, M. J. P. F. G. and Thompson, M. J. 2005. Seismic analysis of the second ionization region of helium in the Sun. I. Sensitivity study and methodology. *MNRAS*, **361**, 1187–96.

Mosser, B., Belkacem, K., Goupil, M.-J., Miglio, A., Morel, T., Barban, C., Baudin, F., Hekker, S., Samadi, R., De Ridder, J., Weiss, W., Auvergne, M., and Baglin, A. 2010. Red giant seismic properties analyzed with CoRoT. *A&A*, **517**, A22.

Mosser, B., Belkacem, K., Goupil, M. J., Michel, E., Elsworth, Y., Barban, C., Kallinger, T., Hekker, S., De Ridder, J., Samadi, R., Baudin, F., Pinheiro, F. J. G., Auvergne, M., Baglin, A., and Catala, C., 2011b. The universal red giant oscillation pattern: an automated determination with CoRoT data. *A&A*, **525**, L9.

Mosser, B., Barban, C., Montalbán, J., Beck, P. G., Miglio, A., Belkacem, K., Goupil, M. J., Hekker, S., De Ridder, J., Dupret, M. A., Elsworth, Y., Noels, A., Baudin, F., Michel, E., Samadi, R., Auvergne, M., Baglin, A., and Catala, C., 2011a. Mixed modes in red giant stars observed with CoRoT. *A&A*, **532**, A86.

Retter, A., Bedding, T. R., Buzasi, D. L., Kjeldsen, H., and Kiss, L. L. 2003. Oscillations in Arcturus from *WIRE* photometry. *ApJ*, **591**, L151–L154. (*Erratum: ApJ*, **596**, L125.)

Rogers, F. J., Swenson, F. J., and Iglesias, C. A. 1996. OPAL equation-of-state tables for astrophysical applications. *ApJ*, **456**, 902–8.

Samadi, R., Georgobiani, D., Trampedach, R., Goupil, M. J., Stein, R. F. and Nordlund, Å. 2007. Excitation of solar-like oscillations across the HR diagram. *A&A*, **463**, 297–308.

Serenelli, A. and Weiss, A. 2005. On constructing horizontal branch models. *A&A*, **442**, 1041–8.

Smith, P. H., McMillan, R. S., and Merline, W. J. 1987. Evidence for periodic radial velocity variations in Arcturus. *ApJ*, **317**, L79–L84.

Soszyński, I., Dziembowski, W. A., Udalski, A., Kubiak, M., Szymański, M. K., Pietrzyński, G., Wyrzykowski, Ł., Szewczyk, O., and Ulaczyk, K. 2007. The optical gravitational lensing experiment. period-luminosity relations of variable red giant stars. *AcA*, **57**, 201–25.

Stein, R. F. 1968. Waves in the solar atmosphere. I. The acoustic energy flux. *ApJ*, **154**, 297–306.

Stello, D., Bruntt, H., Preston, H. and Buzasi, D. 2008. Oscillating K giants with the *WIRE* satellite: determination of their asteroseismic masses. *ApJ*, **674**, L53–L56.

Tabur, V., Bedding, T. R., Kiss, L. L., Giles, T., Derekas, A., and Moon, T. T. 2010. Period-luminosity relations of pulsating M giants in the solar neighbourhood and the Magellanic Clouds. *MNRAS*, **409**, 777–88.

Tassoul, M. 1980. Asymptotic approximations for stellar non-radial pulsations. *ApJS*, **43**, 469–90.

Tassoul, M. 1990. Second-order asymptotic approximations for stellar non-radial acoustic modes. *ApJ*, **358**, 313–27.

Teixeira, T. C., Christensen-Dalsgaard, J., Carrier, F., Aerts, C., Frandsen, S., Stello, D., Maas, T., Burnet, M., Bruntt, H., de Medeiros, J. R., Bouchy, F., Kjeldsen, H. and Pijpers, F. 2003. Giant vibrations in dip. Pages 233–236 of: M. J. Thompson, M. S. Cunha, and

M. J. P. F. G. Monteiro (ed), *Asteroseismology across the HR diagram.* Kluwer Academic Publishers, Dordrecht.

Unno, W. 1967. The stellar radial pulsation coupled with the convection. *PASJ*, **19**, 140–53.

Vandakurov, Yu. V. 1967. The frequency distribution of stellar oscillations. *Astron. Zh.*, **44**, 786–797. (English translation: *Soviet Astronomy A.J*, **11**, 630–638.)

Vorontsov, S. V., Baturin, V. A., and Pamyatnykh, A. A. 1991. Seismological measurement of solar helium abundance. *Nature*, **349**, 49–51.

Xiong, D. R. and Deng, L. 2010. Non-adiabatic oscillations of the low- and intermediate-degree modes of the Sun. *MNRAS*, **405**, 2759–67.

Xiong, D. R., Chen, Q. L., and Deng, L. 1997. Nonlocal time-dependent convection theory. *ApJS*, **108**, 529–44.

Printed in the United States
by Baker & Taylor Publisher Services